Modeling Urban Dynamics

Modeling Urban Dynamics

Mobility, Accessibility and Real Estate Value

Edited by
Marius Thériault
François Des Rosiers

iSTE

WILEY

First published 2011in Great Britain and the United States by ISTE Ltd and John Wiley & Sons, Inc.
Adapted and updated from *Information géographique et dynamiques urbaines* published 2008 in France
by Hermes Science/Lavoisier © LAVOISIER 2008

ISTE Ltd   John Wiley & Sons, Inc.
27-37 St George's Road   111 River Street
London SW19 4EU   Hoboken, NJ 07030
UK   USA

www.iste.co.uk   www.wiley.com

© ISTE Ltd 2011

Library of Congress Cataloging-in-Publication Data

Modeling urban dynamics : mobility, accessibility and real estate value / edited by Marius Thériault,
Francois Des Rosiers.
p. cm.
"Adapted and updated from Information geographique et dynamiques urbaines published 2008 in France
by Hermes Science/Lavoisier."
Includes bibliographical references and index.
ISBN 978-1-84821-268-8
1. Cities and towns--Econometric models. 2. Urbanization--Econometric models. 3. City planning--
Econometric models. 4. Sociology, Urban. I. Thériault, Marius. II. Rosiers, François Des. III.
Information géographique et dynamiques urbaines.
HT151.M5594 2011
307.76--dc22

                                                                              2010042675

British Library Cataloguing-in-Publication Data
A CIP record for this book is available from the British Library
ISBN 978-1-84821-268-8

Printed and bound in Great Britain by CPI Antony Rowe, Chippenham and Eastbourne.

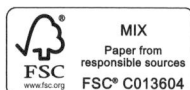

Table of Contents

**Chapter 3. Spatiotemporal Modeling of Destination Choices for
Consumption Purposes: Market Areas Delineation and Market
Share Estimation**.   57
Gjin BIBA and Paul VILLENEUVE

**Chapter 4. Generation of Potential Fields and Route Simulation
Based on the Household Travel Survey**.   83
Arnaud BANOS and Thomas THÉVENIN

Introduction

Modeling Urban Dynamics: Mobility, Accessibility and Real Estate Value

For many decades, aggregate data (pertaining to zones) were used to study urban dynamics, using summary methods that rarely permitted the study of urban phenomena at the spatial process level [CLI 81]. Given the scarcity of data and lack of technology capable of processing data to model the details of the process, urban studies' research was still able to obtain conclusive results with regard to the conceptual aspects of urban dynamics and modeling of interrelationships among a city's basic elements, especially with regard to social and economic aspects. Partly hindered by issues of a modifiable area units problem [OPE 79] and ecological fallacy [ROB 50], this research rarely produced operational tools capable of modeling and simulating the detailed operation of the urban system. This made it difficult to plan for its development following overlapping social, economic and political change at various geographic levels (from the global village to the local neighborhood, and at the national level). Nevertheless, it is indispensible to evaluate the scope and relevance of urban planning and development undertaken by local authorities (cities, higher levels of government associations, etc.) or conducted by the private sector (households, land developers, firms, etc.) [GEE 03].

At the conceptual level, we know that the city can be seen as a group of complex overlapping systems [FOR 69] whose operating mechanisms are the result of the conjunction of, and competition between, individual action processes. These are often intentional, sometimes contradictory, and generally have a legitimate purpose, e.g. improving a person's residential status or reducing daily commute times. Once

---

Introduction written by Marius THÉRIAULT and François DES ROSIERS.

aggregated, such processes may generate counterintuitive results and relatively unhealthy overall consequences for urban agglomeration: pollution, traffic congestion, urban sprawl, inequality, segregation, etc. We then speak of unintentional consequences.

New information technologies, particularly those that process geographic and temporal data, now enable us to study, model and simulate functioning of the various components of a city at different perception levels for elementary decision-making (where, when and how does the action occur?): those of the households, firms and individuals. This technological breakthrough is primarily dependent on the recent progress in information technology and its accompanying information revolution [FRA 01, THI 00]. With the development of:

– spatial technologies (remote sensing, satellite positioning, etc.);

– electronics (increasing calculation and storage capacity);

– networking (wireless communication, ubiquitous databases, spatial data repositories, system compatibility, etc.);

– computer systems (object-oriented programming, encapsulation, heredity, etc.); and

– geomatics (spatial analysis, systemic modeling, geographic information systems, spatiotemporal simulation, etc.),

research in urban studies can now free itself from the constraints of urban phenomena aggregation. These constraints have often caused methodological vulnerability when endeavoring to develop its fields of application (factorial error problems [KIN 97], modifiable area unit problem [GRE 96, OPE 84], autocorrelation and spatial drift [FOT 02, GRI 87], etc.). Could we be witnessing the birth of a paradigm shift longed for by quantitative geographers of the 1970s and the revival of 'housing-transportation-land use' integrator models? This is a dream the Chapel Hill spatial economists had to relinquish during the same period due to lack of means, and which only became possible in a particular scientific and technological context.

The city is a living system that undergoes continual changes: people move about and activities are set up according to logic specific to each area, bringing about changes to urban shape in the long term. The development of the urban fabric is a result of the conjunction of a multitude of individual processes that take place at temporal scales that range from a single hour to several decades. In the medium term, their cumulation brings about a diffusion (or redeployment) of activities in the area. This, with the spatiotemporal configuration of transportation networks (partly subordinated to this redeployment), determines accessibility to urban services in the area (work, shopping, recreation, etc.).

Falling within this progressive trend in urban studies, the creation of this work provides an opportunity for multidisciplinary pooling of interests sparked by the very nature of analytical concepts and processes implemented. The result is the advent of multiple collaborative efforts among many researchers from the fields of social sciences, from geography to economics, and sociology, psychology and transportation studies. The development of urban studies during the past decade bears witness to the potential of this collaboration that targets not only the integration of methodologies, but more specifically the articulation of concepts and theoretical foundations stemming from these various disciplines.

To target the objectives in this book, we voluntarily reduced the urban dynamics thematic to several basic dimensions, while taking care to choose those we believed essential to track and comprehend the structural and functional changes of the urban landscape. These dynamics include concepts of mobility, accessibility, perception and assessment. Most of the chapters place primary importance on the action of persons and households which, using the city as a lifestyle framework, transform it through the cumulation of their decisions. Our collaborative initiative received excellent response, with the 12 chapters being written by 33 collaborators:

– Chapter 1 – the role of mobility in the building of metropolitan polycen-trism, a comparative study of two large agglomerations: Paris and the Montpellier-Avignon-Marseille conurbation in France.

– Chapter 2 – commuting and gender: two cities, one reality? a comparative study of urban contexts and gendered mobility in Brussels (Belgium) and Québec City (Canada).

– Chapter 3 – spatiotemporal modeling of destination choices for consumption purposes: from market area delineation to market share estimation, with an application example for Québec City.

– Chapter 4 – generation of potential fields and route simulation based on a household travel survey, with applications to Lille and Besançon in France.

– Chapter 5 – impacts of road networks on urban mobility, looking at the structural effects of transportation network on perception, behavior, efficiency and modal choice.

– Chapter 6 – daily mobility and urban morphology: abiding landscaping features of activity places used as indicators of environmental values: application to Strasbourg in France.

– Chapter 7 – decision to move house and residential choices upon acquiring a single-family home; a case study in Québec City.

– Chapter 8 – distance, proximity, accessibility and spatial diffusion: a review of concepts.

– Chapter 9 – measuring accessibility to proximity services: an application to servicing areas of poverty in Montreal in Canada.

– Chapter 10 – modeling accessibility to urban services and its impact on residential values: application to Québec City.

– Chapter 11 – hedonic modeling of residential values: measuring urban externalities in Québec City and Brossard in Canada.

– Chapter 12 – hedonic modeling of peri-urban landscapes in the Dijon region residential market in France.

The first seven chapters provide analysis and simulation examples of people's travel behavior in their daily activities, with a focus on how it impacts on urban form in the long term. They investigate the constraints and opportunities afforded by the structure of transportation networks as well as the distribution of urban functions on the landscape. Particular interest is paid to:

– the functional structure of urban contexts, the distribution of activity poles (from monocenterism to polycentrism);

– the comparative analysis of mobility behavior between men and women;

– the destination choice for consumption activities;

– the competition/complementarity of transportation modes in a perspective of intermodality;

– the simulation of individual commutes;

– the structural impact of networks on mobility and the efficiency of transportation services;

– the perception of the surrounding neighborhood;

– the formation of cognitive maps; and, finally,

– residential choices.

The primary objective is to better understand the pulse of urban agglomerations on a daily basis and establish decision-making tools that are capable of informing decision-makers of the possible consequences of their interventions, or to propose various steps that may improve the operation of public and private transport modes.

Beyond the physical infrastructures and the people who live there, a city is also a social construct that is viewed through cognitive filters according to the quality of the environment it offers, its marginal utility and the lifestyle quality it provides for individuals and families who live, work and shop in it. Although often having recourse to quantitative methods to examine the relationship between the city and its perception, Chapters 8 through 12 intend to examine the quality of life associated with urban and periurban environments from a qualitative and critical viewpoint. Perceptions and values will be seen through cognitive filters relating to the mobility of individuals and locations visited. These are sometimes measured by using high-resolution satellite imaging to accurately display the quality of urban environments. These sometimes have an implicit value through residential transactions (sales price of the composite housing good, which is the sum of marginal contributions assigned to the property's physical attributes and urban amenities located in its surrounding area) thanks to hedonic modeling [ROS 74], itself derived from microeconomic utility theory.

In the context of urban sprawl, the thinning of past urban forms and the rapid evolution of networks, Chapter 1 (Berroir, Mathian, Saint-Julien and Sanders) delves into the modes of the emergence of increasingly polycentric urban forms, at different geographical scales. It pays particular attention to the multiscalar process of territorial integration. In this work, the morphology of polycenterism is approached based on the concentration of metropolitan employment, while the functional study is based on residence–work mobility. By comparing two metropolitan structures, we seek to understand how this residence–work mobility contributes to singularly shaping intra-metropolitan polarization patterns, thus helping to reveal new articulations among centers and territories. The first corresponds to the metropolitan region surrounding Paris, while the second focuses on the metropolitan region in Southern France, an expansive urban triangle whose peaks include the urban centres of Montpellier to the west, Avignon to the north, and Toulon to the east. Theoretically speaking, this chapter also aims to clarify the foundations and conceptual bases allowing us to draft a model of the emergence and functioning of polycentric metropolitan territories.

Chapter 2, written by Vandersmissen, Thomas and Verhetsel, studies daily mobility from the viewpoint of the social and professional integration of men and women. It analyzes work-to-residence trips with regard to contact between residence–work spheres. The governing idea is that commutes constitute a decisive form of spatial interaction in understanding urban dynamics. Moreover, an analysis of these commutes raises certain methodological questions that vary from country-to-country and which are discussed in this chapter, with Québec City and Brussels used as examples. In addition to the description of commutes and the possible discovery of differences between men and women, researchers are interested in the factors that influence this mobility and could explain the differences recorded: civil status, type of household, racial or ethnic factors, socioeconomic levels, access to a vehicle, etc. It considers how we model this form of spatial interaction, which also depends on numerous social factors. Lastly, the thematic objective is to draft a Québec City–Brussels comparison with the available data and results, and verify whether these results support the findings obtained for other North American and European cities.

By studying trips made for recreation and shopping purposes, Chapter 3, written by Biba and Villeneuve, analyzes a form of mobility that is less confined in its spatiotemporal framework and that allows more freedom for impulsive or opportunistic decisions. Interest focuses on changes to the urban framework linked to the emergence of the Western consumer society, which triggers a diversification of business and forms of consumption. The most recent forms include power centers and big-box stores (large, stand-alone department stores) that set up on the urban periphery, while the more traditional shopping centers and commercial streets attempt to stay afloat in the more central locations. The authors present a method for determining market areas and competition analysis among the various business forms in the Québec metropolitan area in 2001. The methodology uses disaggregated data within discrete choice models, the specification of which is enhanced by a geographic information system (GIS)-based analysis of the urban context. The objective is to integrate data that are relatively stable in space (location of residences, businesses, transportation network structures), but also relatively unstable over time (business' hours of operation, public transportation service, length of car trips). This is in order to better spatialize the demand for facilities and analyze commuting choices for shopping and recreation. This study illustrates the potential offered by the combination of GIS, spatial analysis, statistical modeling and individual mobility surveys to study territorial competition among businesses. By modeling the overlapping spatiotemporal constraints in the utility function of consumers, the study advances our understanding of business dynamics and, to a point, their systemic connection with urban form and land-use planning policies at the regional level.

Urban sprawl, the scattering of living spaces, the increase in commute distances and the evolution of mobility behavior shaped by the automobile are all transformations that modify people's mobility and repeatedly raise the question of city management. Chapter 4, contributed by Banos and Thévenin, studies these palpitations that liven up a city, from a 'sustainable urbanization' perspective linked to urban planning and the evaluation of consequences on the mobility of people and goods. The authors explain how it is possible to use the French household–commute survey to observe the mobility behavior of various categories of people for all transportation modes at the urban level. However, the use of data provided by the survey must be adapted to meet the risks and issues generated through the evolution of urban mobility. This requires new analysis methods and representation of individual behavior, and also overall movement occurring at the agglomeration scale. The chapter provides the link between mobility studies and multiagent simulation approaches by introducing three key steps:

– the creation of a virtual city within a GIS and the definition of subjacent potential fields;

– the simulation of likely itineraries for trips using the road network;

– the use of geo-visualization methods to reveal simulated individual trips, on one hand, and daily or seasonal urban space "palpitations" on the other.

Introducing three application examples, Chapter 5, (written by Foltête, Genre-Grandpierre and Josselin) focuses on the impact of network structures on the organization of mobility and the potential competition among transportation modes: walking, regular and demand-responsive public transit and the personal car. These modes are studied from the point of view of their characteristics, competition and complementarity in terms of access to urban facilities. They are also studied from an environmental perspective, making reference in particular to the principles of spatial syntax. The first example deals with pedestrian trips in Lille and aims to verify whether the density and connectivity of the road network, synthesized by a calculation of its accessibility potential, has an influence on walking habits. The second example studies the relationships between the configuration of road networks and the potential effectiveness of a demand-responsive transit service in Besançon. The final study conducted in Avignon, Valabrix and in the Doubs Central, focuses on the relationships between network metrics, accessibility and modal competition which, in the long term, generates urban sprawl. The chapter proposes novel avenues of research that may improve competition among active and public transportation modes, by reconfiguring transportation networks.

Identifying urban forms associated with daily mobility is an important issue in city planning management as, in the long run, it helps to identify their contribution to the choice of an itinerary over a myriad of possibilities associated with each mode of transportation. A second issue deals with analyzing the structuring elements of mobility by drawing on the sociocognitive dimension of accessibility, a dimension that is based on the social legibility of space, that is variable from one person to another. This approach presupposes calling upon available knowledge and know-how in geography, psychology and sociology, without isolating the effects of location, social groups and mobility. Chapter 6, authored by Ramadier, Petropoulou, Haniotou, Bronner and Énaux, presents a study of an analysis procedure that has been developed according to this approach in order to resolve two interdependent methodological issues:

1) difficulties related to the recording and reflective sharing process of cognitive maps in relation to urban morphology; and

2) the establishment of links with the spatial mobility practices of individuals and the identification of their living spaces.

The methodological objective is to identify constants within landscapes frequented and represented, then to define the relationship between these two types of landscape. This in turn allows us to identify the environmental values objectified simultaneously (or singularly) by the landscapes and (or) by the landscapes represented, but also by the relationships observed among them. Preliminary results obtained in Strasburg, France, enable us to clarify the process and evaluate its application potential for urban studies.

The situation regarding households evolves according to different rates that correspond to lifecycles that have a major impact on the evolution of residential location aspirations and strategies. In Chapter 7, Kestens, Thériault and Des Rosiers study the reasons for moving and the selection criteria for neighborhoods and residences as declared by new single-family homeowners. Using data obtained by a telephone interview conducted in Québec City among families that had purchased a single-family home between 1993 and 2001, the selection criteria are analyzed according to the household's attributes. The results stress the importance of the relationship between the lifecycle and residential choices. They shed light on location strategies, particularly with regard to the perception of the neighborhood and the feeling of belonging that occurs over the long term. It thus echoes the preceding chapter, with a statistical analysis of the relationship between the choice of location in the city and the criteria mentioned by households during the study that motivate their choices.

Distance, proximity, centrality, accessibility and mobility are terms that refer to the relationships between the formal structure of cities (urban fabric, segregation of activities, land use, communications network) and the use of facilities by residents. These residents condition urban dynamics in the short and long term, laying the foundation for the future modeling of urban operation and evolution. The exact signification of the terms may vary from one study to another, even resulting in semantic intersections where we confuse proximity and accessibility, accessibility and centrality, or even mobility and centrality potential. To instil order in this garbling of connected concepts, in Chapter 8 Dumolard reviews the basic concepts of distance(s), spatial accessibility, temporal accessibility and diffusion. The author discusses various problems related to the operational definition of distance, proximity and accessibility, notably those that are connected to their basic properties (validity scale, impacts on urban dynamics). These are compared to the dimensionality of geometric objects required to locate phenomena in a metric space, as well as the inherent difficulties in terms of precision and validity. The chapter then focuses on the intrinsic link between concepts of accessibility and spatial diffusion by noting the contributions of geography. These include the consideration of rugosity (also called impedance in simulation tools) that varies according to transportation and communication technologies; functional hierarchy (at the supraurban and intraurban levels); and socioeconomic inequalities.

Chapter 8 then delves into various approaches to operationalize the concept of accessibility by considering:

– representation modes of the territory (continuous versus discontinuous space, and representation by imaging – rugosity matrix – or by objects – network vector graph);

– classical solutions (operational research);

– studies of itineraries and optimization in network graphs;

– multicriteria analysis; and

– artificial intelligence simulations (cellular automation, multiagent systems, and genetic algorithms).

The inherent difficulties of these approaches are discussed, especially with regard to the quality of flux data (high spatiotemporal variability), their acquisition, and the difficulties encountered in modeling the consistencies (recurring behaviors) and temporal singularities (emergent behaviors).

Accessibility to public services and facilities is a crucial issue for socially disadvantaged groups who are less mobile and often without a car, given their financial insecurity. Moreover, for disadvantaged households, low accessibility to public resources contributes to exacerbating their economic handicap, while the reverse scenario partially compensates for the lack of individual resources. The main objective of Chapter 9, written by Apparicio and Séguin, is to describe how GIS has enabled us to evaluate accessibility to proximity services for populations living in Montreal's disadvantaged neighborhoods. With regard to methodology, the assessment of accessibility to urban resources requires that we apply parameters to four elements:

1) the spatial unit of reference to which the population is attached;

2) the method of aggregation;

3) the measurement or measurements of accessibility;

4) the type of distance used to calculate the measurements of accessibility.

The methodology presented is illustrated by an application example of the population living in subsidized housing in the Montreal region, in order to compare their accessibility to five categories of facilities: cultural, health, education, sports and recreation, and banking services. In complementary studies, other types of urban facilities, especially subway stations and supermarkets, are also considered.

Accessibility entails more than the result of the structure and the net capacity of communication networks (variable in space and time). It also reflects the time available to citizens to make trips (schedule constraints), their willingness to accept the trip duration (variable depending on the type and duration of the targeted activity) and their perception of added value (or interest) of each attainable point of service. In many cases, we thus cannot measure accessibility solely according to physical factors (for example, the closest service point), but must also include social and economic factors relating to preferences, either stated (during a survey, for example) or revealed through action.

In Chapter 10 for the Quèbec City region, Thériault, Voisin and Des Rosiers present an example of the modeling of accessibility that combines trip simulations in reference to network graphs with impedance constraints on nodes and links. Preferences are revealed by analyzing significant differences in mobility behavior (according to person/household/activity type) drawn from an origin-destination mobility study. Theoretically, this method is based on the principles of Hägerstrand's [HÄG 70] time geography. It is validated according to a process based on the utility theory, notably through the integration of various accessibility indicators obtained in modeling their marginal contribution to the shaping of residential values, as measured in a hedonic model [ROS 74]. This last approach uses multiple regression techniques to estimate the marginal contribution of accessibility in the sales price of single-family homes. However, the standard *ordinary least-square* multiple regression approach is unsuitable to estimate robust coefficients because of the multicollinearity of indicators on one hand, and the spatial autocorrelation of residuals on the other. The solution to the methodological problem involves an implementation, either of an autoregressive approach to estimate the global coefficients (stationary hypothesis of coefficients), or a geographically-weighted regression approach when effects display significant spatial drift. These hedonic modeling concepts are reviewed in detail at the end of the book, in Chapters 11 and 12.

Real estate transactions involve a complex commodity whose attributes are difficult to isolate and whose value is arrived at by the sum of advantages relating to the commodity's structural attributes (size of lot, size, quality and condition of buildings, appurtenances, etc.) and externalities that the location provides it with. The objective evaluation of the price of each of these qualitative attributes is tricky. In a free market, assessing value is based on the perception of the desirability (preferences) of each attribute during negotiations between sellers and buyers (supply and demand), where the negotiated price is aggregate (selling price) and some attributes form associations (e.g. luxury items: pool, garage, high-quality finishing). Chapter 11 presents the principles of an approach called hedonic modeling that allows us to determine the marginal contribution of specific and environmental attributes. Using multiple regression techniques, Des Rosiers, Dubé and Thériault show the hedonic approach faces the usual problems of multicollinearity, heteroskedasticity, autocorrelation of residuals and coefficient nonstationarity.

Aside from the inherent characteristics of each property, there are generally two types of spatial effects that contribute to setting real estate prices, both residential and commercial:

1) the effects of positive externality that are seen as advantages (e.g, good accessibility, pleasant view, well-developed neighborhood) and that contribute to increasing their value; and

2) the effects of negative externality that are seen as disadvantages (e.g. air pollution, unpleasant view, noise, unsafe neighborhood) and that contribute to decreasing the value.

Moreover, proximity effects can act in phases or in opposition to the preceding and normally respond to observable facts in relatively close proximity to the property. When markets are efficient, these positive or negative externalities are capitalized in the price of properties at their time of sale, which allows their variations to be measured. The authors present three examples of the modeling of externalities on single-family residential markets by using the hedonic approach:

1) the proximity effect of shopping centers in Québec City;

2) the impact of visible hydroelectric transmission lines in the Montreal region; and

3) the internalization of property characteristics in Québec City residential neighborhoods.

The last example suggests a nonstationary phenomenon of coefficients linked to a heterogeneous appreciation of implicit prices according to the sociological characteristics of the residents.

With everything else being equal, are households willing to pay for a house that is more expensive than another because of the view that it affords? If yes, at what cost does this view come? In Chapter 12, Brossard, Cavailhès, Hilal, Joly, Tourneux and Wavresky answer these questions. The economic value of landscapes is estimated through the combined processes of spatial modeling (digital elevation models, remote sensing, inter-visibility area calculation) and economic appraisal methods, particularly the hedonic approach. The contribution of geography helps to characterize the landscape seen by an observer walking the boundaries of his/her residence based on basic information: the terrain and land-based objects, all modeled thanks to digital elevation models and 3D spatial analysis processes. The extent of the viewshed is thus quantified in order to produce indicators that describe various aspects of the surrounding landscape. Then there is the economical analysis, using the hedonic price method based on 4,050 real estate transactions in the region of Dijon, France. These transactions show that various aspects of the landscape have a significant price and that, consequently, the attraction of the peri-urban lifestyle is a factor in residential choice. The approach adopted stems from a theoretical bias: it aims for maximum objectivity by introducing quantitative models and reproducible analytical methods.

Acknowledgements

The completion of this volume was made possible through the excellent collaboration of 33 coauthors. Funding for the *programme de soutien aux équipes de recherche, du Réseau Villes, Régions, Monde* was received from the *Fonds Québécois de la Recherche sur la Société et la Culture* and the Social Sciences and Humanities Research Council. The publication's coeditors would especially like to thank Marion Voisin and Jean Dubé for their invaluable editing assistance. We would like to express our gratitude to Professor Pierre Dumolard, who offered to produce this work for the Traité IGAT series, and to Chantal Ménascé and Alexandra Toulze from Hermès Science Publishing and ISTE-Wiley, who helped guide us through the editing process. Lastly, the coeditors wish to acknowledge the efficient collaboration of Gordon Cruise for the English translation and proof reading of chapters in this book.

Bibliography

[CLI 81] CLIFF A.D., ORD J.K., *Spatial Processes. Models and Applications*, London, Pion, 1981.

[FOR 69] FORRESTER J.W., *Urban Dynamics*, Waltham (Mass.), Pegasus Communications, 1969.

[FOT 02] FOTHERINGHAM A.S., BRUNSDON C., CHARLTON M.E., *Geographically Weighted Regression: The Analysis of Spatially Varying Relationships*, Chichester, Wiley, 2002.

[FRA 01] FRANK A.U., RAPER J., CHEYLAN J.P. (eds.), *Life and Motion of Socio-economic Units*, London, Taylor & Francis, 2001.

[GEE 03] GEERTMAN S., STILLWELL J. (eds.), *Planning Support Systems in Practice*, Berlin, Springer, 2003.

[GRE 96] GREEN M., FLOWERDEW R., "New evidence on the modifiable areal unit problem", in: P. LONGLEY, M. BATTY (eds.) *Spatial Analysis: Modelling in a GIS Environment*, Cambridge, GeoInformation International, pp. 41-54, 1996.

[GRI 87] GRIFFITH D.A., *Spatial Autocorrelation: A Primer*, Washington D.C., Association of American Geographers, 1987.

[HÄG 70] HÄGERSTRAND R.T., "What about people in regional science?", *Papers of the Regional Science Association*, vol. 4, pp. 7-21, 1970.

[KIN 97] KING G., *A Solution to the Ecological Inference Problem*, Princeton University Press, 1997.

[OPE 79] OPENSHAW S., TAYLOR P., "A million or so correlation coefficients: Three experiments on the modifiable area unit problem", in: N. WRIGLEY (ed.), *Statistical Applications in the Spatial Sciences*, London, Pion, pp. 127-144, 1979.

[OPE 84] OPENSHAW S., "The modifiable areal unit problem", in: *Concepts and Techniques in Modern Geography*, Norwich, Geobooks, 1984.

[ROB 50] ROBINSON W.S., "Ecological correlation and the behavior of individuals", *American Sociological Review*, vol. 15, pp. 351-357, 1950.

[ROS 74] ROSEN S., "Hedonic prices and implicit markets: Product differentiation in pure competition", *Journal of Political Economy*, vol. 82, pp. 34-55, 1974.

[THI 00] THILL J.C. (ed.), *Geographic Information Systems in Transportation Research*, Amsterdam, Pergamon, 2000.

Chapter 1

The Role of Mobility in the Building of Metropolitan Polycentrism

1.1. Introduction

With the rapid evolution of transportation and communication networks, the distances, areas and volumes of the population and activity at stake in the process of urban sprawl have been growing by leaps and bounds for 40 years, in contrast to the simpler growth processes of the largest European metropolises before that time. In response to the necessity of developing network functioning between the different parts of metropolitan spaces, increasingly complex mobility patterns have emerged and new forms of centrality have appeared.

We may wonder about the articulations between the new forms of mobility and the characteristics of metropolitan centralities that are emerging. It is a question of showing how the ties between urban structures and mobility behavior contribute to the formation of increasingly polycentric metropolitan areas, from both morphological and functional viewpoints. The hypothesis has been formulated that the organization and intensity of the flows between places of residence and work depend on the settlement pattern and on the context of urbanization; complementarily. We consider that the spatial organization of the flows enters fully into the constitution of the existing centers of activity.

Chapter written by Sandrine Berroir, Hélène Mathian, Thérèse Saint-Julien and Lena Sanders.

The comparison of the processes at work in two French areas, the Parisian and the Mediterranean metropolises, is central to our discussion. The Mediterranean metropolitan area[1] is the perfect example of an initial context of morphological polycentrism that has its roots in a *de facto* divided-up regional territory, as each locality is attracted by one and only one major center. Thus, the global level of spatial integration of that region was initially quite low. The transition towards functional polycentrism should draw it toward a model in which the relationships, particularly between the major centers, are intensified. In comparison with this, in the metropolitan area of Paris[2] we begin with an initial context of strong spatial integration that is built according to very high dependence on a single, powerful center. Therefore, its evolution towards a functional polycentrism should correspond to a re-arrangement of the directions and intensities of the links between the center, and to the construction of a less dissymmetric model of relationships. Consequently, we need to verify that similar processes of spatial integration are acting in the two metropolitan areas, and are simply adapting to the various local and regional contexts.

This study is organized into two parts, the first part being methodological in nature. It presents a method for identifying the centers that structure a metropolitan space that is meant to be generic and therefore reproducible. The stakes of such an identification exceed those of the single theoretical goal of this chapter and also address a major planning problematic, in terms of transportation, accessibility, location choices of public amenities and equity, for example. Empirical and comparative in scope, the second part of this study analyzes the links upheld between the commuting of active persons and centralities of jobs, in the two metropolitan areas. It shows how the densities and directions of the links refer, in each of the areas, to various states of completion of the territorial processes leading to metropolitan integration.

1.2. Identification of centers and sub-centers

The first step in this work consists in defining and delineating the centers and sub-centers in the two regions. This step cannot be taken without raising a certain

---

1 The French Mediterranean metropolitan area had a population of approximately 4.2 million inhabitants and 1.5 million jobs in 1999. It forms a vast urban triangle of which the peaks correspond to the urban area of Montpellier to the west, Avignon to the north and Toulon to the east.

2 The metropolitan region of Paris, which corresponds to the urban area of Paris defined by the INSEE (French National Institute of Statistics and Economic Studies), had a population of 11.2 million inhabitants and 5.1 million jobs in 1999. It is also termed Ile-de-France, often referred to as the *Francilian* region.

number of issues that are both conceptual and empirical. First, we will summarize the state of previous work before presenting the details of the steps of the method.

1.2.1. *A most widespread morphological approach*

From a theoretical point of view, the notion of a center refers to the idea of places of a certain hierarchical level, characterized by a heavy concentration of economic activity, which allows them to generate agglomeration economies. Thus, it is a question of central places that possess a strong capacity of attraction on the surroundings and which, consequently, attract jobs and services from the whole of the region they dominate. As shown by Boudeville [BOU 72], center attractiveness rests on two major characteristics: the relationships among the centers; and the hierarchy of those relationships leading to dissymmetric flows.

From an empirical point of view, the identification of those centers and sub-centers implies a set of choices, both in terms of the scales of observation and the measurement criteria. The various works that have been undertaken in this domain have always clearly highlighted the difficulty of this task, which is increasingly complex under varied, dense and contiguous urban contexts. The usual methods define centers according to a certain degree of concentration of jobs. They envision a center as a concentration of firms large enough to have a significant effect on the urban spatial structure, in terms of spatial distribution of the population, jobs and land-bid rents. The sub-centers are then located as local density peaks. Several empirical methods have been proposed to delineate them.

One method, examined in particular by McDonald [MCD 87], consists of estimating a function linking the density of jobs with distance to the center and using it for identifying sub-centers as positive residuals from the estimated gradient. The methods of adjustment have become increasingly sophisticated, as can be seen, for example, in [CRA 01].

Another common method was formulated by Giuliano and Small [GIU 91]. It combines three criteria:

– a job density higher than a given threshold (e.g. 10 jobs/acre);

– a job density higher than the surrounding places; and

– a minimum number of jobs (e.g. 10,000 jobs).

This approach has often been used with thresholds adapted to the local context, particularly in studies of American cities (for example, see McMillen [MCM 03] for the definition of 32 "sub-centers" in Chicago in 2000, or McMillen and Smith [MCM 04] for a comparison of several large American cities).

The weight of the economic function in relation to the residential population (the ratio between the number of jobs and number of workers living in the area) often complements those criteria of global concentration of jobs. Beckouche, Damette and Vire [BEC 96] defined *Francilian* centers by combining a significant volume of jobs and a high employment rate, weighted by the population of workers not working in Paris. Similarly, in recent work Huriot [HUR 03] defined employment centers in the Ile-de-France region by combining a minimum threshold of jobs (7,000) with the ratio of the number of residents to the number of jobs. This was computed at the *commune*[3] level and weighted by a reference value computed at the scale of the department.

Lastly, in most empirical applications, the measurement of the concentration rests on two main criteria: employment density and the rate of employment, considered separately or in combination. In the various cases, the methods prove to be sensitive to the choices of thresholds and size of the spatial units, which both influence the number and delineation of the centers.

These studies raise the issue of a possible generalization of a method of delineation to varied contexts of urbanization. Indeed, the variation in thresholds for the various indicators constitutes a subjective way of taking into consideration different local contexts that are relatively dense or were formerly urbanized (variation in thresholds of employment densities, as well as relatively functionally homogeneous (variation in employment rate thresholds). In some extremely heterogeneous metropolitan contexts, integrating both historic centers and suburbs or sub-centers caught up in the process of peri-urbanization, the variation in thresholds is not sufficient and authors most often rely on criteria of various types selected according to the type of urban environment. Thus, in works pertaining to the *Francilian* metropolis, taking into account overlapping activities and the contiguity of economic activity in the dense heart of the agglomeration, the centers of this sector (Paris and some *communes* of the "*Petite Couronne,*" that is, the ring closest to the center) are subject to specific treatment based on the differentiation of the economic profiles of the *communes* [BEC 96, HUR 03]. Our study does not escape those issues.

The quasi-totality of empirical research, whereby employment centers are defined according to the levels of concentration, refers to morphological dimension of polycentrism. While the method proposed in this work supports this essential dimension, it also aims at integrating a complementary and more functional

---

3 The smallest administrative entity in France is the *commune*, characterized by its very small average size in comparison with other countries' administrative grids (altogether there are 36,000 *communes* in France). *Commune* is the French equivalent of local municipality.

dimension of polycentrism that considers the capacities and scales of attraction of these centers.

The chosen method can be broken down into two steps, each of which belongs to a geographical level. The objective of the first step is to identify *kernel units* from indicators of concentration and attraction observed at the finest scale, which is that of the *commune*. The second step consists of building *multi-commune* clusters, in order to take into account the spatial proximity of certain *kernel units* in the definition of the main center (central business district) and sub-centers.

1.2.2. *Identification of kernel units*

The first step consists of identifying units that have the potential to be centers among the set of *communes* of each metropolitan area. The selection simultaneously accounts for two fundamental and complementary dimensions of centrality: concentration and attraction. These follow two measurement logics: raw and relative. Several criteria are combined in order to encompass various aspects of the centrality concept (see Figure 1.1). The selected *communes* are called kernel units.

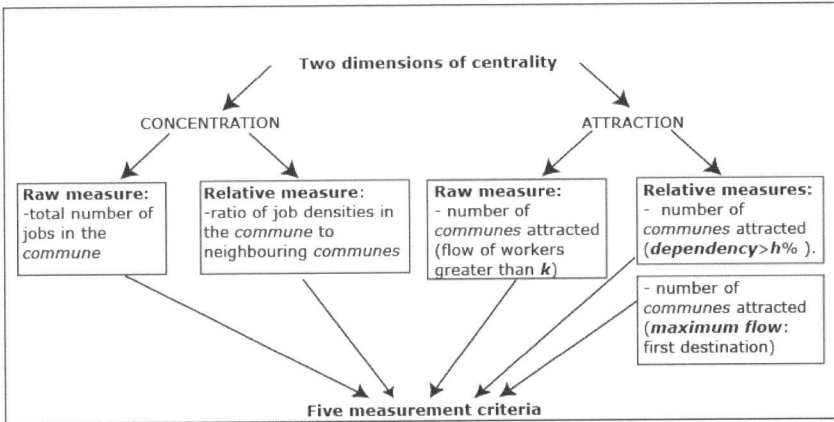

Figure 1.1. *Definition of the five criteria identifying the kernel units of a metropolitan area*

Criteria of concentration: the main criterion for grasping this concept is the gross number of jobs. However, the same number of jobs does not have the same meaning in different urban contexts. Thus, while rather trivial in the first Parisian ring, a concentration of 10,000 jobs in a city located in the distant periphery implies an influence exerted on the organization of the surrounding environment. This raw criterion has been complemented by a second criterion that describes the relative aspect of the concentration. This has been estimated on the propensity of a *commune* to generate a density peak in relation to its local environment, reflecting its capacity to structure the surrounding areas.

Criteria of attraction: the second dimension highlights mechanisms of attraction. The capacity of a *commune* to constitute a center is thus determined by the attraction it exerts on the surrounding areas, that is by the intensity and diversity of the flows that converge toward it (here, the flow of commuters). On one hand, the raw attraction capacity of each *commune* is computed based on the number of *commune*s that send a significant flow of commuters to it. On the other hand, its ability to exert a privileged attraction (highest outgoing flow) is assessed as well as the level of dependency for surrounding *commune*s.

Finally, a group of five indicators is selected, each of which highlights one or other of the properties[4] that we expect from a center. While the two phenomena of concentration and centrality clearly coincide in a certain number of cases, empirical observation shows that those different dimensions may be incompletely superimposed. The whole process consists in identifying kernel units based on the sum of their ranks based on the five criteria. Thus, the *commune*s selected correspond to those found at the top of the hierarchy for most criteria, but rarely do they dominate all of them. This method allows for the creation of compensation mechanisms between the types of logic induced by the various criteria, which include the concentration or attraction of raw or relative effects.

The application of this method[5] leads us to identify 149 kernel units in the Parisian metropolitan area and 66 in the Mediterranean metropolitan area (see Table 1.1). Examining their respective size illustrates the different meanings that a given number of jobs can have according to the urban context. The threshold of 30,000 jobs for the Parisian metropolitan area and 10,000 jobs for the Mediterranean metropolitan area characterize the respective levels of the urban hierarchy for which the various criteria converge sufficiently so that there is no ambiguity when defining a *commune* as a center. In contrast, below that, there is a diversity of cases:

– *commune*s concentrating a significant number of jobs but without having much influence on their environments (essentially in the first ring of the Parisian suburbs, e.g. Boulogne); and

---

4 The employment rate was not taken into consideration in this application, because in the French metropolitan areas the economic and residential spaces often closely overlap. It is not uncommon for a *commune* with a well-developed residential function to also display a concentration of jobs typical of an employment center.
5 The values selected in this application are $k = 10$ commuters for the size of the flow and $h = 10\%$ for the degree of dependency.

– some *communes* that benefit from economic weight that is much less, but possesses a clear attraction potential at the local level (for example, Etampes or even Coulommiers in the Parisian metropolitan area).

Such properties are very unusual at the lower levels of the urban hierarchy and are thus less often taken into account, whereas despite their relatively weak economic weight, such centers play a significant role in structuring the fringes of metropolitan areas. One of the advantages of this method is to highlight these areas. After having thereby identified some kernel units that display a certain level of diversity, the constitution of the centers associated with them still remains consistent.

Number of jobs	Parisian metropolitan area		Mediterranean metropolitan area	
	Number of *communes*	Number of kernel units	Number of *communes*	Number of kernel units
<5,000	1,407	22	788	28
5,000–10,000	89	38	19	16
10,001–30,000	68	50	16	16
>30,000	39	39	5	6
Total	1,603	149	829	66

Table 1.1. *Distribution of communes and kernel units according to the number of jobs. Source: Census 1999 (INSEE, the French National Institute of Statistics and Economic Studies)*

1.2.3. *Building multi-commune clusters*

The geographical distribution of kernel units shows numerous spatial proximities. The issue of possible links between contiguous kernel units arises, which leads us to search for centers at an aggregation scale larger than *communes*. Mobility patterns reveal the existence of forces of attraction that are built up around centers comprised of a set of *communes*. These *multi-commune clusters* operate then at the *supra-communal* level, which corresponds better to their modes of functioning than the *commune* level.

This issue is crucial for developing a theoretical approach to the processes of polycentric structuring in dense metropolitan contexts. Under such an approach, the multiscalar dimensions of those structures are handled directly, without limiting the application of this approach to other fields of study, like territorial action. It is instrumental to the analysis of policies as well as practices of intercommunality, which are particularly complex in dense metropolitan regions. Lastly, it is of particular significance for the authorities responsible for the management and regulation of mobility in those areas.

In the process of building multi-commune clusters, the multiple contiguities observed lead us to wonder what relationships these contiguous *commune*s have with one another. The aim is to go beyond the *commune* limits that are not appropriate for handling commuter flows. Indeed, commuter patterns can appear as being split at the *commune* level while they would converge at a higher level. For example, an activity zone could overlap several *commune*s, as is the case around Roissy, Les Ulis or Saint-Denis in the Parisian area or around Avignon or Étang-de-Berre in the Mediterranean region. There are also known complementarity effects or associations between contiguous kernel units. Some of them have links, which find all of their meaning in the definition of higher-level attractions. These relationships may come from complementarity induced by spatial proximity (Nanterre-Rueil, Tarascon-Beaucaire, etc.), or from inter-*commune* planning policies in connection with shared economic development strategies.

1.2.4. *Aggregation criteria*

The spatial proximities of these kernel units have been investigated according to the interactions observed. The goal was to group together the kernel units that had strong interactions, thus justifying that they be considered as belonging to a single, higher-level center. Different elements underlie these interactions, among which are resemblance and complementarity. The first element concerns areas of attraction: their relative positioning allows us to formulate hypotheses on the type of relationship between various kernel units. For example, one interaction linked to a multi-*commune* activity zone involves attraction areas that either overlap or are included within one another.

Figure 1.2 provides four examples of various types of overlapping areas. The second element is based on the existence of *intense* and *symmetrical* exchanges between the contiguous kernel units. The simultaneous consideration of those two aspects, namely the *relative position of main areas of attraction and that of the exchanges between kernel units* allows us to measure both the interaction between pairs of kernel units and the result of that interaction. The methodology used is based on the successive observation of contiguous or very close kernel units[6].

---

6 Here "close" means either contiguous or located less than 7 km apart; the "broad" rule of proximity is not meant to eliminate the strong relationships between non-contiguous centers that would still function together. In the latter case, an analysis of the gap between the *commune*s is conducted in the second step.

Type of overlap	Example in the Parisian area
a) Disjoint	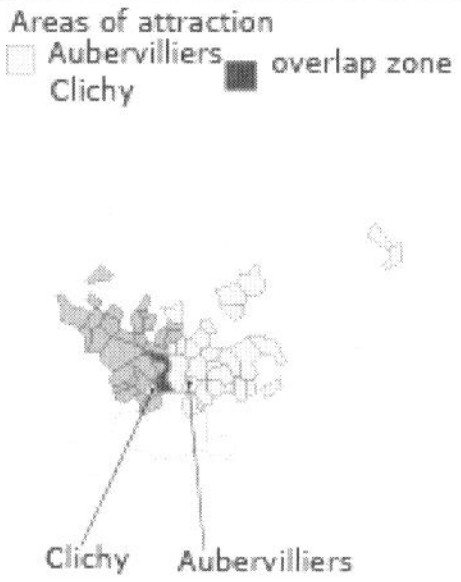
b) Intersection (partial overlap)	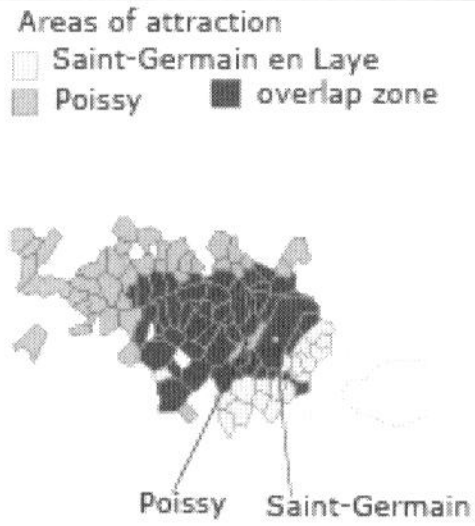
c) Inclusion	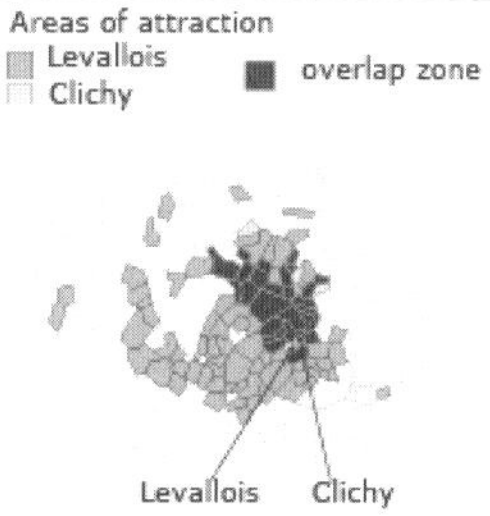
d) Spatial coincidence	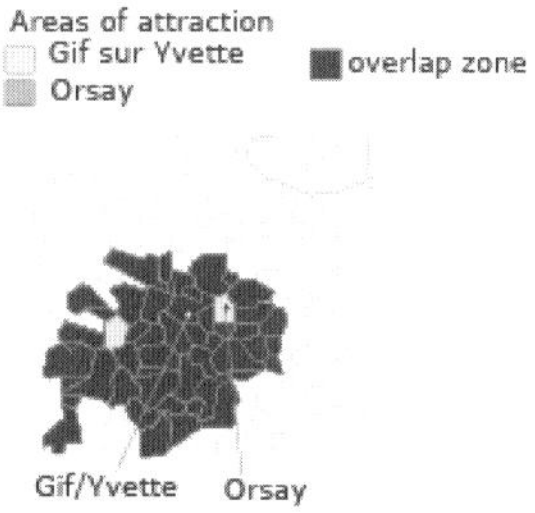

Figure 1.2. *Types of attraction area overlapping in the Paris metropolitan area*

Areas of attraction are defined on the basis of main flows originating from the *commune*s contributing to the labor market of the kernel units (flows of more than five commuters representing more than 1% of the workers living in the *commune* of origin). With this initial criterion of the *relative position of the major areas of attraction*, the pairs of kernel units with which the areas of attraction have strong intersections (cases b, c and d in Figure 1.2) can be systematically identified. A closer analysis of the attraction areas on a case-by-case basis (shapes, directions, etc.) then follows in order to check the consistency of the aggregation.

Exchanges between the kernel units allow us to identify the pairs of *close kernel units* that have intense and symmetric relationships. The aim is to identify mutual dependencies within the pairs considered. Conversely, a one-way dependency of one *commune* in relation to another would correspond to a hierarchical dependency.

1.2.5. *Aggregation of kernel units into clusters: a three-step approach*

The aggregation of the kernel units is carried out in three steps that gradually integrate them into *multi-commune clusters* of variable sizes.

– **First step:** identification of strong links between close kernel units:

- *overlapping of areas* of attraction: kernel units for which the labor markets correspond to the same attraction areas are grouped together. This is often the case, for example, in the context of a *multi-commune* zone of activity. The threshold selected is an overlap of at least 30% of the total area of attraction,

- *interaction between the kernel units*: two kernel units are associated if symmetrical exchanges of commuters can be observed. The indicators used to make this selection simultaneously take into account the relative importance of the flows originating from each of the two kernel units. The thresholds chosen follow two principles: symmetrical dependency links of at least 2% or belonging to one of the first five flows;

This first step provides a list of pairs of contiguous kernel units that are linked by *strong relationships* (i.e. that verify the two criteria simultaneously).

– **Second step:** combined together, these links generate geographical sub-sets at the higher level. The second step aims to verify the consistency of these sub-sets, considering the criteria already used in the first step. This ensures the existence of sufficient links based on either one or the other logic between the pairs of *communes* (weak links) or both logics simultaneously (strong links). This helps us to delineate consistent entities while avoiding the chain effects that gradual aggregations might bring about. A preliminary representation of *multi-commune clusters* is displayed in Figure 1.3a for Versailles and Trappes-Guyancourt (western part of the Parisian metropolitan area).

– **Third step:** finally, a systematic analysis of the neighborhoods of the preliminary multi-*commune* clusters is performed in order to identify *commune*s that, although they do not necessarily appear in the selection of the kernel units, belong to a multi-*commune* cluster based on symmetrical exchanges and overlapping of attraction areas (see Figure 1.3b). This last operation yields the final definition of multi-*commune* clusters.

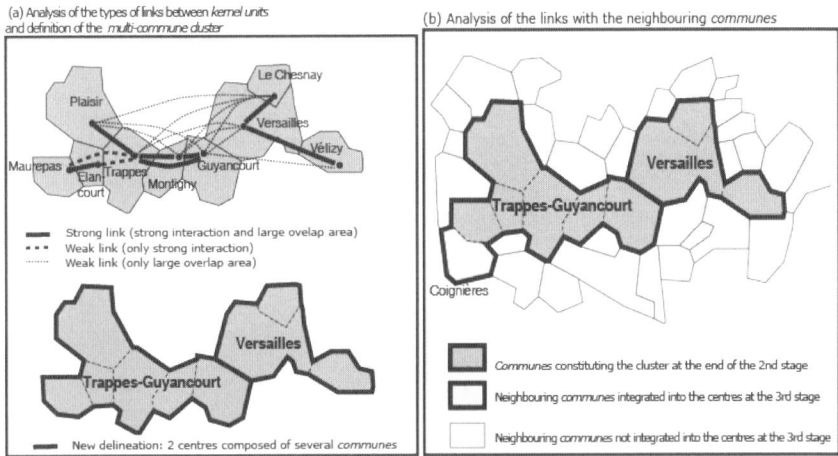

Figure 1.3. *Analysis of the weak links and neighboring links for defining multi-commune clusters*

Steps/operations	Indicators	Nature of geographical objects
(1) Delineation of metropolitan areas	- jobs - flows of commuters	- 1,600 *commune*s (Parisian area) - 800 *commune*s (Mediterranean area)
(2) Identification of kernel units	- concentration - attraction	- 149 kernel units for the Parisian metropolitan area - 66 kernel units for the Mediterranean metropolitan area
(3) Aggregation of *commune*s to build centers	- proximity - exchanges - attraction areas	- 67 multi-*commune* clusters for the Parisian metropolis (175 *commune*s) - 50 multi-*commune*-clusters for the Mediterranean metropolis (67 *commune*s)

Table 1.2. *From communes to multi-commune clusters: summary of methodology*

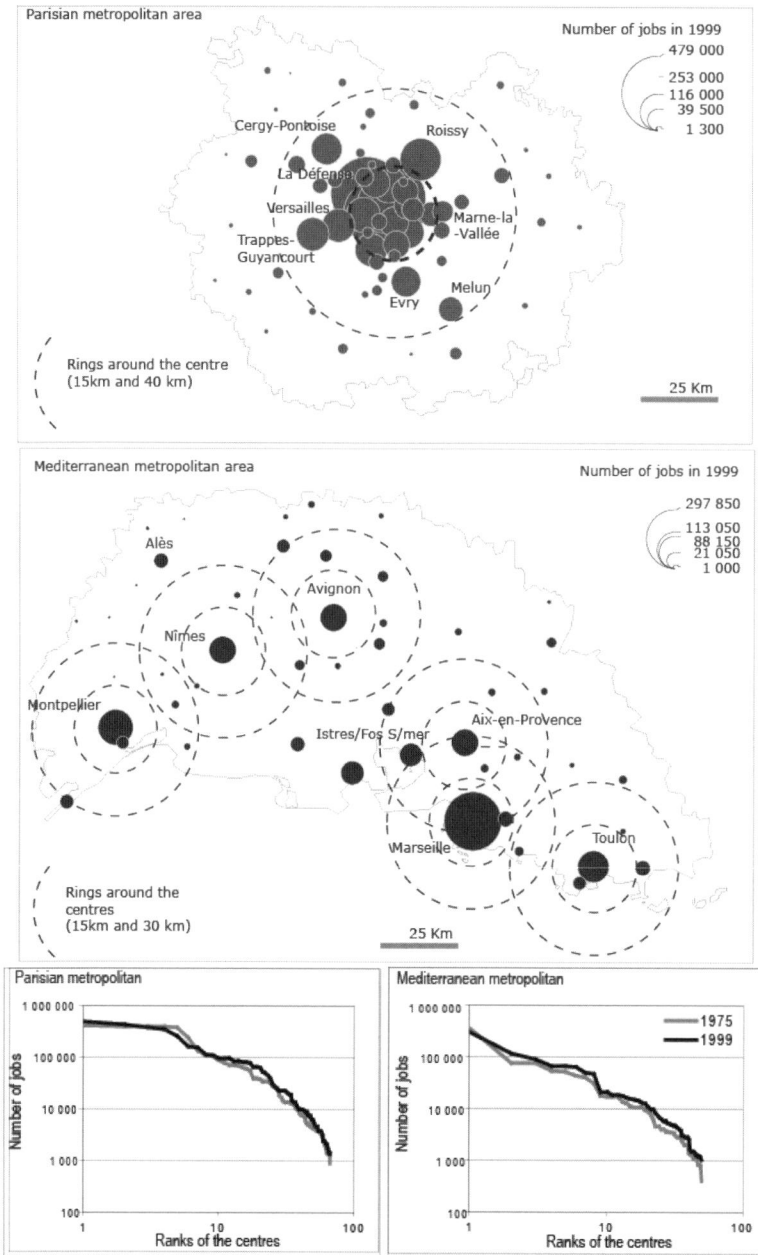

Figure 1.4. *The multi-commune employment clusters*

The key steps in the method used for defining centers are summarized in Table 1.2, while results of this *supra-communal* analysis are displayed for each of the two metropolitan regions in Figure 1.4. As can be seen, the density of centers is particularly high in the Parisian metropolis, even more so as we approach the main center. It is much weaker in the Mediterranean metropolitan area, even if the same phenomenon of densification in proximity to the largest centers also applies, although at another level. Thus in each of the two areas, main centers and sub-centers can be differentiated. The main centers of the two areas correspond to a very concentrated spatial pattern in the case of the Parisian area, with several contiguous centers, while it is very dispersed in the case of the Mediterranean area where the six main centers are regularly spaced out over the entire region.

1.3. Polycentric functioning in two metropolitan contexts

The consistency of the definition and delineation of the centers in the Parisian and Mediterranean metropolitan areas is tested through the analysis of the links between workers commuting and the centralities. In so doing, the processes that enhance polycentric metropolitan integration are identified.

1.3.1. *Morphological evolutions*

The reinforcement of morphological polycentrism translates into a weakening of the hierarchy based on the size of the centers, which clearly corresponds to an increase in the relative weight of the sub-centers in relation to the main centers. Most probably the hierarchical weakening, or at least the simplification of the hierarchies, generates lower dissymmetry in the interdependencies between the centers. Starting with initial settlement patterns that are morphologically quite different, we now show that the two metropolitan areas are clearly engaged in a similar process of polycentric evolution.

In the two metropolitan areas, one very distinct threshold separates the main centers from the other sub-centers whose size decreases, first slowly then more rapidly. In addition, whereas in the two regions the average size of the main centers is substantially higher than that of the first ring centers (with a ratio of 14:1 in the Mediterranean area and close to 7:1 in the Parisian area), the tendency of the centers to decrease in size with increasing distance to the main center is specific to the Parisian area, where the gradient linked to strong monocentrism is only weakly influenced by the peripheral centers. In the Mediterranean area, the gradients around the dispersed main centers are relatively weak. In the spaces lying between these centers, sub-centers are mostly similar in size, as the central attraction is weak (see Table 1.3).

Rings around the main centers[7]	Mediterranean metropolitan area				Parisian metropolitan area			
	No. of centers in 1999	Average number of jobs 1975	Average number of jobs 1999	Rate of variation 1975–1999 (%)	o. of centers in 1999	Average number of jobs 1975	Average number of jobs 1999	Rate of variation 1975–1999 (%)
Main centers	6	106,254	115,848	9	6	365,599	339,190	-7
First ring	6	7,098	10,753	51	16	54,402	53,337	-2
Second ring	21	7,690	10,396	35	22	22,368	38,821	74
Third ring	17	6,200	7,524	21	23	6,244	7,992	28
Total	50				67			

Table 1.3. *Size of the sub-centers according to their relative position toward the main centers*

Indeed, a period of 15 years may not be long enough to adjust these two models. Yet the regional tendencies are generally convergent. On one hand, in both regions the distributions of center sizes according to their rank have a more curved shape in 1999 (see Figure 1.4). On the other hand, as attested by the rates of variation in the average size of centers according to their distance from the main centers, the process of urban sprawl that underlies the emergence of more polycentric structures is at work in both metropolises. Much more advanced in the Parisian metropolitan region, the process there takes the form of a slight decrease in the size of the main centers and those of the first ring. Inversely, center size displays a very strong growth in the second ring, while it is much less pronounced in the third ring. During the 1975–1999 period, besides the fact that the main centers of the Mediterranean area have again recorded a slight growth, the highest rates of growth occur in the centers of the first ring; rates then decrease with distance, although they still maintain quite high levels. In the third ring, the observed growth rates are even closer to those recorded in the Parisian area, considering a comparable peripheral location.

1.3.2. *Evolving mobility: from local to metropolitan integration*

A diversity of mobility patterns based on the relative positions of centers within the metropolitan area is associated with the morphological diversity that characterizes the centers of the two metropolitan areas. A first way to apprehend this

---

7 In relation to the main centers, the rings outline the spaces included between 0 and 15 km (first), 15–40 km (second) and beyond 40 km (third) for the Parisian metropolitan area; in the Mediterranean area, the second and third rings encompass spaces within 15–30 km and beyond 30 km, respectively.

diversity and its evolution from 1975–1999 is to jointly analyze the rates of employment[8] and the indexes of turbulence[9] of the centers. A second way consists of analyzing the attraction patterns of the centers based on the spatial configuration of the *communes* from which the commuters who work there originate.

1.3.2.1. Autonomy and turbulence in a dynamic of polycentric metropolitan integration

From a theoretical point of view, two diametrically opposite models may be distinguished:

– *The "autonomous" center model*, whereby all of the jobs are filled by workers living in the center while, reciprocally, all active center residents have a job there. This theoretical situation corresponds to an employment rate of one and a turbulence index of zero.

– *The "turbulent" center model*, whereby all of the jobs are filled by workers living outside the center while all workers living in the center work elsewhere. Such a case reflects the total absence of local adequacy between employment supply and demand. The employment rate corresponding to this theoretical situation may well be high or low but the turbulence is at its maximum rate and must be higher than one.[10]

The two metropolitan areas under analysis provide examples of each of these extreme cases. The "autonomous"-type centers were quite common in the Mediterranean metropolis in 1975 (all the turbulence indexes there are lower than one, with the minimum standing at 0.14 in Marseille). At the other extreme, we find centers very close to the "turbulent" type in the Parisian metropolis in 1999 (all the turbulence indexes stand between one and two). In the case of centers based on the "autonomous" model, on average seven jobs out of 10 are filled by "residents", whereas the ratio is barely one out of three for the alternative model.

By magnifying this process, we could say that the establishment of polycentric functioning is accompanied by the shift from the first to the second model, each center becoming a potential place of employment for any worker, regardless of where the worker lives in the metropolis. Territorial cohesion is then expressed according to the scale of the metropolis and no longer solely at the local level.

---

8 Here, the employment rate is calculated as the rate between the number of jobs in the center and the number of workers living there.

9 The turbulence index is measured as the ratio of the total number of commuters (sum of the incoming and outgoing commuters) to the number of jobs in the center.

10 It will in fact be equal to the inverse of the employment rate plus one, the number of commuters entering being equal to the number of jobs and the number of commuters leaving being equal to the number of resident workers.

Given the constraints imposed by metropolitan transportation, the logics of proximity remain preponderant today, and such a model is still in its infancy. Here, the idea is to use these extreme patterns, interpreted as the limit cases, in order to describe the position and evolution of the two metropolitan areas under study. The centers of these two areas record, in fact, parallel evolutions from the first ("autonomous") to the second ("turbulent") type, but starting from different initial positions and evolving at different paces (see Table 1.4).

Rings around centers	Parisian metropolitan area					Mediterranean metropolitan area				
	No. of centers	Turbulence index		Employment rate (%)		No. of centers	Turbulence index		Employment rate (%)	
		1975	1999	1975	1999		1975	1999	1975	1999
Main centers	6	1.1	1.0	3.1	3.1	6	0.3	0.6	1.1	1.4
First ring	16	1.5	1.5	1.1	1.1	6	0.7	1.1	1.1	1.4
Second ring	22	1.3	1.4	0.9	1.0	21	0.5	0.9	2.0	2.0
Third ring	23	0.7	1.2	1.1	1.1	17	0.5	0.8	1.1	1.3

Table 1.4. *Rates of employment and levels of turbulence in the centers according to location*

The employment rates are stable in the Parisian metropolis, whereas they are increasing in the Mediterranean metropolis. The turbulence indexes are the highest for the centers of the first ring around the main centers, in both the Mediterranean and Parisian metropolises, while they are weakest for centers located in the distant periphery. Thus, the inner-ring centers stand out with a more "open" job market. They escape the hierarchical logics and are inscribed into a pattern of overlapping flows in various directions. Finally, it can be observed that the level of turbulence increases for all Mediterranean centers (from main centers to remote sub-centers), but only in those of the most distant ring for the Parisian metropolis. The centers of the first Parisian ring have evolved little between 1975 and 1999, as if a balance has been reached.

The result is that, in 1999, the Mediterranean main centers came close to the Parisian centers, although they did not reach the same level of turbulence. In all, beginning with different initial situations, the two metropolises underwent the same evolution with a reinforcement of turbulence. This was particularly heavy for all centers in the Mediterranean metropolis and for the centers of the more distant rings in the Parisian metropolis. We easily perceive the effects of a diffusion of this phenomenon from the inner rings to the most distant rings.

1.3.2.2. *Attraction patterns vary according to urban contexts*

Here again, two opposite models are useful for describing the nature of the attraction area of a center. There is a "classical" model in which each center attracts the workers from surrounding areas over a range that is reasonably proportionate to its size. The attraction areas of the whole set of centers thus induce a partition of the space. The inverse scheme, which could be referred to as "integrated," corresponds to the case in which each center attracts commuters from the entire metropolitan area. It is hypothesized that during the process of metropolitan integration, the commuting patterns evolve from the first model to the second. In order to compare the actual situations in the two metropolitan areas with these two extreme models, the raw and relative attraction[11] of the centers according to their relatively central or peripheral position within the metropolitan area have been computed and compared (see Table 1.5).

The attraction areas of the centers in the two metropolitan areas reveal a complex organization. In a certain way, each of them refers to the "classical" and "integrated" models, according to the measurement level selected. Thus, the set of *communes* that sends at least 10% of their workers to one center forms, in general, a concentric and compact area around that center. On the other hand, if the whole set of *communes* that send out a flow of at least 10 commuters is considered, the image of a very large, although split and discontinuous, attraction area is obtained (see Figure 1.5). Those two measurement levels are only partially correlated: thus, a center that exerts a raw attraction on a high number of *communes* (e.g. 300 in the case of the Mediterranean centers of Alès and Fos-sur-Mer Istres) may display a dependency area that is extensive (82 *communes* for Alès) or weak (13 *communes* for Fos-sur-Mer Istres) and *vice versa*[12]. The former is more prevalent in the peripheral rings, where low-density urban environments are found; while the latter dominates in inner rings, in a dense urban context. The comparison of four centers of the Parisian area – namely Orly, Versailles, Saint-Quentin-en-Yvelines and Cergy-Pontoise – suggests that other differentiating factors exist. With a similar hierarchical level, the area of raw attraction of each of the centers covers two-thirds of the metropolitan space, although their areas of relative attraction encompass five, 31, 107 and 152 *communes*, respectively. This illustrates the exceptional capacity of "new towns" (Saint-Quentin-en-Yvelines and Cergy-Pontoise) to structure wide proximity areas.

11 The reader is reminded that raw attraction is measured by the total number of *communes* that produce a flow of commuters toward the center which is at least equal to 10, while relative attraction is measured by the number of dependent *communes*, inasmuch as 10% of its workers work in the center.

12 The example of Orly and Cergy-Pontoise in the Parisian metropolis is illustrative: both have a comparable hierarchical level, and they both have an area of raw attraction that includes approximately 1,000 *communes*. The area of dependency of the former is limited (five *communes*), however, whereas that of the latter is among the most significant in the metropolis (152 *communes*).

The extent of those two types of attraction areas in increasing everywhere (see Table 1.5) with a single notable exception: the centers of the first ring in the Parisian metropolis. Indeed, whereas raw attraction is reinforced with increasingly wide areas of influence being integrated, relative attraction decreases reflecting the emergence of split and overlapping areas of attraction. Such a decrease reveals some dismantling of the local areas of influence to the benefit of a multiplication in commuting destinations, thereby reflecting functional integration at a higher territorial scale.

Metropolitan area of Paris							
Rings around main centers	No. of centers	No. of communes sending >3 commuters 1975	No. of communes sending >3 commuters 1999	Average distance between dependent commune and center 1975	Average distance between dependent commune and center 1999	No. of dependent communes 1975	No. of dependent communes 1999
Centers	6	828	1,373	33.4	41.1	72.7	32.7
0–15 km	16	343	849	21.3	33.1	6.0	2.7
15–40 km	22	225	671	20.2	31.0	16.6	30.7
>40 km	23	91	300	21.9	30.5	19.2	22.5
Mediterranean area							
Rings around main centers	No. of centers	No. of communes sending >3 commuters 1975	No. of communes sending >3 commuters 1999	Average distance between dependent commune and center 1975	Average distance between dependent commune and center 1999	No. of dependent communes 1975	No. of dependent communes 1999
Centers	6	177	455	43.0	65.0	45.0	76.0
0–15 km	6	37	138	25.0	41.0	2.5	3.8
15–30 km	21	50	199	25.0	38.0	8.0	12.2
>30 km	17	42	146	25.0	35.0	10.0	15.2

Table 1.5. *Various measurements of the sprawl of the centers' areas of attraction*

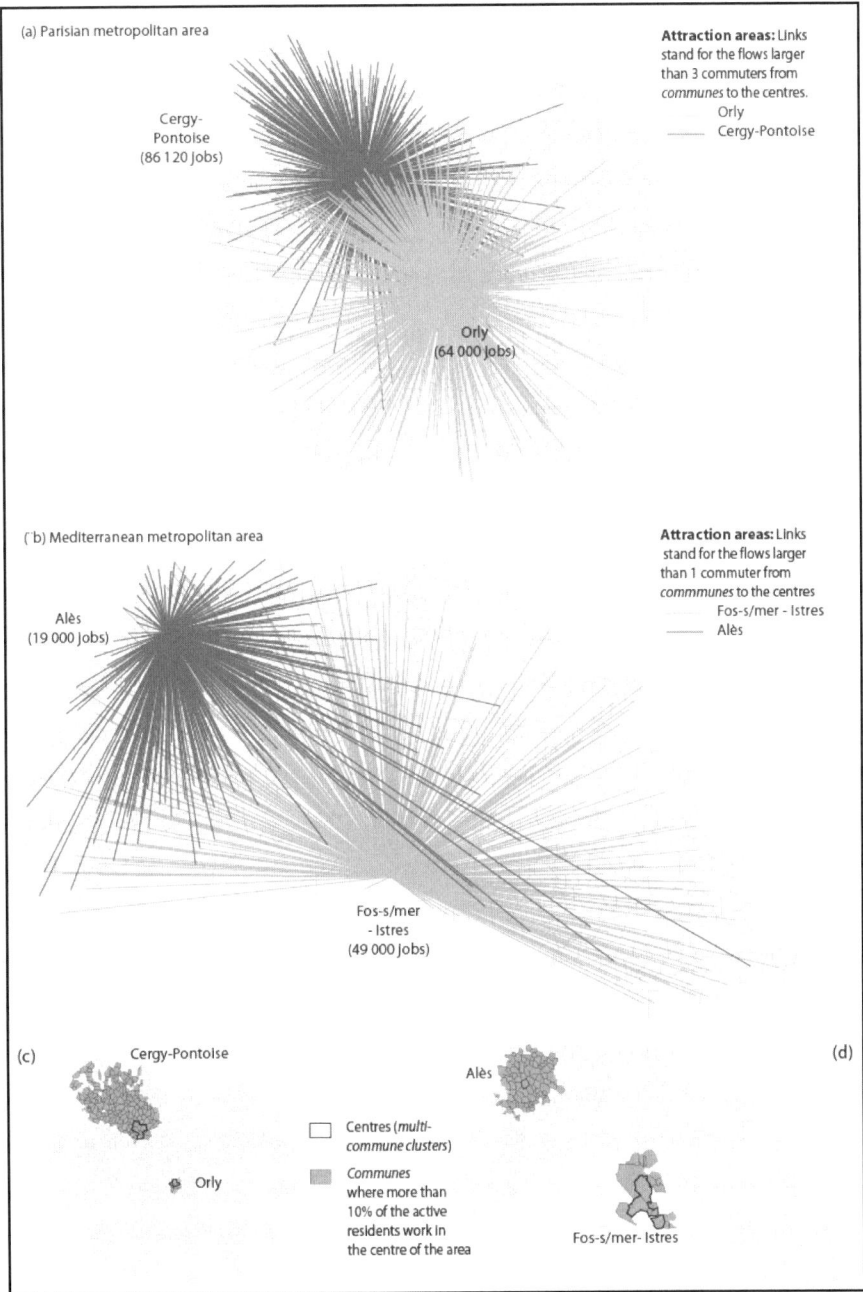

(a) Parisian metropolitan area

Attraction areas: Links stand for the flows larger than 3 commuters from *communes* to the centres.
Orly
Cergy-Pontoise

Cergy-Pontoise
(86 120 jobs)

Orly
(64 000 jobs)

('b) Mediterranean metropolitan area

Attraction areas: Links stand for the flows larger than 1 commuter from *commmunes* to the centres
Fos-s/mer - Istres
Alès

Alès
(19 000 jobs)

Fos-s/mer
- Istres
(49 000 jobs)

(c) Cergy-Pontoise

Alès

(d)

Orly

Centres (*multi-commune clusters*)

Communes where more than 10% of the active residents work in the centre of the area

Fos/s/mer- Istres

Figure 1.5. *Two measurement levels of the area of attraction of a center*

1.3.3. *Pace of metropolitan integration*

The hypothesis is that polycentric metropolitan integration combines two processes. The first is the process of concentration/dispersal, which contributes to the reinforcement of morphological polycentrism. It refers to a morphological evolution of the spatial structures, which is, as previously shown, clearly at work in the two regions between 1975 and 1999. The second process corresponds to the establishment of a network operating among centers. On one hand, it is brought out through a strengthening of the flows of commuters on several levels, in particular between centers of the same hierarchical level, or between centers which, although operating at different hierarchical levels, belong to traditionally separate areas of attraction. To this increase in the multidirectionality of the flows at all levels of a metropolitan region that enhance the links and channels of interaction between places, a lessening of the dissymmetry of the exchanges is added. This confirms the emergence of a shift away from integration based on quite hierarchical models, towards urban forms based on increasingly polycentric models.

In order to validate these hypotheses, it is considered that at the intra-metropolitan level the evolutions of directions of flows emitted by the various centers are good indicators of this shift.

The evolutions in the distributions of commuter flows from the *commune*s of origin are very convergent from one metropolitan area to another in terms of main directions. These directions show the traces of network dynamics and the emergence of intra-metropolitan polycentrism (see Figure 1.6).

The rise in the multidirectionality of the flows is clear in the two regions. First, it takes the form of a decrease in internal flows to each of the centers, from 63% to 51% in the Mediterranean area and from 13% to 11% in the Parisian area, where the center expansion process is both older and much more advanced. Second, it increases the flows linking any sub-center with the sub-centers from another ring in the same main center area. Third, it is obvious in the reinforcement of the links between centers in the area and *commune*s of the hinterland, as this part was multiplied by two in the Mediterranean area and by 1.4 in the Parisian area. Beyond that, the increase in multidirectionality of flows and the slow erosion of their dissymmetry is more a function of the initial form of the region and stage of the polycentric integration process through the establishment of a commuting network.

In the Parisian area, this phenomenon is accompanied by a relative weakening of the flows directed towards the main centers of the inner ring, which had become relatively less attractive, in favor of the other centers of the same ring. In the Mediterranean area, two tendencies seem to emerge. The first is based on the densification of links within each main center's attraction area. On one hand, there is an increase in the flows linking the different centers of the area to its main center, while on the other a relative intensification of links between the centers of the same ring is emerging. The second tendency is reflected in the rise in interactions between the various areas of that metropolitan space, which was quite compartmentalized in the past, under the then-prevailing regional logic that used to keep these centers relatively autonomous from each other. Thus, the establishment of network dynamics implies an increase in the relative weight of commuting flows between:

– the centers and the main centers of areas other than their own;

– the centers and sub-centers attached to other main centers.

These trends are clear signs that this metropolitan area is evolving toward a spatially more integrated restructuring.

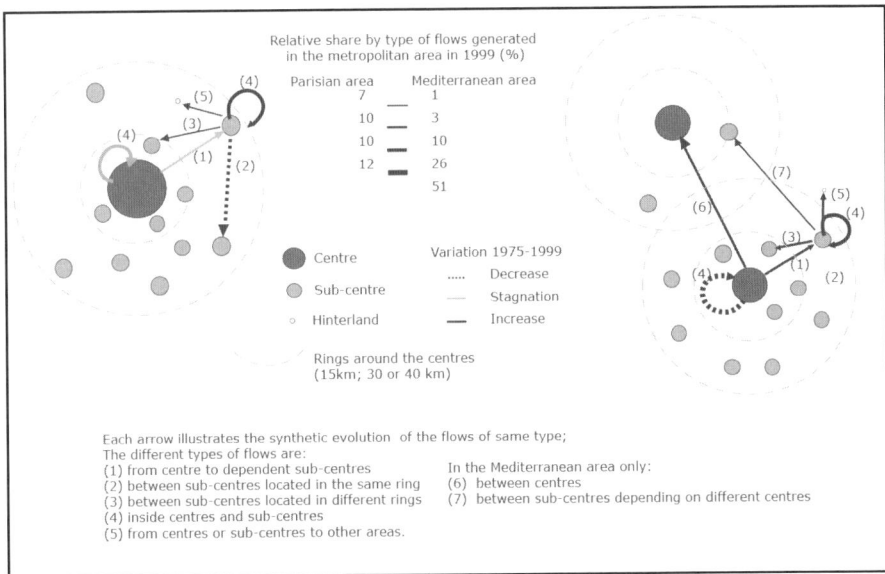

Relative share by type of flows generated
in the metropolitan area in 1999 (%)

Parisian area	Mediterranean area
7	1
10	3
10	10
12	26
	51

Centre

Sub-centre

Hinterland

Variation 1975-1999
..... Decrease
— Stagnation
— Increase

Rings around the centres
(15km; 30 or 40 km)

Each arrow illustrates the synthetic evolution of the flows of same type;
The different types of flows are:
(1) from centre to dependent sub-centres
(2) between sub-centres located in the same ring
(3) between sub-centres located in different rings
(4) inside centres and sub-centres
(5) from centres or sub-centres to other areas.

In the Mediterranean area only:
(6) between centres
(7) between sub-centres depending on different centres

Figure 1.6. *The main directions of flows originating from the centers*

The links of dependency between the various centers of each of the two metropolitan areas (see Figure 1.7) clearly show the paths towards a single integrating polycentrism. In the Paris region, polycentrism is built in the form of a star around the city center, outside of which the density of reticulation between centers increases. In numerous directions, including those of "new towns", articulations with remote centers are built step-by-step. In the Mediterranean area, the dependencies quite strongly trace connected infra-regional networks, each of which is itself centered on a polycentric structure that is supported, for example, by Montpellier, Nîmes and Alès to the west, and Avignon or even Aix-en-Provence, Marseille and Toulon to the east. Moreover, these three infra-regional systems are linked together by the dependencies of the sub-centers among themselves and the links between the main centers. These reinforcing connections are indicative of slow polycentric integration that is enveloping the whole of the metropolitan area.

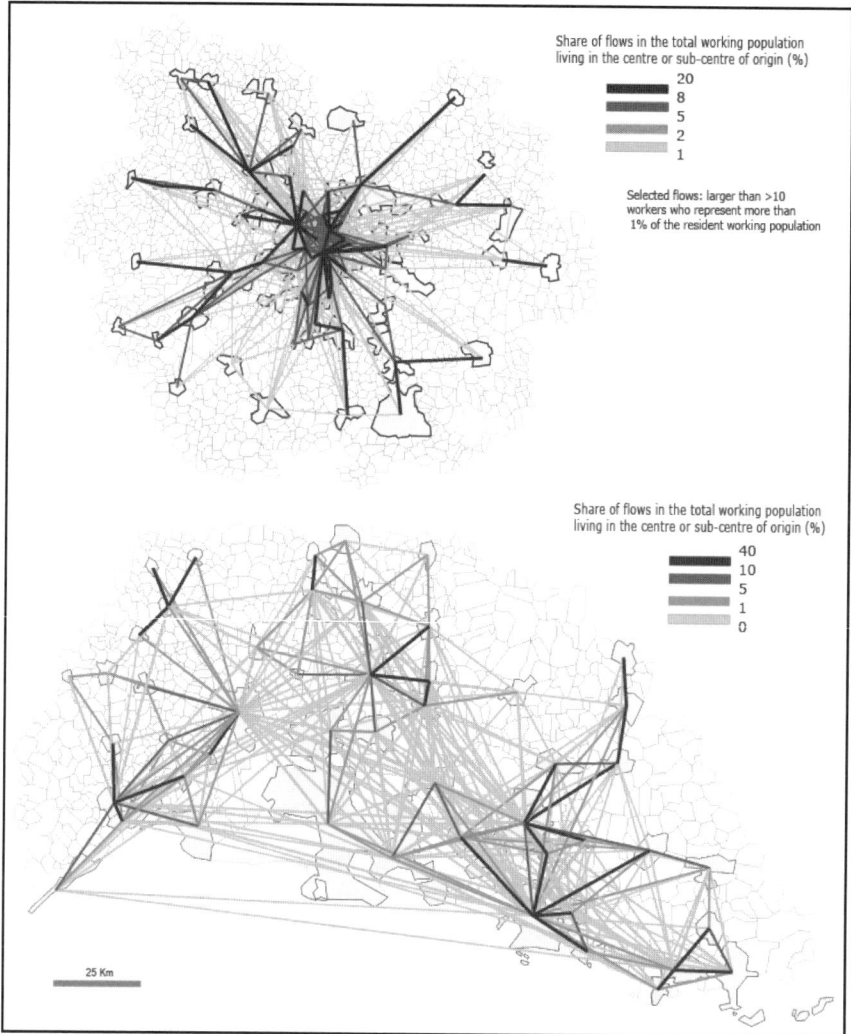

Figure 1.7. *Towards polycentric metropolitan structures: the reticulation of dependency links*

1.4. Conclusion

The contribution of this research to knowledge is at once theoretical, methodological and thematic. Polycentrism has been defined as a process of re-territorialization in that the new forms of territorial integration that it bears redefine not only the shapes of places but also their relationships; hence their functions as well.

The complex process of polycentric territorial integration plays out on several levels, as was demonstrated and emphasized through the above approach used for defining the centers. The method was developed from a generic perspective after its transferability from one metropolitan area to another was tested. This explains the focal importance given to it in this chapter.

The comparison of the processes of polycentric territorial integration at work in the two metropolitan regions has allowed us to show that a single process drives its evolution. In the Mediterranean context, where a morphological polycentrism has long been associated with strong functional compartmentalization of the region, the evolution clearly followed a path that led toward a multiscale and multidirectional strengthening of the links between centers. In particular, the relationships between the main centers are being reinforced. Although based on an initial context of a very high dependency to the main center, the metropolitan area of Paris follows a parallel path. Here, the transition takes place towards a polycentrism that is both morphological and functional. It passes through a redistribution of the directions and intensities of links between the centers, contributing to the emergence of a new model of relationships. The forces of integration have become much more multidirectional in the center, as well as over quite a wide peri-central ring, where the newest and most developed forms of functional polycentrism have appeared. In the peripheral ring, the reinforcement of the sub-centers takes the form of a greater individualization of dependency systems with regard to a sub-center in a continuous environment.

In conclusion, beyond its theoretical and methodological scope, this comparative analysis may bring about new ideas for improving territorial planning policies and strategies. Quite clearly, these processes of re-territorialization of metropolitan spaces through polycentrism now stand at the very core of debates about the development of the new planning scheme of the Ile-de-France region.

1.5. Acknowledgements

This chapter is based in part on the results of the following study: S. BERROIR, H. MATHIAN, Th. SAINT-JULIEN and L. SANDERS, 2004, *Mobilités et Polarisations:*

Vers des Métropoles Polycentriques. Le Cas des Métropoles Francilienne et Méditerranéenne, a report written for the *Plan Urbanisme Construction Architecture,* within the framework of the research program *mobilités et territoires urbains.* Details are available at: http://www.parisgeo.cnrs.fr/publications /mobilités2004.

1.6. Bibliography

[ANA 98] ANAS A., ARNOTT R., SMALL K.A., "Urban spatial structure", *Journal of Economic Literature,* vol. 36, no. 3, pp. 1426-1464, 1998.

[ASC 95] ASCHER F, *Metapolis ou l'Avenir des Villes,* Paris, Odile Jacob, 1995.

[BAU 00] BAUMONT C., LE GALLO J., "Les nouvelles centralités urbaines", in: BAUMONT C., COMBES P.P., DERYCKE P.H., JAYET H. (eds.), *Economie Géographique: Approches Théoriques et Empiriques,* Paris, Economica, 2000.

[BEC 97] BECKOUCHE P., DAMETTE F., VIRE, E, *Géographie de la Région Parisienne,* Direction Régionale de l'Équipement de l'Ile de France, 1997.

[BER 98] BERROIR S., Concentration et Polarisation. Vers une Nouvelle Organisation des Espaces Urbanisés, PhD Thesis, University of Paris I, 1998.

[BER 03] BERROIR S., Saint-JULIEN T., SANDERS L., "Spécialisation fonctionnelle et mobilité: les pôles d'emploi de l'aire urbaine de Paris", *Données Urbaines 4,* Paris, Anthropos, pp. 169-181, 2003.

[BER 07] BERROIR S., MATHIAN H., SAINT-JULIEN T., SANDERS L., "Les pôles de l'activité métropolitaine", in: SAINT-JULIEN T., LE GOIX R. (eds.), *La Métropole Parisienne. Centralités, Inégalités, Proximités,* Paris, Belin, pp. 11-38, 2007.

[BER 07] BERROIR S., MATHIAN H., SAINT-JULIEN T., SANDERS L., "Navettes et disjonction sociale dans une métropole multipolaire", in: SAINT-JULIEN T., LE GOIX R. (eds.), *La Métropole Parisienne. Centralités, Inégalités, Proximités,* Paris, Belin, pp. 89-109, 2007.

[BOU 72] BOUDEVILLE J.R, *Aménagement du Territoire et Polarisation,* Marie-Thérèse Genin, Libraries Techniques, Paris, 1972.

[CER 97] CERVERO R., WU K.L., "Polycentrism, commuting and residential location in the San Francisco Bay area", *Environment and Planning A,* vol. 29, pp. 865-886, 1997.

[COF 01] COFFEY W.J., SHEARMUR R., "The identification of employment centers in Canadian metropolitan areas: The example of Montreal 1996", *The Canadian Geographer,* vol. 45, no. 3, pp. 371-386, 2001.

[CRA 01] CRAIG S.G., PIN T.N., "Using quantile smoothing splines to identify employment subcenters in a multicentric urban area", *Journal of Urban Economics,* vol. 49, p. 100-120, 2001.

[FOU 98] FOUCHIER V, "Le polycentrisme: du concept au concret", *Urbanisme,* no. 301, pp. 53-59, 1998.

[GAR 91] GARREAU J., *Edge City: Life on the New Frontier*, New York, Anchor Books, Doubleday, 1991.

[GAS 02] GASCHET F., LACOUR C., "Métropolisation, center et centralité", *Revue d'Economie Régionale et Urbaine,* no. 1, pp. 49-72, 2002.

[GIU 91] GIULIANO G., SMALL K.A., "Subcenters in the Los Angeles region", *Regional Science and Urban Economics*, vol. 21, pp. 163-182, 1991.

[GLA 01] GLAESER E.L., KAHN M.E., "Decentralized employment and the transformation of the American city", *Harvard Institute of Economic Research Paper*, no. 1912, 2001.

[GUE 00] GUÉROIS M., LE GOIX R., "La multipolarité dans les espaces métropolitains: Paris, Lyon, Marseille et Lille", *Données Urbaines 3*, Paris, Anthropos, pp. 235-250, 2000.

[HAL 04] HALBERT L., "The decentralization of intra-metropolitan business services in the Paris region: Patterns, interpretation, consequences", *Economic Geography*, vol. 80, pp. 381-404, 2004.

[HUR 03] HURIOT J.M. (ed.), Services aux Entreprises et Nouvelles Centralités Urbaines, Rapport Final, DARES-PUCA, 2003.

[IAU 03] IAURIF-INSEE, *Atlas des Franciliens*, Volume 4, INSEE, Paris, 2003.

[MCD 87] McDONALD J.F., "The identification of urban employment subcenters", *Journal of Urban Economics*, vol. 21, no. 2, pp. 242-258, 1987.

[MCM 03] McMILLEN D.P., "Employment subcenters in Chicago: past, present and future", *Economic Perspectives*, QII, pp. 2-14, 2003.

[MCM 04] McMILLEN D.P., SMITH S., "The number of subcenters in large urban areas", *Journal of Urban Economics*, vol. 53, pp. 321-338, 2004.

[SCH 01] SCHWANEN T., DIELMAN F.M., DIJST M., "Travel behaviour in Dutch monocentric and polycentric urban systems", *Journal of Transport Geography*, vol. 9, pp. 173-186, 2001.

Chapter 2

Commuting and Gender:
Two Cities, One Reality?

2.1. Commuting, gender and urban dynamics

Many studies in both North America and Europe have focused on the analysis of urban mobility behavior (see Chapters 1, 3, 4 and 6). These studies have focused primarily on travel from home to the workplace (commuting) for various reasons. First, this type of travel represents approximately one-third of all daily travel and is concentrated in time (rush hours), with the resulting well-known urban congestion. Second, people tend to travel farther to go to work than for any other reason [HAN 04]. The increase in daily travel distances is thus closely linked to peri-urbanization [SUL 07] and, according to socioeconomic contexts, to the difficulty in finding employment.

This heavy and highly concentrated demand on transportation has resulted in household travel surveys in many large cities, in order to better understand and accommodate this demand. These surveys have provided detailed analyses of the commuting behavior and resulted in a vast array of scientific literature. There is a growing interest on the part of geographers introduced to the issues of mobility, since the studies on time geography conducted by Hägerstrand in 1970 [HÄG 70].

Chapter written by Marie-Hélène VANDERSMISSEN, Isabelle THOMAS and Ann VERHETSEL.

Commuting is probably the human activity that best establishes the spatial and symbolic distinction between the public and private domains, this distinction being a typical paradigm in Western urbanism. Commutes link production and reproduction locations, the workplace and home. Home and work are not only considered as two spatially-distinct spheres, but also as being differentiated by gender: women (home, private and reproductive space) versus men (work, public and production space). Thus, the first women's urban studies quickly criticized the underlying principles of transportation planning that hid the differential relationship between gender and space, particularly in the context of women's increased participation in the labor market and its implications on transportation planning [GIU 79, ROS 78]. Studies on commuting conducted in the 1970s identified major and significant differences between genders, relating to distances, durations and transportation modes [AND 78, BRO 02, FAG 83, HAN 81, PAS 84][1].

The past few decades have witnessed much social and economic change that has transformed urban cities in the Western world: women's increased participation in the labor market, tertiarization of the economy, social polarization and the relocation of work places [FUJ 99, ROS 93]. During this same period, social relationships between men and women changed dramatically. This can be seen as the primary trigger for change in the labor market, in the structure of households and home locations that characterize most North American cities. In this changing social environment, daily travel underwent considerable change, especially as far as women are concerned.

As mentioned by Coutras [COU 96], mobility does not occur in a neutral space with no significance. In every urban agglomeration, we can identify social groups that effectively control their mobility and others that do not. Mobility has thus become an issue in reaching equality between the sexes. In fact, mobility is essential in daily life, as it is means combining both domestic and professional activities [COU 93], in spite of the emergence of telecommuting and information and communication technologies (ICT). In addition to this, in the context of modern Western cities, mobility is a fundamental condition for the integration of women, just as much as for men, into the labor market. Thus, mobility reflects social and spatial admission, and we can expect that expanding their activity sphere should help women obtain better employment and higher salaries, and result in a decrease in social gender inequality, bringing about a better sharing of power in society.

Commuting, gender and urban dynamics are tightly interrelated. Commuting is a decisive form of spatial interaction in understanding urban dynamics, especially those pertaining to the accessibility of work places. The first objective of this chapter is to show how the effective processing and use of geographical information

---

1 A partial overview of these studies is presented in section 2.1.1.

can allow us to expand our understanding of urban dynamics linked to daily mobility. In order to achieve this, the methodological tools used to analyze the commutes and their fluctuations are listed. Two study areas are used in view of the type of available data for each location: Brussels (Belgium), which is a "North American-style" city with regard to residential location [BRO 02, GOF 00, THO 99]; and Québec City (Canada), which is probably the most European of all North American cities. Here, we are not trying to systematically compare both cities, but rather to show the diversity and potential of several geographical methods.

This chapter contains four sections. The first section presents a brief review of the literature on mobility differences by gender (2.1.1) and various methodological aspects (2.1.2). Section 2.2 pertains to Belgium and section 2.3 concerns Québec City. The last section (2.4) concludes the work.

2.1.1. *Commuting and gender: state of the art*

Gender differences in daily mobility have been repeatedly studied since the 1970s. In spite of the apparent homogeneity of the findings, the large diversity of sources, size and space that they refer to and the calculation methods of some indicators (distance and duration of commuting trips) must be acknowledged. However, some trends are evident. This chapter briefly introduces some of these trends, drawn essentially from the social and geographical fields. The interested reader will find additional bibliographical information of a more economical and econometric nature in [BRO 02].

Generally, women travel less frequently than men, use public transportation more often, and their commuting trips are shorter than those for men [BAC 96, COU 97, GAL 98, HAN 85, MAD 81, MCL 97, RUT 88, TUR 97, WYL 98]. The factors explaining women's lower mobility rate at the urban level reflect their position in the labor market and the constraints associated with their double role in society. Their status in the labor market would first explain the shorter trips: part-time employment and low salaries do not encourage traveling long distances to work [BLU 95, JOH 95]. Women's double role reduces their potential mobility and, indirectly, access to the labor market in order to have more time to care for children and household responsibilities. This results in women reducing their travel time by choosing to work closer to home [BAC 97, MLA 97, PRE 93]. However, the difficulty in measuring family and household responsibilities produces contradictory results with regard to the influence of children and marital status [BLU 95, BRO 02, CAM 96, JOH 92, JOH 95, MLA 97, WYL 98]. In many cases, indeed, the presence of young children has been associated with longer commutes for families living in the suburbs and on the outskirts of the city [THO 98, VAN 01].

The third explanatory factor regards women's reduced access to automobiles, given their lower salaries, the fact that fewer women have driving permits, and the tendency for men to use the family car [BLU 95, COU 97, ROS 89, SEG 97, VAN 04]. With the spatial redeployment of activity places, access to workplaces can be difficult for women with no access to automobiles, regardless of their place of residence. The spatial context also plays a role in commuting, with women's mobility depending on where the family lives and on the location of the main workplaces for women in the city.

More recent research suggests a growing similarity between mobility profiles for both men and women, especially with regard to commuting distance [VAN 07a, VER 09]. However, some differences do remain: although women's commutes are longer now than in the past, they are still shorter than those for men, from 15% across France to 25% in the urban zone of Paris [BAC 96, COU 97, GAL 98]. Similar values are observed for Belgium [VER 09]. Women still use public transportation more than men, while becoming bigger users of the personal automobile [VAN 97, VER 09]. In the Québec City area, work-residence distances increased for both men and women between 1977 and 1996, while travel time increased for men and decreased for women [VAN 01]. Thus, a number of women appear to have abandoned public transportation in favor of the automobile, resulting in reduced travel time. In 1996, however, women's commutes were still shorter than those of their male counterparts, and in 2001 the gender difference in the duration of the trip was still significant, although steady [THÉ 05]. Similar results were obtained for Belgium between 1991 and 2001 [VER 09].

2.1.2. *Some methodological issues*

The main characteristics of commuting trips are the transportation modes used and distance between the place of residence and the workplace (metric distance and time). Distances and duration of travel are unfortunately not systematically included in the databases that describe these trips. Researchers must then reconstruct commuting trips based on the geo-referencing of departure and arrival locations, which may be very precise (at the building level) or approximate (centroid of the spatial unit where work and home are located). A Euclidian approximation of distance is therefore used [BLU 90, BLU 95, LEV 94, VIL 88]. In some cases, the Euclidian distance can be corrected by a curve factor [CAM 96]. Euclidian distances can present a linear relationship rather close to true distances [PEE 00, VIL 00b], but are still approximations, and they do not allow travel times to be estimated. Time is the variable that most influences travel behavior [GOR 91, MAK 98]. Travel time estimates, whether self-declared or calculated according to departure and arrival times, are subject to much bias, including daily variations due to traffic, variations from one person to another, or according to the transportation mode used [FAG 86, GOR 91, HAN 95].

Another way to calculate commuting distance is to simulate travel on a digital road network having characteristics as similar as possible to the actual network (length of segments, maximum allowable speed, wait time at intersections, etc.). This digital modeling of travel was conducted at the *Centre de Recherche en Aménagement et Développement* (CRAD) using a geographical information system and a transportation data management system [THÉ 99] (also see Chapter 10). The simulation is far from perfect as it is based on a given number of hypotheses (for example, the shortest route between two points on a graph) and approximations (fastest legal speed versus true speed), but it also allows the duration of trips to be estimated. These approximations are more difficult when it comes to simulating commuting trips made by bus. Although average trip times can be calculated from established bus company schedules, waiting times for bus correspondence must be simulated. Walking distances/times from home to the closest bus stop (access time) must also be added to the trip time, as must egress time (walking time to destination). Lastly, the simulation process should integrate bus traffic frequency, boarding, disembarkation and transfer locations, which may be rather difficult given the fact that bus companies do not necessarily store archives on the development of their network.

The modeling process loses part of its effectiveness and accuracy when it is not applied to precise departure and destination locations but to spatial units' centroids. In spite of its imperfections, it allows us to estimate homogeneous values that are comparable from one person to another, while avoiding biases of perception and "cultural" variations inherent in self-declarations. Ideally, it would be best to have recourse to both series of values. This modeling process was used in Québec City in order to obtain an overview of the evolution of mobility between 1977 (date of the first computerized origin destination or OD study) and 1996 [VAN 01]. The relationship between the length of trips by automobile and by bus was satisfactory. The average public transportation trip times obtained with the 1977 OD survey based on the difference between departure and arrival times were also similar to the average times obtained through modeling [VIL 00a]. The modeling process developed at CRAD is currently being improved, notably with regard to the estimation process for bus commutes, in order to more faithfully reproduce operational data of the actual service (stop locations, schedule, route frequency, etc.).

Distance can also be estimated by means of the duration of the trip and speed of the travel mode [WYL 96]. Some censuses, including those in Canada and Belgium, use spatial units to compute the number of workers who work in the municipality of their residence or in another municipality. This is another way to estimate whether people work far from their residence [BAR 07, VER 09], although the diversity of the size and shape of entities may pose typical methodological problems in the quantitative spatial analysis (spatial granularity, geographical scale, MAUP, see Chapter 1).

2.2. Commuting and gender in Belgium

2.2.1. *Spatial data*

The population census is the most widely-used database for conducting a detailed spatiotemporal analysis. Whereas in the past only spatial unit means were available, in 2001 researchers were provided with individual variables. Trip distances and duration were then available for each worker; they corresponded to the average situations declared by respondents and were able to include stops or be biased by subjective perception phenomena [VER 09]. Despite this, spatial analyses show that the length of commutes is closely linked to the urban, economic and social geography, at both national and city levels [VER 09]. The advantage is that such average trip data can be used in relationship with multiple characteristics of the person and his or her dwelling.

Several alternative studies exist in Belgium (such as MOBEL and OVG-Vlaanderen), but they are often limited in size and thus in spatial representiveness. Moreover, it is difficult to compare these studies temporally and they cannot be used in terms of accurate geographical differentiations. They do, however, have the advantage of being based on trip diaries completed over a precise time period; thus, activity chaining can be traced and each stop geo-referenced. It must also be noted that biases had to be corrected, such as well-known problems with answers rounded off to multiples of five replacing precise distances [RIE 99].

2.2.2. *Assessing distance decay with survey data*

Analyzing differential distance decay in individual surveys in an exploratory and quantitative manner and according to gender raises the issue as to whether space is seen differently by men and women and, if so, whether such differences can be measured at all. For the Belgian part, our research is confined within the limits of the national MOBEL survey [HUB 02]. The interested reader will find results based on other surveys as well as methodological developments in [HAM 03]. MOBEL is based on diaries kept by 3,064 households for a weekday in 1999, including 21,096 trips made by 7,037 persons. This assessment allows us to rediscover a well-known geographic technique: measuring distance decay [FOT 81, TAY 75].

The exponential model was the best suited to analyzing and measuring distance decay (see [HAM 03] for a full theoretical justification). The model, often referred to as the distance decay function [TAY 75], takes the following form:

$$I_{ij} = ae^{-bD_{ij}}$$
[2.1]

where I_{ij} is the interaction between location i and location j, D_{ij} the distance, a and b are estimated parameters, while e is the natural logarithm base (Neper number). The value of the parameters is obtained by regression; the b parameter, called the distance gradient, measures the intensity of this interaction by distance unit and its pace of decrease with increasing distance.

Fotheringham [FOT 81] presents several empirical results that illustrate the relationships between the parameters of distance decay function and spatial patterns. Initially, as accessibility to a point of origin decreases, the absolute value of the negative parameter increases. While a constant parameter should be expected in a relatively homogeneous urban society, this is not what is actually observed [FOT 81]. It would also seem that positive estimates for this parameter are occasionally obtained. This suggests that interactions may increase with distance, which only occurs in very accessible zones. Note also that the most accessible points of origin have higher average trip lengths and high and negative parametric values. Lastly, the magnitude of the estimated parameters can also be an indication of relationship: strong, negative estimates indicate that the distance is seen as a major constraint to interaction; while it is less of a constraint where negative estimates exhibit smaller magnitudes.

Estimating this model can be done in two different ways: by using nonlinear estimation techniques *such as* maximum likelihood, the nonlinear least-squares method, or by the ordinary least-squares method with a logarithmic transformation in order to obtain a classic linear model.

Hammadou *et al.* [HAM 03] investigate whether the distance decay parameter b, varies significantly with the characteristics of trips and tour chaining, the trip purpose, transportation modes, household composition, type of urbanization of the place of residence and that of the destination, etc. A maximum likelihood test is used.

The difference between men and women proved to be statistically significant: distance decay is higher for women (see Table 2.1), suggesting that men and women react differently in space. It must be noted, however, that other household characteristics interact here (family composition, socioeconomic status, etc.).

Interested readers should refer to the full paper for more details [HAM 03]. The method proved to be very useful in quantifying differences in spatial practices.

Sex	B	T	R^2
Men	0.146	18.44	0.69
Women	0.181	24.21	0.75

Table 2.1. *Distance decay parameters for men and women. Source : [HAM 03]*

2.2.3. *A model for Brussels based on the 1991 census*

In 1991, population census data were available exhaustively in the form of averages by spatial entity (statistical areas, municipalities (*communes*[2])) throughout Belgium, and also in the form of a sample of individual values drawn randomly. The sample allows for detailed econometric, although not spatial, analyses. Broze *et al.* [BRO 02] proposed an exploratory analysis of Brussels based on these data.

It must be remembered that economic literature has shed a different light on discrimination issues in the labor market and/or urban segregation [BRO 02]. In all models proposed, however, the direct link between these discrimination issues is ambiguous, except in models with poor spatial matching (*spatial mismatch*), whose objective is to explain the high rates of poverty and unemployment among racially-discriminated groups in the downtown core of American cities. The influencing mechanism is the relocation of businesses at the outskirts of cities (a well-known trend in the US). This is reinforced by the discriminatory attitude of landowners from wealthy suburbs. Segregation on the housing market front limits potential residential choices and hinders access to employment.

In European cities, patterns are often less evident than in the US. In Paris, for example, the downtown/suburb distinction is less evident due to the fact that the city has a rather polycentric structure (see Chapter 1), even if the poorest sections are found on the city's outskirts. The situation in Brussels is quite similar to the one found in American cities, since the best-paid residents live in spacious homes on the outskirts of town, where there are few economically precarious immigrants. Inversely, it is at the city center – where housing is of poorer quality – that the highest concentration of immigrants, low-wage workers and the unemployed can be found [VAN 07b]. Regarding jobs, Brussels does not fit the pattern of very large American cities in terms of polycentrism: jobs remain in the central zones [RIG 07].

---

2 In Belgium, *communes* are a close equivalent to local municipalities in the US.

By "Brussels," we mean the extended urban agglomeration – both the city and its inner suburbs [DUJ 07]. The sample analyzed includes approximately 20,000 citizens living in 9,000 households, drawn randomly. Sub-populations of men and women were created and put in perspective with other forms of segregation. A descriptive analysis shows the importance of each discriminatory variable in table form [BRO 02], and includes a tentative multivariate explanatory model. It confirms observations often made for other cities: 55% of the sample's households are couples (married or not), as opposed to 9% for single-parent family households. Among the latter, 83% are comprised of a single woman with children. As expected, "traditional" families live further towards the outskirts than single-parent families or singles (one-person households).

Location	Immigrants		Belgians		Total	
	Men	Women	Men	Women	Men	Women
Center	29	43	9	14	13	17
Outskirts	n.s.	n.s.	4	10	5	10
Total	29	42	7	13	10	15

Table 2.2. *Unemployment rate for Belgians and immigrants (%)*
(n.s. – less than 50 persons). Source: [BRO 02]

Table 2.2 shows that the unemployment rate for immigrants is much higher than for Belgians and that, regardless of the category chosen, women are more affected by unemployment than men. Unemployment is also higher in the city center than on the outskirts. This conforms to the theory of *spatial mismatch* [THO 99]. Women are not uniformly affected by unemployment. Single mothers and women living in the city center experience the highest rates of unemployment. Also, part-time work is more common on the outskirts than in the city center and among women living as part of a couple. This corresponds to the *spatial entrapment* of women.

In 1991, 60% of people surveyed used a car to get to work, either as the driver (90%) or passenger (10%). While 22% used public transportation, 8% walked to work and 7% used a combination of transportation modes (car and public transportation). Cars were used more by those living on the outskirts, while urban dwellers favored public transportation. Women comprised the majority of both public transportation users and car passengers, especially for combined transportation modes, which resulted in longer travel time. Men primarily drove their car. The median work-to-residence trip was 30 minutes (total trip, including stops) for all subjects, slightly more than the residence-to-work trip, with a well-defined center-periphery spatial structure regardless of gender.

A Weibull plot was traced to explain the duration of commuting (see Table 2.3). While this model is more complex to estimate than the regression model based on the maximum likelihood method, it is better adapted to analyzing travel lengths [BRO 02]. For interpretive purposes, this estimation tells us that, all other things being equal, immigrants experience shorter trip duration than native Belgians and that married people make longer trips than single people or unmarried couples. People with higher education also have shorter trip durations. Homeowners make slightly longer trips, although this trend is not very significant (there is a low propensity to move in Belgium due to the taxation system). Married women have longer trip durations than married men.

Variable	Parameter	Standard Error	Significance	p
Constant	-7.84	0.100	**	0.000
Immigrant	-0.24	0.078	**	0.002
Married[1] female	0.08	0.036	*	0.024
Married[1] male	0.06	0.032	*	0.049
Higher educated	-0.07	0.031	*	0.016
Homeowner	0.03	0.028	n.s.	0.138
b	1.87	0.021	**	0.000
LogL	-20,748.94			

Table 2.3. *Results of the Weibull model estimation on commuting duration (Variable: duration of commuting; 1married or living as a couple; significance: ** 0.001; * 0.002–0.05; n.s.: >0.05.). Source: [BRO 02]*

Of course, this model is only a first attempt, *is* several years old and isolated in context. It does however show that at the level of commuters the effects of social, economic and spatial characteristics on the journey to work. Current trends in spatial econometrics will no doubt soon lead us to continue the modeling process for Brussels and Québec City, keeping in mind the numerous possible biases, such as spatial autocorrelation or endogeneity [DUJ 06, DUJ 08].

2.2.4. *Trips to Brussels according to the 2001 census*

In 2001, data from the population census was anonymously produced; this database includes information about each household in Belgium. A detailed geographical analysis of commuting activity is to be found in [VER 09]. Some of the elements are reported here as a reminder of the virtues of the statistical map.

We know that jobs in Belgium are always concentrated in the urban centers, and that this concentration is higher than that of the population [RIG 07]. Thus, commuting is a Belgian reality confirmed by the latest census. In Belgium, only a fifth of the active population works in their *commune* of residence. On average, the length of the journey to work is 22.1 km or 28 minutes (regardless of the transportation mode); in 1991, these measures were 20.2 km and 25.8 minutes, respectively. Brussels is the primary employment center, with nearly 630,000 persons stating they work there; half of these jobs are held by people living outside the urban agglomeration.

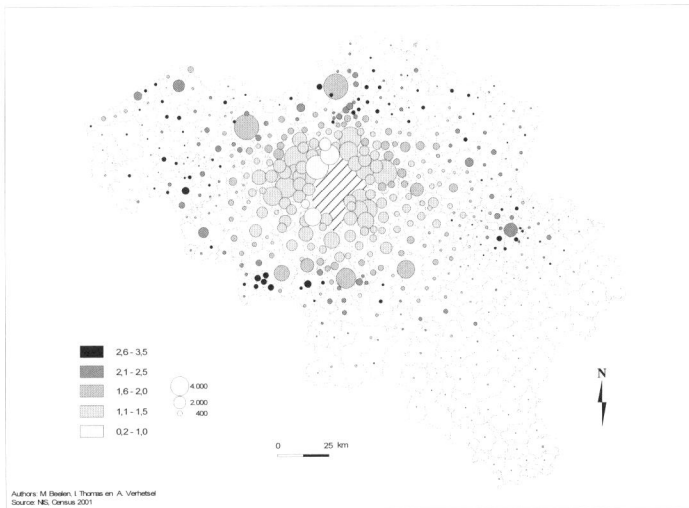

Figure 2.1. *Number of commuters to the "extended Brussels urban center" and the ratio between men and women. Source: [VER 09]*

Figure 2.1 summarizes the spatial distribution of 303,385 persons commuting to the extended Brussels urban center. Note that the morphological agglomeration used here is defined according to various international criteria [DUJ 07]. It is a symbol map wherein the area of each circle is proportional to the number of commuters to Brussels. Each circle is centered on the *commune* of residence and colored according to the ratio of the number of men to women commuting to Brussels. Where the ratio is less than 1, the number of women is larger than the number of men, and *vice versa*.

This map confirms the results obtained in section 2.1: the size of the circles decreases as the distance to Brussels increases. The regularity of the decrease is punctually ruptured by the proximity to a large-sized city, as predicted by the gravity model: the number of commuters to Brussels in a *commune* of residence *j*, is directly dependent on the number of jobs available in *j* and is inversely proportional to the distance to Brussels [DUJ 01].

The circle color provides useful information: the lighter the color, the smaller the gender ratio. This confirms the fact that women generally make shorter commuting trips (in time or distance). It should be noted that in some peri-urban municipalities (*communes*) of mid-sized cities, this ratio can be higher than 3. In other words, women working in Brussels live near Brussels and gender influences geographical mobility, the choice of employment and also that of residence. This confirms the results of prior studies [BRO 02, SCH 02, VAN 01]. The determinants of this difference are studied later in this chapter. They include: lower salaries, labor-market discrimination, the need to combine family and employment, etc.

2.3. Commuting and gender in Québec City

2.3.1. *Evolution of transport modes, trip durations and distances*

The analysis of mobility in the Québec metropolitan srea (QMA) is primarily based on the use of databanks compiled from OD surveys conducted jointly by the *Ministère des Transports du Québec* and the *Réseau de Transport de la Capitale* (RTC). These studies characterize the commutes made during a weekday by all persons in a representative sample of households living in the QMA. They allow us to obtain data on trip purpose, transportation mode, places of origin and destination, commuter and household profiles (age, gender, occupation, etc.) as well as on the resources at their disposal (e.g. number of automobiles). It must be mentioned that the QMA is an average-sized urban region (670,000 inhabitants). Like many North American cities of this size, Québec City has a modern highway network complemented by a public transportation system limited to bus service.

The first series of studies covers the period from 1977 to 1996. The duration and distances of commutes are computed (see section 2.1.2) between the centroids of the 248 public transport zones for both surveys [VAN 01, VAN 03]. A second series of studies covers the period from 1991 to 2001. The Euclidian distance is used to estimate commuting distances between postal codes [VAN 07a].

Regarding the transportation modes used, statistical tests show that there are many significant differences between men and women in both 1977 and 1996. While less than 40% of women's trips were made as the driver of a vehicle in 1977, this figure climbed to 70% in 1996. Women's access to a dual-income family's sole vehicle is clearly lower. However, in 1996 women drove the family vehicle for only 44.2% of their trips to work. Average commute times rose significantly for women between 1977 and 1996, regardless of the point of departure and transportation modes (except by bus, from the second suburban ring). Thus, average trips for men and women appear to have become similar, especially when the origin is located in the urban central zone. Commuting distances also increased for women between 1977 and 1996, but are still significantly less than for men, especially in the central zone and the first suburban ring (the reader may learn more in [VAN 01]).

	1991		2001		1991		2001	
	M	F	M	F	M	F	M	F
	A. One automobile/with children				B. One automobile/no children			
Driver	7.40	6.54**	7.16	6.46*	7.26	5.84**	6.68	6.82
Passenger	7.49	7.03	6.88	7.32	6.44	6.64	7.28	8.35
Bus	7.65	7.43	7.86	7.14	5.93	5.43	7.62	5.93**
Average distance	7.07	6.26**	6.64	6.16*	6.53	5.44**	6.15	6.13
	C. More than one automobile/with children				D. More than one automobile/no children			
Driver	8.02	7.24**	8.28	8.30	8.08	7.65	8.01	8.51*
Passenger	8.34	7.61	6.73	9.23	11.22	8.22	8.56	8.77
Bus	7.48	6.52	7.51	7.31	7.93	7.21	7.35	8.43
Average distance	7.95	7.12**	8.12	8.15	8.00	7.49	7.87	8.31

** significant difference at threshold α: 0.01; * significant difference at threshold α: 0.05

Table 2.4. *Commute distance (km) of two-worker families according to modal choice, presence of children and number of vehicles in the household in QMA. Source: [VAN 07a]*

The reduction in gender difference between 1991 and 2001 supports the idea of an improvement in female mobility. Results are more subtle when the transportation mode, the level of household motorization and the presence/absence of children in the household are controlled for (see Table 2.4). In 2001, in those households with two adults working and with children, and possessing more than one vehicle, there is no significant difference between the length of commutes made by men and women. Differences still exist, however, for two-worker households with children but only one vehicle. Moreover, trips were slightly shorter in 2001 than in 1991 for both men and women, which is not the case for households with more than one vehicle.

2.3.2. *Evolution of activity areas*

Centrographic analysis techniques were applied to job locations in order to compare activity spaces[3] for men and women and their evolution between 1991 and 2001. The centrographic analysis generates a group of indicators that permits the identification and measurement of the overall characteristics of spatial distribution of a phenomenon associated with the frequency count [KEL 81]. It corresponds to the equivalent of dispersion and central tendency measures adapted to a 2D geographical space. This technique is applied as much in economic geography [GRE 91] as it is in urban economy [THÉ 95], and even to very specific fields such as road security [LEV 95]. For a group of points (workplaces) whose geographical coordinates (x_i, y_i) and frequency counts (f_i) are known, the spatial statistics (first and second moments) are:

– the center of gravity $\left(\overline{x}_{fcg}, \overline{y}_{fcg}\right)$, or weighted average center of distribution of the workplaces, representing the barycenter of the job market:

$$\left(\overline{x}_{fcg}, \overline{y}_{fcg}\right) = \left(\frac{\displaystyle\sum_{i=1}^{n} f_i x_i}{\displaystyle\sum_{i=1}^{n} f_i}, \frac{\displaystyle\sum_{i=1}^{n} f_i y_i}{\displaystyle\sum_{i=1}^{n} f_i}\right) \qquad [2.2]$$

– the standard distance or squared-mean deviations of weighted points to the center of gravity:

$$dt_{cg} = \sqrt{\frac{\displaystyle\sum_{i=1}^{n} f_i\left(x_i - x_{cg}\right)^2 + f_i\left(y_i - y_{cg}\right)^2}{\displaystyle\sum_{i=1}^{n} f_i}} \qquad [2.3]$$

---

3 In spite of given conceptual links, our estimation of activity spaces is based on a simpler method than that used to establish "action space" (see [CHA 01, DIJ 03, KWA 04], among others).

Spatial distributions are generally anisotropic and require particular indexes to illustrate their shape. The dispersion ellipse is used to represent the uneven direction of the scatter points around the centroid. Calculating dispersion ellipses provides several indexes (direction, relative length, area); in this chapter though, only the area is considered for comparing the geographical extent of employment pools, based on frequency-weighted workplaces. Activity spaces differentiated by gender were analyzed first according to the transportation mode, but only female spaces are presented here (see Figure 2.2). The evolution of activity spaces between 1991 and 2001 reveals opposite trends according to gender. The activity space of women commuting by car (driver or passenger) increased, while that of men decreased. The inverse situation defines the activity spaces resulting from bus commutes. Male activity spaces are, however, larger than those for women, regardless of the transportation modes or year.

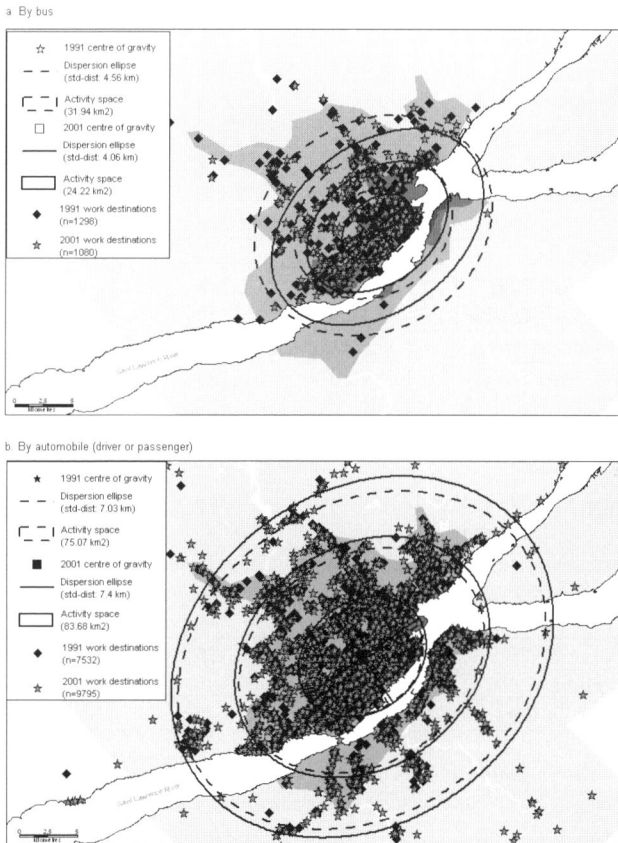

Figure 2.2. *Centrographic analysis of women's workplaces in the QMA using: a) bus; and b) automobile. Source: [VAN 07a]*

The influence of children and number of automobiles on the activity space of workers was also studied using centrographic analysis (see Table 2.5). For households with only one vehicle gender differences decreased, particularly for homes with no children. When children were present, activity spaces for men were still larger in 2001 than for women. For households with more than one vehicle, the differences were reduced: in 2001, female activity space covered the same area as that for men. This is far from the case for multivehicle homes with children: although the difference decreased between 1991 and 2001, it still remained higher than for households with children and only one vehicle.

Household type	N	Standard distance (m)	First ellipse area (km^2)
A. Household with two workers and one automobile			
Men/no children 1991	1,004	7,989	92.70
Women/no children 1991	946	5,978	53.90
Variation		*33.63%*	*71.97%*
Men/with children 1991	1,339	7,546	85.05
Women/with children 1991	1,118	5,806	52.27
Variation		*29.97%*	*62.71%*
Men/no children 2001	902	6,927	71.90
Women/no children 2001	895	6,350	61.52
Variation		*9.09%*	*16.87%*
Men/with children 2001	1,297	7,333	82.14
Women/with children 2001	1,271	6,378	62.66
Variation		*14.98%*	*31.10%*
B. Household with two workers and more than one automobile			
Men/no children 1991	1,089	9,123	121.40
Women/no children 1991	1,154	7,411	81.92
Variation		*23.10%*	*48.19%*
Men/with children 1991	1,987	8,874	115.95
Women/with children 1991	1,841	7,362	83.45
Variation		*20.54%*	*38.95%*
Men/no children 2001	1,393	7,778	93.43
Women/no children 2001	1,557	7,686	90.94
Variation		*1.19%*	*2.75%*
Men/with children 2001	2,416	8,726	116.66
Women/with children 2001	2,583	7,584	87.69
Variation		*15.05%*	*33.05%*

Table 2.5. *Centrographic analysis of workplaces reached by automobile, according to gender, presence of children and number of vehicles in QMA households. Source: [VAN 07a]*

The addition of a vehicle in households with no children reduces the difference between male and female activity spaces. When households have children, women cover shorter activity space than men, regardless of the number of vehicles in the household. Here, the addition of a vehicle allows activity spaces to be expanded, but more so for men than for women.

2.3.3. *Evolution of mobility determinants*

The methods presented in the preceding sections describe the mobility associated with residence-work trips, but tell us little about the determining factors. Managing given variables (see Table 2.4) is the first step to this end, but it does not allow us to estimate the marginal effect of a variable when other variables are held constant, which regression does. Thus, we designed cross-sectional models (see Table 2.6) in order to compare the social and spatial determinants of the duration of commutes in the Québec City metropolitan area, in 1977 and 1996, for women and men living in dual-income households [VAN 03].

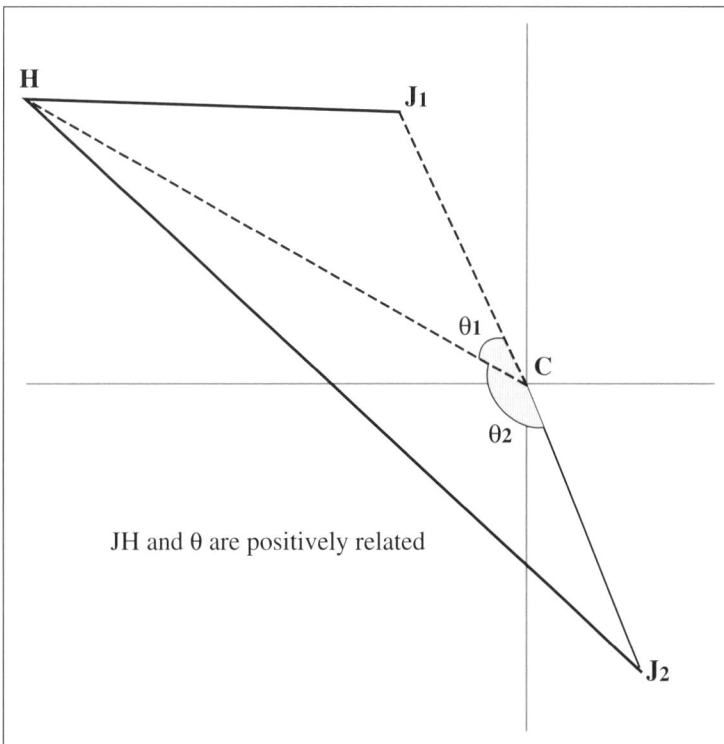

Figure 2.3. *Angles formed at the center of town (C) between workplaces (J1 and J2) and residence (H). Source: adapted from [VAN 03]*

Spatial determinants were operationalized in three 3D. The first considers the influence of the distance to the downtown central axis[4] and is expressed by two variables: the distance between the place of residence and the central axis of the city[5]; and the distance between the workplace and the central axis of the city. The second dimension concerns the estimation of a directional bias, in order to reflect the relative location of the residence and work in relationship to the central business district[6] and to test the concept of the evolution of Québec City towards an increasingly less monocentric urban structure. It is a trigonometric process that measures the angle θ formed by two straight lines: job-to-center Euclidian distance (JC) and home-to-center Euclidian distance (HC) (see Figure 2.3). If, over the years, the city evolved towards a polycentric shape, and more particularly, if a non-radial transportation network developed, the effect of θ on travel time should be reduced. For each commute, the cosine of θ is calculated and normalized (see the direction variable in Table 2.9). The third dimension of urban structure focuses more on the relative location of the residence and workplace, and can be measured by accessibility indices. Based on the job-residence equilibrium, living in a job-rich zone and working in a residence-rich zone should equate to shorter commutes [LEV 98]. Moreover, living in a residence-rich zone and working in a job-rich zone should result in longer commutes, due essentially to competitive employers in the case of jobs, and competitive workers in the case of residences. The following accessibility indices express these relationships:

$$A_{iJm} = \sum_j \left(J_j * f\left(c_{ijm} \right) \right) \tag{2.4}$$

and:

$$A_{iHm} = \sum_j \left(H_j * f\left(c_{ijm} \right) \right) \tag{2.5}$$

where:

– A_{iJm} is the accessibility to employment in zone i by transportation mode m;

– A_{iHm} is the accessibility to residences in zone i by transportation mode m;

– J_j is the number of jobs in zone j;

---

4 Preliminary analyses of these studies show that using a central axis rather than a single center point is preferable to outline the urban shape of the QMA.

5 Québec City has two main central activity areas (the historical center in the Old Town and the new central business district in Sainte-Foy); therefore we measure accessibility to the center using the distance to the central axis linking both main activity centers.

6 In this case, directional bias was calculated based on the historical center of the city, located at the north-west extremity of the central axis.

– H_j is the number of residences in zone j; and

– $f(c_{ijm})$ is a function of impedance in terms of cost (duration of trip) between zones i and j.

These accessibility indices were estimated separately according to commutes made by car and by bus. The number of residences and jobs in each travel zone were registered according to weighted OD surveys using expansion factors (sample adjustment derived from census data). Impedance functions were calculated using the clustering procedure [LEV 98] with five-minute classes. The reader will find details of these functions as well as the accessibility index chart in [VAN 03]. For a large-scale, monocentric urban region that becomes polycentric, these accessibility indices should be better predictors of the duration of commuting than the variables formulated by the first two dimensions. In smaller urban zones, maximum acceptable commuting distances can make the whole urban region accessible if both workplaces and dwellings are more markedly decentralized. The result would be a decrease between 1977 and 1996 of the accessibility indices' influence on the duration of commutes.

	C1	C2	C3
% proportion of variance	31.00	23.42	19.95
Indices loadings[*]			
AiEc	**0.535**	0.083	-0.014
AiEb	**0.788**	0.017	0.027
AjRc	0.021	0.293	**0.845**
AjRb	0.037	-0.001	**0.918**
AjEc	0.064	**0.940**	0.161
AjEb	0.078	**0.942**	0.095
AiRc	**0.889**	0.092	0.044
AiRb	**0.879**	0.094	0.049

[*] A is the accessibility index from the living place i or job location j to employment E or residences R, either by bus (b) or by automobile (c)
Figures in bold represent the highest loadings on components

Table 2.6. *Overview of the principal components analysis with the Varimax rotation of accessibility indices of the QMA in 1977. Source [VAN 03]*

In order to avoid multicollinearity issues, a principal components analysis allows us to summarize the information in three orthogonal components. The results for 1977 are presented in Table 2.6 (the general pattern is the same for 1996). Generally, zones having good accessibility by car for a given type of activity also have good accessibility by bus for the same type of activity, even if the accessibility levels differ. The first component (*C1*) measures a worker's level of access to city employment and residences by car or bus from his or her place of residence. The second component (*C2*) measures each worker's level of access to city employment from his or her place of work. The third component (*C3*) indicates the level of access of workplaces to metropolitan residences. These components will be used in the explanatory model presented in Table 2.7.

The attributes of commutes, commuters and households were then used to select control variables which, according to previous studies, have an influence on trip duration (gender, age, occupational status, motorization, size and composition of household). The transportation mode is introduced as a binary variable; it is a causal factor in the duration of commutes [TRA 78]. A cross-sectional model is calculated for each year (1977 and 1996); the coefficient of determination is identical for each model (63%). Interested readers should review [VAN 03] for a description of each model, as only the difference between 1977 and 1996 is discussed here. As expected, the transportation mode has the strongest impact on the duration of commuting. Some variables, such as the number of workers and vehicles in the household or occupational status are only significant in 1977; while others, such as the presence of children younger than six and from six to 15 years old are only significant in 1996. Also, for women there is a significant reduction in commuting duration (negative coefficient), but the effect is weaker in 1996.

Among the six variables measuring the dimensions of the urban structure (distances, direction), only one is not significant in 1977, while they all are in 1996, sometimes with unexpected signs. The distance between residence location and the central axis has a major positive and increasing influence on the commuting duration, while the distance between the workplace and the central axis has a negative effect, decreasing over time. Directional bias greatly influences the commuting duration, but as expected its effect decreases over time. Two accessibility indices have the same level of negative impact in 1977 and 1996: living in a job-rich and residence-rich zone decreases the commuting duration, as does living in a zone well served by job opportunities. It must be mentioned that these results differ from those obtained by Levinson in Washington [LEV 98], probably due to Québec City's smaller urban agglomeration and the relatively short commuting lengths.

Dependent variable: *Ln* commuting duration (in minutes)	B	t	Sig.
Constant	1.683	69.335	0.000
Year (1977 = 0; 1996 = 1)	**0.045**	**4.255**	**0.000**
Household characteristics			
Number of workers	-0073	-16.176	0.000
Number of workers × Year	**0.040**	**7.905**	**0.000**
Number of automobiles	0.038	7.006	0.000
Number of automobiles × Year	**-0.022**	**-3.754**	**0.000**
Single person	-0.024	-3.828	0.000
Single person × Year	0.010	1.439	0.150
With children <6 years old	0.006	1.102	0.270
With children <6 years old × Year	**0.017**	**2.928**	**0.003**
With children 6–15 years old	0.010	1.824	0.068
With children 6–15 years old × Year	**0.013**	**2.295**	**0.022**
Worker characteristics			
Age	-0.018	-3.643	0.000
Age × Year	0.006	1.170	0.242
Female	-0.044	-7.821	0.000
Female x Year	**0.023**	**3.492**	**0.000**
Manager-Professional	0.014	2.466	0.014
Manager-Professional × Year	-0.008	-1.280	0.201
Employee (services)	-0.026	-4.855	0.000
Employee (services) × Year	0.000	-0.086	0.931
Transportation mode			
Bus	0.647	92.844	0.000
Bus × Year	**0.015**	**2.747**	**0.006**
Spatial factors			
Residence–central axis distance	0.263	38.692	0.000
Residence–central axis distance × Year	**0.044**	**6.162**	**0.000**
Work–central axis distance	-0.072	-9.975	0.000
Work–central axis distance × Year	0.007	1.362	0.173
Direction	0.330	54.913	0.000
Direction x Year	**-0.060**	**-9.408**	**0.000**
Accessibility from home base – C1	-0.116	-18.709	0.000
Accessibility from home base × Year	**0.039**	**7.579**	**0.000**
Accessibility from workplace to employment – C2	-0.162	-20.870	0.000
Accessibility from workplace to employment × Year	**0.090**	**15.562**	**0.000**
Accessibility from workplace to residences – C3	0.000	0.058	0.954
Accessibility from workplace to residences × Year	**0.017**	**3.277**	**0.001**

Note: Figures in bold indicate variables for which the standardized coefficients (B) differ significantly for 1996 *versus* 1977 (5% level)
N = 15,605; R^2 adjusted: 0.638; standard error: 0.486; F: 1,447.31 p <0.001

Table 2.7. *Changes in determinants of commuting duration, QMA 1977–1996. Source: [VAN 03]*

In order to test for the statistical significance of changes in the determinants of the length of commutes between 1977 and 1996, the categorical variables approach was adopted so as to test for the statistical significance of the differences between the y-intercepts and regression coefficients [GUJ 88, MAD 92]. To do so, both sets of observations were consolidated to build a sample of 15,605 commutes to work (8,368 in 1977; 7,237 in 1996). To test for the difference between y-intercepts, a primary categorical variable was generated: the year ($Yr = 0$ for observations in 1977 or $Yr = 1$ for observations in 1996). The difference between the regression coefficients for each variable was tested by creating as many interaction variables as the number of independent variables with the variable Yr as the interactive term, after having centered the interval scale variables in order to avoid multi-collinearity [JAC 90]. The coefficients obtained can be interpreted as categorical variables and standard interaction.

In Table 2.7, interaction terms measuring the changes between 1977 and 1996 directly follow the source variables, which measure level effects. First, Yr has a significant and positive coefficient, indicating an overall increase in the level of commuting duration between the two dates, which differs from the results obtained in some North American metropolitan regions where households seem to experience shorter trip durations [LEV 94]. The explanation for this discrepancy is also based on the commuting duration, which in Québec City is always less than that for larger US cities. Various individual and household characteristics have a significantly decreasing effect; this is the case for the number of workers, vehicles and gender. Other attributes have more of an increasing statistical effect: the transportation mode and the presence of children younger than six and those six to 15 years old. Finally, some characteristics do not show any change: living alone, age, and the two occupational status indicators.

All spatial determinants (excluding the distance between the workplace and central axis) display significant changes. Commuting duration is more influenced by the residence-central axis distance in 1996 than in 1977, but less affected by the directional bias established from the historical center. The negative effect of the two first accessibility indices is lower in 1996 than in 1977, while the positive effect of the third accessibility index is significantly higher in 1996. We interpret these changes in the commuting duration determinants as the result of the QMA's particular urban structure evolution. The key element of this evolution is the gradual adaptation of work and living places to the context resulting from the construction of an extensive highway network at the start of the 1970s. If the creation of suburban activity centers had maintained a given balance between jobs and residences, the accessibility indices would have had a negative effect on the duration of commutes at least as strong in 1996 as in 1977. On the contrary, these indices indicate a higher dispersion of residences in relation to the agglomeration's central axis.

2.4. Québec City and Brussels: two cities, one reality?

Commutes are a determining measure of spatial interaction in the understanding of territories. This chapter revealed how the proper use and processing of geographical data can allow us to expand our understanding of some of the urban dynamics relating to daily commuting. To do so, we made use of the work conducted in two areas: Québec City and Brussels. Different aggregated databases are used, ranging from individual data samples to official census tables. All contain data relevant to the spatial analysis of residence-work trips. Analysis tools must be adapted to the characteristics of the database and its limitations. Some are given in this chapter as examples: exploratory analysis, statistical tests, statistical cartography, distance decay parameter, Weibull model, centrographic analysis, or multivariate regression models.

Based on the results obtained, we can judge whether the expression "two cities, one reality" is relevant. In Brussels, it is mostly women who use public transportation, are passengers in cars and use combined transportation modes; while men primarily drive their own car. The Weibull model shows that family status plays a significant role, but only among married couples: married women have significantly longer commutes than married men. When analyzing activity chaining, we see a significant difference between men and women: distance decay is higher for women, thus men and women react differently in space. The spatial distribution of employees who work in Brussels but do not live there indicates that women make shorter commutes (in time or distance), with gender thus limiting geographical mobility as well as the choice of both employment and residence.

In Québec City, observations are relatively similar to those recorded in Brussels, in spite of the difference in city size and analysis tools used. Women appeared to travel more by car as passengers and by bus than men, although their use of the car as a driver increased between 1977 and 1996. Commuting times and distances for women increased significantly during these years, with distances remaining less than for men, while travel time appeared to get closer to that of men. It must be noted, however, that in dual-income households with children and only one vehicle, there are still significant gender differences. An analysis of activity spaces shows that owning more than one vehicle for households with no children reduces the difference between male and female activity spaces. When households have children, women's activity spaces are smaller than for men, regardless of the number of vehicles in the family. Thus, adding a vehicle allows activity spaces to be expanded, but more so for men than for women.

The diversity of techniques, time frames and territories analyzed herein limit the scope of our conclusions. It appears that the trends observed on either side are similar for both Brussels and Québec City: gender limits geographical mobility, however to a lesser degree at the start of this 21st Century. Comparative studies for Québec City and Brussels will be conducted in the near future. With the quality of spatial disaggregation constantly increasing (individual data), these studies will definitely be supported by models whose spatial component will be increasingly taken into account. Their results will be linked with urban dynamics (social, economic, environmental, administrative and/or political) and could lead to effective decisions regarding urban management tools. Once again, transportation is seen as a unique geographical application field, where several geographical aspects interact to produce a promising sustainable development space.

2.5. Acknowledgements

The authors would like to thank the Département de Géographie and the Center de Recherche en Aménagement et Développement (Laval University), the Fonds National de Recherche Scientifique (Catholic University at Louvain-la-Neuve) and the Departement Transport en Ruimtelijke Economie (Antwerp University) for sharing their resources in order to produce this chapter.

2.6. Bibliography

[AND 78] ANDREWS H.F., "Journey to work considerations in the labour force participation of married women", *Regional Studies*, vol. 12, no. 1, pp. 11-20, 1978.

[BAC 96] BACCAÏNI B., "L'évolution récente des navettes en Île-de-France", *L'Espace Géographique*, vol. 1, pp. 37-52, 1996.

[BAC 97] BACCAÏNI B., "Les navettes des périurbains d'Ile-de-France", *Population*, vol. 2, pp. 327-364, 1997.

[BAR 07] BARBONNE R., VILLENEUVE P., THÉRIAULT M., "La dynamique spatiale des marchés locaux de l'emploi au sein du champ métropolitain de Québec: 1981-2001", *The Canadian Geographer/Le Géographe Canadien*, vol. 51, no. 3, pp. 303-322, 2007.

[BLU 90] BLUMEN O., KELLERMAN A., "Gender differences in commuting distance, residence and employment location: metropolitan Haïfa 1972 and 1983", *The Professional Geographer*, vol. 42, no. 1, pp. 54-71, 1990.

[BLU 95] BLUMEN O., "Gender differences in the journey to work", *Urban Geography*, vol. 15, no. 3, pp. 223-245, 1995.

[BRO 02] BROZE L., STEINAUER M., THOMAS I., "Discrimination spatiale des femmes et ségrégation sur le marché du travail: l'exemple de Bruxelles", *Espace, Populations, Sociétés*, vol. 3, pp. 323-345, 2002.

[CAM 96] CAMSTRA R., "Commuting and gender in a lifestyle perspective", *Urban Studies*, vol. 33, no. 2, pp. 283-300, 1996.

[CHA 01] CHARDONNEL S., "La time-geography: les individus dans le temps et l'espace", in: SANDERS L., (ed), *Modèles en Analyse Spatiale*, Hermès Science, 2001.

[COU 93] COUTRAS J., "La mobilité des femmes au quotidien. Un enjeu des rapports sociaux de sexe?", *Les Annales de la Recherche Urbaine*, pp. 59-60, 1993.

[COU 96] COUTRAS J., *Crise Urbaine et Espaces Sexués*, Paris, Colin et Masson, 1996.

[COU 97] COUTRAS J., "La mobilité quotidienne et les inégalités de sexe à travers le prisme des statistiques", *Recherches Féministes*, vol. 10, no. 2, pp. 77-90, 1997.

[DIJ 03] DIJST M., "ICTs and accessibility: An action space perspective on the impact of new information and communication technologies", in: BEUTHE, M., HIMANEN, V., REGGIANI, A. (eds), *Transport Developments and Innovations in an Evaluating World*, Springer, 2003.

[DUJ 01] DUJARDIN C., "Effet de frontière et interaction spatiale. Les migrations alternantes et la frontière linguistique en Belgique", *L'Espace Géographique,* vol. 30, no. 4, pp. 307-320, 2001.

[DUJ 06] DUJARDIN C., The role of residential location on labor-market outcomes. The urban agglomerations of Lyon and Brussels, Doctorate thesis, Département de géographie, UCL, Louvain-la-Neuve, Belgium, 2006.

[DUJ 07] DUJARDIN C., THOMAS I., TULKENS H., "Quelles frontières pour Bruxelles? Une mise à jour ", *Reflets et Perspectives de la Vie Economique*, vol. 46, no. 2-3, pp. 155-176, 2007.

[DUJ 08] DUJARDIN C., SELOD H., THOMAS I., "Unemployment and urban structure for young adults. The case of Brussels", *Urban Studies*, vol. 45, no. 1, pp. 89-113, 2008.

[FAG 83] FAGNANI J., "Women's commuting patterns in the Paris region", *Tijdschrift voor Economische en Sociale Geografie*, vol. 74, pp. 12-24, 1983.

[FAG 86] FAGNANI J., "La durée des trajets quotidiens: un enjeu pour les mères actives", *Économie et Statistique*, vol. 185, pp. 47-55, 1986.

[FOT 81] FOTHERINGHAM A.S., "Spatial structure and distance-decay parameters", *Annals of the Association of American Geographers*, vol. 71, pp. 425-436, 1981.

[FUJ 99] FUJITA M., KRUGMAN P., VENABLES A., *The Spatial Economy. Cities, Regions and International Trade*, MIT Press, Cambridge (Mass.), 1999.

[GAL 98] GALLEZ C., ORFEUIL J.P, POLACCHINI A., "L'évolution de la mobilité quotidienne: Croissance ou réduction des disparités?", *Recherche Transports Sécurité*, vol. 56, pp. 27-41, 1998.

[GIU 79] GIULIANO G., "Public transportation and the travel needs of women", *Traffic Quarterly*, vol. 33, pp. 607-616, 1979.

[GOF 00] GOFETTE-NAGOT F., THOMAS I., ZENOU Y., "Structure urbaine et revenu des ménages", in BEAUMONT, C., COMBES P.P., DERIJCKE P.H., JAYET, H. (eds), *Économie Géographique. Les Théories à l'Épreuve des Faits*, Paris, Economica, pp. 276-302, 2000.

[GOR 91] GORDON P., RICHARDSON H., JUN M. J., "The commuting paradox evidence from the top twenty", *Journal of the American Planning Association*, vol. 57, no. 4, pp. 416-420, 1991.

[GRE 91] GREENE R., "Poverty concentration measures and the urban underclass", *Economic Geography*, vol. 67, no. 3, pp. 240-252, 1991.

[GUJ 88] GUJARATI D.N., *Basic Econometrics*, New York, McGraw-Hill, 1988.

[HÄG 70] HÄGERSTRAND T., "What about people in regional science?", *Papers of the Regional Science Association*, vol. 24, pp. 7-21, 1970.

[HAM 03] HAMMADOU H., THOMAS I., VAN HOFSTRAETEN D., VERHETSEL A., "Distance decay in activity chains analysis. A Belgian case study", in: DULLAERT, W., JOURQUIN B., POLAK J (eds), *Across the Border. Building upon a Quarter Century of Transport Research in the Benelux*, Antwerpen, De Boeck, pp. 1-26, 2003.

[HAN 81] HANSON S., HANSON P., "The impact of married women's employment on household travel patterns: A Swedish example", *Transportation*, vol. 10, no. 2, pp. 165-183, 1981.

[HAN 85] HANSON S., JOHNSTON I., "Gender differences in work-trip length: Explanations and implications", *Urban Geography*, vol. 6, no. 3, pp. 193-219, 1985.

[HAN 95] HANSON S., *The Geography of Urban Transportation*, New-York, The Guilford Press, 1995.

[HAN 04] HANSON S., "The context of urban travel: Concepts and recent trends", in: HANSON, S., GIULIANO, G., (eds), *The Geography of Urban Transportation* (3rd Edition), The Guilford Press, 2004.

[HUB 02] HUBERT J.P., TOINT P., *La Mobilité Quotidienne des Belges*, Namur PUN, 2002.

[JAC 90] JACCARD J., ROBERT T., CHOI K.W., *Interaction Effects in Multiple Regression*, Newbury Park, CA, Sage Publications Inc., 1990.

[JOH 92] JOHNSTON–ANUMONWO I., "The influence of household type on gender differences in work trip distance", *The Professional Geographer*, vol. 44, no. 2, pp. 161-169, 1992.

[JOH 95] JOHNSTON-ANUMONWO I., McLAFFERTY S., PRESTON V., "Gender, race and the spatial context of women's employment", in: GARBER J.A., TURNER R.S., (eds), *Gender in Urban Research: Urban Affairs Annual Review*, SAGE Publications 42, 1995.

[KEL 81] KELLERMAN A., *Centrographic Measures in Geography*, CATMOG # 32, Geobooks, Norwich, 1981.

[KWA 04] KWAN M.P., "Geovisualization of human activity pattern using 3D GIS: A time-geographic approach", in: GOODCHILD M., JANELLE D.G. (eds), *Spatially Integrated Social Science*, Oxford University Press, 2004.

[LEV 94] LEVINSON D., KUMAR A., "The rational locator: Why travel times have remained stable?", *Journal of American Planning Association*, vol. 60, no. 3, pp. 319-332, 1994.

[LEV 95] LEVINE N., KIM N.E., NITZ L.H., "Spatial analysis of Honolulu motor vehicle crashes I: Spatial patterns", *Accident Analysis and Prevention*, vol. 27, no. 5, pp. 663-674, 1995.

[LEV 98] LEVINSON D.M., "Accessibility and the journey to work", *Journal of Transport Geography*, vol. 6, pp. 11-21, 1998.

[MAD 81] MADDEN J.F., "Why women work closer to home?", *Urban Studies*, vol. 18, pp. 181-194, 1981.

[MAD 92] MADDALA G.S., *Introduction to Econometrics*, 2nd Edition, New York, MacMillan Publishing Company, 1992.

[MAK 98] MAKIN J., HEALEY R.G., DOWERS S., "Simulation modelling with object-oriented GIS: A prototype application of the time geography of shopping behaviour", *Geographical Systems*, vol. 4, no. 4, pp. 396-430, 1998.

[MCL 97] MCLAFFERTY S., PRESTON V., "Gender, race and determinants of commuting: New-York in 1990", *Urban Geography*, vol. 18, no. 3, pp. 192-212, 1997.

[PAS 84] PAS E., "The effect of selected sociodemographic characteristics on daily travel-activity behavior", *Environment and Planning A*, vol. 16, pp. 571-581, 1984.

[PEE 00] PEETERS D., THOMAS I., "Choosing a measure of distance for applied location-allocation models: Simulations based on the *lp*-distance and the *k*-median model", *Geographical Systems*, vol. 6, no. 2, pp. 167-184, 2000.

[PRE 93] PRESTON V., MCLAFFERTY S., "Gender Differences in Commuting at Suburban and Central Locations", *Revue Canadienne des Sciences Régionales*, vol. 16, no. 2, pp. 237-259, 1993.

[RIE 99] RIETVELD P., ZWART B., VAN WEE B., VAN DEN HOORN F. "On the relationship between travel time and travel distance of commuters, Reported versus network travel data in the Netherlands", *Annals of Regional Science*, vol. 33, no. 3, pp. 269-288, 1999.

[RIG 07] RIGUELLE F., THOMAS I., VERHETSEL A., "Urban polycentrism: a measurable reality. The case of Brussels", *Journal of Economic Geography*, vol. 7, pp. 193-215, 2007.

[ROS 78] ROSENBLOOM S. "The need for study of women's travel issues", *Transportation*, vol. 7, no. 4, pp. 347-350, 1978.

[ROS 89] ROSENBLOOM S., "Differences by sex in the home-to-work travel patterns of married parents in two major metropolitan areas", *Espace Populations Sociétés*, vol., 1, pp. 65-76, 1989.

[ROS 93] ROSE D., VILLENEUVE P., "Work, labour markets and households in transition", in: BOURNE L.S., LEY D.F., (eds), *The Changing Social Geography of Canadian Cities*, McGill-Queen's University Press, 1993.

[RUT 88] RUTHERFORD B.M., WEKERLE G.R., "Captive rider, captive labor: Spatial constraints and women's employment", *Urban Geography*, vol. 9, no. 2, pp. 116-137, 1988.

[SCH 02] SCHWANEN T., DIJST M., DIELEMAN F.M., "A microlevel analysis of residential context and travel", *Environment and Planning A*, vol. 34, no. 8, pp. 1486-1507, 2002.

[SEG 97] SÉGUIN A.M., BUSSIÈRES Y., "Household forms and patterns of mobility: The case of Montreal metropolitan area", in: STOPHER P., LEE-GOSSELIN M., (eds), *Understanding Travel Behaviour in an Era of Change*, New York, Pergamon, 1997.

[SUL 07] SULTANA S., WEBER J., "Journey-to-work patterns in the age of sprawl: Evidence from two midsize Southern metropolitan areas", *The Professional Geographer*, vol. 59, no. 2, pp. 193-208, 2007.

[TAY 75] TAYLOR P.J., *Distance Decay Models in Spatial Interactions*, Norwich Geo-Abstracts, 1975.

[THÉ 95] THÉRIAULT M., DES ROSIERS F., "Combining hedonic modelling, GIS and spatial statistics to analyze residential markets in the Québec urban community", *Proceedings of the Joint European Conference and Exhibition on Geographical Information*, The Hague, The Netherlands, vol. 2, pp. 131-136, 1995.

[THÉ 99] THÉRIAULT M., VANDERSMISSEN M.H., LEE-GOSSELIN M., "Modelling commuter's trips length and duration within GIS: Application to an O-D survey", *Journal of Geographic Information and Decision Analysis,* vol. 3, no. 1, pp. 41-55, 1999.

[THÉ 05] THÉRIAULT M., DES ROSIERS F., JOERIN F., "Modelling accessibility to urban services using fuzzy logic", *Journal of Property Investment & Finance*, vol. 23, no. 1, pp. 22-54, 2005.

[THO 98] THOMAS C., VILLENEUVE P., "Les navettes à Québec: genre, famille et résidence", *Espace géographique*, vol. 3, pp. 285-316, 1998.

[THO 99] THOMAS I., ZENOU Y., "Ségrégation urbaine et discrimination sur le marché du travail: le cas de Bruxelles", in: CATIN M., LESUEUR J.Y., ZENOU Y., (eds), *Emploi, Concurrence et Concentration Spatiales*, Paris, Economica, pp. 105-127, 1999.

[TRA 78] TRAIN K.E., "The sensivity of parameter estimates to data specification in mode choice models", *Transportation*, vol. 7, pp. 301-309, 1978.

[TUR 97] TURNER T., NIEMEIER D., "Travel to work and household responsibility: New evidence", *Transportation*, vol. 24, pp. 397-419, 1997.

[VAN 97] VAN BEEK P., KALFS N., BLOM U., "Gender differences in activities and mobility in the Netherlands, 1975 to 1990", *Transportation Research Record: Journal of the Transportation Research Board*, vol. 1607, pp. 134-138, 1997.

[VAN 01] VANDERSMISSEN M.H., VILLENEUVE P., THÉRIAULT, M., "L'évolution de la mobilité des femmes à Québec, entre 1977 et 1996", *Cahiers de Géographie du Québec*, vol. 45, no. 125, p. 211-243, 2001.

[VAN 03] VANDERSMISSEN M.H., VILLENEUVE P., THÉRIAULT M., "Analyzing changes in urban form and commuting time", *The Professional Geographer*, vol. 55, no. 4, pp. 446-463, 2003.

[VAN 04] VANDERSMISSEN M.H., VILLENEUVE P., THÉRIAULT M., "What about effective access to cars in motorized households?", *The Canadian Geographer/Le Géographe Canadien*, vol. 48, no. 4, pp. 488-504, 2004.

[VAN 07a] VANDERSMISSEN M.H., "Évolution récente de la mobilité à Québec: Qu'en est-il des différences selon le genre?", in LANNOY P., RAMADIER T. (eds), *La Mobilité Généralisée. Formes et Valeurs de la Mobilité Quotidienne*, Académia-Bruylant, Louvain-la-Neuve, Belgium, pp. 47-63, 2007.

[VAN 07b] VANNESTE D., THOMAS I., GOOSSENS L., *Le Logement en Belgique*, Bruxelles, SPF-Economie, P.M.E., Classes moyennes et énergie, 2007, available at: http://statbel.fgo v.be/fr/modules/digilib/marche_du_travail_et_conditions_de_vie/0733_le_logement_en_b elgique.jsp.

[VER 09] VERHETSEL A., VAN HECKE E., THOMAS I., BEELEN M., HALLEUX J., LAMBOTTE J., RIXHON G., MERENNE-SHOUMAKER B., "Le mouvement pendulaire en Belgique. Les déplacements domicile-lieu de travail. Les déplacements domicile-école", in: *Monographies Enquête Socio-économique*, Brussel, SPF Economie en Politique Scientifique Fédérale, 2009.

[VIL 88] VILLENEUVE P., ROSE D., "Gender and the separation of employment from home in metropolitan Montreal, 1971-1981", *Urban Geography*, vol. 9, no. 2, pp. 155-179, 1988.

[VIL 00a] VILLENEUVE P., VANDERSMISSEN M.H., THÉRIAULT M., "Comparing self-reported and computed trip-lengths", *Canadian Regional Science Association Annual Meeting*, Toronto, 2000.

[VIL 00b] VILLENEUVE P., VANDERSMISSEN M.H., THÉRIAULT M., "Urban form, gender and work trip length in the Québec metropolitan area", *Canadian Regional Science Association Annual Meeting*, Toronto, 2000.

[WYL 96] WYLY E.K., "Race, gender, and spatial segmentation in the twin cities", *The Professional Geographer*, vol. 48, no. 4, pp. 431-444, 1996.

[WYL 98] WYLY E.K., "Containment and mismatch: Gender differences in commuting in metropolitan labor markets", *Urban Geography*, vol. 19, pp. 395-430, 1998.

Chapter 3

Spatiotemporal Modeling of Destination Choices for Consumption Purposes: Market Areas Delineation and Market Share Estimation

3.1. Introduction

Over the past few decades, Western nations' headlong rush into becoming consumer societies has resulted in a multitude of commercial structures and consumption areas. The most recent forms are no doubt the power centers and big box stores that generally open up in the suburbs, while the traditional shopping malls and streets attempt to keep their doors open. The purpose of establishing the location of a point of sale is to maximize potential market share, and is done according to the profile of the targeted clientele, location of other businesses, and consumers' mobility behavior. Deciding on the best place to locate a business depends on classical marketing studies, which only provide the decision maker with partial information. A major challenge lies in the ability to predict individual consumer choices and estimate the attractiveness of retail and sales outlets, keeping in mind the spatiotemporal instability arising from the competitive environment.

In order to predict the impact competition has on new outlets, in addition to identifying targeted customer profiles and their place of residence (or work),

Chapter written by Gjin BIBA and Paul VILLENEUVE.

entrepreneurs seek information on the location of competing businesses and consumers' propensity to travel in order to consume [THR 02]. Attempting to model the acceptability of a trip by a consumer and the choice of destination involves a series of theoretical and practical difficulties that spatial analysis and the use of geographic information systems (GIS) can help to resolve [CLA 98]. On one hand, for the theoretical aspect, utility theory and mobility behavior theories must be combined to assess three key dimensions:

– the subjective nature of the perception of space and time [KIM 03];

– the cumulative effect of spatial cognition (perception–action–production) in formulating destination choices that subsequently affect the development of urban form [POR 04]; and

– the potential combination of individual consumption needs leading to multipurpose trips [ROY 01].

On the other hand, this intrinsic complexity of modeling choices and behavior is multiplied by the necessity of combining, in a singular operational framework, phenomena that are relatively stable in space (location of residences, businesses, transportation network structures), but also highly unstable over time. Such instabilities include business' hours of operation [BAK 00], public transportation services and the frequentation of each outlet. This brings about notable difficulties, both for the representation and analysis of individual decisions, and for the microsimulation of urban processes [BUL 04].

This chapter shall illustrate how using new methods and data sources allows us to delve deeper into the spatial analysis of retail businesses. It has a double objective. First, through empirical observation we seek to develop theoretical statements concerning market areas and the spatial behavior of consumers. Second, we seek to better understand the impact that the development of business structure has on land planning and the urban dynamics at the intra-regional level.

This chapter comprises two sections. Section 3.2 deals with the foundations and evolution of analytical methods of spatial and economic competition among types of businesses. In reviewing the attributes and limitations of the various methods of delineating market areas and modeling consumer behavior, we outline the main challenges of the spatial analysis of business activities and stress the contribution of GIS in this area. Section 3.3 presents a methodology used to study competition between commercial streets, shopping malls and big box stores in the Québec metropolitan region in 2001. Using data on the individual mobility behavior of consumers drawn from a large origin-destination studies (OD) mobility study, the analysis employs a regional GIS to model commutes made by road network in order to take into account the accessibility perturbations linked to transportation infrastructures. The spatial delineation of market areas and the statistical modeling of consumer behavior enable us to study competition among types of retail outlets and draw up an appraisal of the overall effect of retail dynamics on other urban activities and, to a certain degree, on urban form.

3.2. Main approaches to the spatial analysis of retail activity

Spatial analysis of markets provides an indication of the attractiveness, economic and strategic value of a business or group of businesses. It is widely used by stakeholders. Spatial influence (market area) is measured based on the frequency (or probability) of commutes by consumers (more often than not residents) to the point of sale. The scope of the market area may vary considerably, according to whether the influence is measured based on the extent of the supply basin of daily consumer products, departure point (work, residence, etc.), frequency of use by neighboring populations of various types of urban businesses or various means of transportation. The increase in the speed of commutes and reduction in the relative importance of proximity relations between service providers and potential consumers [DES 01] have resulted in a decrease in the relevance of some of the methods used for delineating zones of influence. Thanks to the development of geomatics and statistics, new and more accurate analytical tools and methods are offering new perspectives, not only to test existing theories, but also to discover new explanatory models [YEA 01].

A review of existing work allows us to identify three approaches to the spatial analysis of retail activity:

– Traditional approaches based primarily on the economic theory of market areas and on central-place theory. Here the criterion of minimization of distance, the concept of rent charges and the principle of least differentiation play an essential role in explaining both the spatial distribution of the supply of goods and services and the choices of consumers.

– Behavioral approaches that attempt to identify the decision-making processes of consumers and, in particular, the factors that influence consumer location choices.

– Microsimulation approaches, which set out to model the time or distance of commutes through GIS.

Each of these approaches has its own theoretical foundations, favored methods, advantages and limitations, which we analyze in the following sections.

3.2.1. *Traditional approaches*

Initial studies attempted to understand how businesses with a given location share the market spatially. This required identifying the mechanisms producing the geographical extension of the market and the patronage characteristics of a point of sale [FOT 88]. Traditionally, three main approaches were used to define and analyze the spatial influence of a business:

– radial approaches;

– gravitational models;

– central-place theory.

3.2.1.1. *Theory of radial market areas*

The *radial approach* is inspired by the von Thünen model[1] [HUR 94] and constitutes an attempt to link economics and geography. By considering a point of sale in a spatial monopolistic situation where the unit cost of transportation is identical in all directions, this approach demarcates its market area in the form of a hyperbole [FET 24]. The maximum radius of the market area is the result of the relationship between the maximum price the consumer is prepared to pay to acquire a given item, minus the actual sale price and divided by the unit cost of transportation [HYS 50]. In spite of its inherent interest, this approach relies on strong assumptions that may, however, be relaxed. We only have to introduce a margin of indifference to price among consumers or even a difference in the cost of transportation to have the market area boundary established by this approach become vague and irregular.

3.2.1.2. *Gravitational models*

Under *gravitational models* (or spatial interaction models) the market area is established by considering the spatial distribution of all sales points and by evaluating their relative attractiveness. Formulated by Reilly [REI 31], this approach works by analogy with the universal law of gravity. It considers that the attractiveness of a store is directly proportional to its size and inversely proportional to its distance, raised to a power that measures its degree of "friction" (see Chapter 8). Differing from the radial approach, which is based on price and distance, the Reilly model invokes the size of the business and the distance. This approach has been criticized for its overly-mechanical conception of spatial behavior that does not place enough emphasis on the individual as a decision-maker.

---

1 Von Thünen considers an isolated city center, surrounded by homogeneous agricultural land, served by an isotropic communications system, wherein transportation cost is uniquely proportional to the Euclidian distance, and a market having set prices. The result is an organization of farm specializations in concentric circles around the city-market [HUR 94].

To counter the gaps of traditional models, new processes stemming from utility theory[2] have been developed [HUF 64]. What makes the Huff model stand out is that it adds preference factors to location factors. Nakanishi and Cooper [NAK 74] generalize the Huff model, including business attributes (interior appearance, number and type of services, parking availability, etc.) and consumers' sensitivity to these variables.

3.2.1.3. *Central-place theory*

This theory describes the spatial organization of various types of services according to two fundamental parameters [BER 58]: the range of a good (maximum distance the consumer agrees to travel to buy this good); and the entry threshold (the demand volume in order for a point of sale for a given good to be viable). According to this theory, in a homogeneous space, merchants with the same threshold will tend to group together to benefit from agglomeration economies. This consolidation permits the emergence of a hierarchical system of places occupying the centers of hexagon-shaped market areas. Each hierarchical level offers a specific selection of goods and services [CHR 33] or a combination of goods and services [LOS 54].

Widely used, especially during the 1950–1960 period to identify and classify the components of urban business structure, the theory has been criticized because of its overly-simplified hypotheses [PAR 95]. In seeking to establish a theoretical base to represent the spatial organization of business activities, it does not adequately encompass all aspects of consumers' spatial behavior or all interactions between businesses that sell similar products in the same commercial area [EPP 94]. Some of the main elements for which this theory is less adapted to the current business context are:

– the consumer does not systematically choose the center closest to home. He or she is more likely to make multipurpose trips to simultaneously purchase goods and services of a lower quality at higher quality shopping centers, thus compensating for the longer commute by means of a more diversified offering;

– the hexagonal and regular market areas proposed by Christaller are inappropriate, as their configuration is based on the hypothesis that the customers' distribution is uniform;

---

2 According to strict utility theory [LUC 59], an individual is able to categorize the various alternatives in a coherent and unambiguous manner, and then choose the alternative that maximizes utility. This is a deterministic representation of the choice process which considers that the probability of choosing a given store can be expressed by the relationship of this store's utility value with the sum of utilities of the other stores considered [HUF 64].

– in the current context, where trip duration takes precedence over physical distance [BON 00], the organization of the urban business structure tends to elude central-place theory.

3.2.2. *Modeling consumer behavior choices*

Beginning in the 1970s, in response to the classical approaches and their logic of commute minimization, we have seen an increase in theoretical and empirical studies that propose a new analysis of the spatial behavior of consumers [THI 92]. This new approach is much less deductive and aggregated, since it relies greatly on empirical observations made at the individual level [BRO 92]. The behavioral approach resulted in many contributions that can be grouped under two categories: one based on utilitarianism, and the other on the cognitive dimension of the consumer [MER 03].

3.2.2.1. *Utilitarian approach*

Consumer spatial behavior analysis was widely based on approaches that integrated the utility-function concept [HAN 04, TIM 82]. Contrary to gravitational models, those developed beginning in the 1970s were not based solely on utility, but on a random utility function [MAN 77]. In these, the consumer is led to choose an alternative (*i*) from a set (*C*) of possible alternatives, basing his or her decision on the hypothesis of utility maximization. Given the heterogeneity of individual preferences, however, measurement errors, the omission of certain variables and approximate estimates, the utility measure includes an uncertainty dimension that makes it impossible to accurately predict a consumer's choice. Thus, the utility function (U_{in}) of alternative *i* for an individual *n*, can be defined as being composed of a deterministic part (V_{in}) and a stochastic part (ε_{in}). The deterministic component includes all of the independent variables that describe the alternative and, if possible, characteristics of the individual (X_{ink}). The marginal effects of the variables, X_{ink}, on the utility function are provided by the coefficients (β_{ik}).

$$U_{in} = V_{in} + \varepsilon_{in} \qquad\qquad [3.1]$$

$$V_{in} = \sum_{k=1}^{K} \beta_{ik} X_{ink} = \beta_i' X_{in} \qquad\qquad [3.2]$$

The error term (ε_{in}) results from random variables for which distribution hypotheses are formulated. In the consumer destination choice models, where the dependent variable (the alternative choice) is discrete, the most widely-used hypothesis in the multinomial logistic regression context is the identical and independent distribution (IID) hypothesis [BEN 85, MCF 81]. This type of modeling – widely used in spatial, transportation, marketing analyses, etc. – allows us to estimate the probabilities of choosing between several alternatives by using variables pertaining to the profiles of individuals, environmental characteristics (socioeconomic and spatial) and the attributes of alternatives [LOU 00]. Equation [3.3] shows the common form of the multinomial logit (MNL):

$$P_n(i) = \frac{e^{\mu(\beta_i X_{in})}}{\sum_{j=1}^{J} e^{\mu(\beta_i X_{in})}}, i \in C_n, j \in C_n \qquad [3.3]$$

The probability of an individual n choosing alternative i is expressed as the utility ratio offered by this alternative ($\beta_i' X_{in}$) over the sum of the utilities offered by all other alternatives in the group, C_n. The expression μ is a scale parameter inversely related to the variation of error terms (stochastic utilities). Estimating the coefficients (β_i) is done using the maximum likelihood procedure which, according to attributes X_{in}, allows us to estimate β_i parameters so as to maximize the probability of obtaining the choice made by the consumer [LOU 00].

In spite of its wide utility, the MNL is criticized for its postulate of independence from irrelevant alternatives (IIA). This property signifies that the introduction of a new alternative in a given set of alternatives will modify the hard-and-fast market shares, but will not affect the relative alternative shares when considered in pairs. The IIA assumption is quite plausible for numerous applications, however, and empirical experience has shown that the MNL is relatively robust[3] [HAU 84]. This allows us to recognize the operational validity of using this model to analyze consumers' behavioral choices, and thus the competition among the various alternatives of consumption.

3.2.2.2. *Cognitive approach*

The development of the cognitive process in analyzing the spatial behavior of consumers is founded on the concept that consumers make decisions based on imperfect information and a subjective evaluation of opportunities [BRO 92]. The consumer's view of the alternatives plays an essential role in his or her choices

---

3 Various methods are available to test the validity of the IIA hypothesis. Hausman and McFadden [HAU 84] suggest that if a sub-part of the group of possible choices is in fact not relevant, its omission from the model will not fundamentally change the parameter estimates. Moreover, including these choices will be ineffective, but will not make the model non-significant. If the likelihood of choosing a given alternative is not truly independent from these other choices (i.e. the IIA hypothesis is unverified), however, the parameters estimated once these choices are eliminated will not be significant.

[SHE 80]. Empirical studies based on the cognitive approach generally borrow from tools and methods used in psychology [POT 82]. Through the construction of cognitive maps or the comparison of real and perceived distances, these studies have shown that consumers have a rather limited information field (the alternatives they are aware of) and a more restricted action space, which only includes the places they actually frequented. Potter [POT 82] also points out that these fields depend on the consumer's residential and work locations and that they expand with the increase in socioeconomic level of the individual/household. Moreover, marketing studies have placed emphasis on the cognitive aspects of consumer buying behavior by attempting to identify the product and commercial outlets differentiation process [LEO 00]. In spite of the interest in this approach in order to learn more about consumer behavior, its use remains generally limited. The high cost of collecting relevant empirical data and the difficulties respondents have in correctly answering the questionnaires are major stumbling blocks. As we will see in the following section, however, the development of GIS helps us to implement this approach and better take into consideration the empirical results dealing with the spatial behavior of consumers.

3.2.3. *Microsimulation of trip duration and distance within a GIS*

For the past few decades, researchers have been attempting to construct increasingly disaggregate models, in order to better consider individual decision processes [HOL 05]. Primarily based on the economic theory of human behavior, these processes are represented through the construction of simulation models of individuals' behavior. Hägerstrand [HÄG 53] was the first to focus on the spatiotemporal prism (*time-geography*) that influences interaction among individual subjects, resources and constraints. By emphasizing both the location and duration of actions, he formulated the theoretical bases of spatial microsimulation.

The conceptual framework of time-geography can be described as follows [HOL 05]. Actions stem from the behavior of individual subjects (people, households, firms). The subjects, resources and constraints are located in space. This has an effect on the succession of events; actions and events are influenced by the individual characteristics of subjects, by the environmental context, and by the actions of other subjects. The succession of events is influenced by a random factor.

Hägerstrand has shown that individuals' spatial choices are guided by constraints and that they are not independent decisions made by spatially and temporally autonomous individuals. In the case of commercial activities, understanding the context is made easier by GIS-associated tools.

The use of GIS in the analysis of retail activity has greatly increased over the past few years [LON 05]. It is used particularly in the spatial organization analysis of retail outlets and their competitive environment [HER 99, JON 01, SLI 02]. For example, by using a computerized road network and keeping in mind given speed limits [THE 99, THE 05], GIS allows us to model the time (and/or distance) of commutes made by consumers from their starting point to their destination, and thus identify the market area and customer profile for each retail outlet. By making the analysis of individuals' spatial decisions easier, GIS allows us to understand an individual consumer's reactions *vis-à-vis* the various types of retail outlets, notably by considering their accessibility level.

3.2.4. *GIS contribution to the spatial analysis of retail activity*

The widespread use of personal vehicles has greatly contributed to the complexification of Western cities' business structure. Thomas and Bromley [THO 02] have reviewed the impacts of this transformation on the traditional business hierarchy of cities in Britain, notably on the decrease in trips to small, local markets and central urban areas. Municipal authorities have tried to remedy this local development problem by promoting the revitalization of commercial streets in the downtown core. These steps must, however, be taken in the context of lively competition among businesses, rurbanization and the fiscal consolidation of national and multinational chain stores.

With the shopping phenomenon in hubs located in outlying areas of metropolitan regions becoming more widespread over the past few decades [GOT 03], there is a need for revitalization strategies aiming to increase the competitive level of traditional mid-sized centers – those most affected by competition from the big box stores that have opened in the outlying areas [THO 02]. We must also be able to study the competitive mechanisms among the various retail types within the specific context of each metropolitan region, in order to target public intervention and, once revitalization initiatives are introduced, to track their evolution over several years in order to evaluate their relevance and effectiveness. This analytical objective is fully compatible with the development of methodologies targeting the optimization of specific retail business positioning [ZEN 01]. It is more complex, having to analyze competition at the metropolitan level and establish methods that enable us to track the evolution of customers' purchasing power (consumer pools and market areas).

To succeed, it is necessary to properly align theoretical approaches and sophisticated spatial analysis tools [THR 02, ZEN 01]. Thus, a better consideration of the size of businesses and agglomeration economies associated with the various components of the urban business structure provides a more realistic representation of market areas and a clearer identification of spatial impact [PAR 95]. Moreover, regarding questions pertaining to land use and mobility behavior, other authors [MAA 05] conclude that the densification of urban areas does not necessarily result in a consequent reduction in the length of city commutes.

To better analyze consumers' mobility habits, it is necessary to introduce a combination of the utility theory (economics) and the cognitive approach (psychology). The authors note that behavioral adjustments of urban dwellers in terms of destination choices are not only related to a reduction in the length of a single commute, but rely increasingly on optimizing multiple-destination commutes [ROY 01] in the short term and housing choices in the long term [KES 04]. Thus, accessibility to urban services competes with the quality of the social and natural (e.g. vegetation) environment of the chosen neighborhood.

Empirical studies conducted in the Québec City region [BIB 06, DES 03, DES 05, THE 04b] have shown how the joint use of GIS and an OD study is useful to model the spatial extent of market areas specific to each shopping center. Thériault and Des Rosiers [THE 04a] have shown that average commute lengths vary significantly according to:

– the reason for the commute (work, school, grocery shopping, other shopping, leisure, restaurant, health services);

– commuters' gender; and in some cases

– according to the household composition (single person, couple with no children, with children, etc.).

Moreover, Thériault [THE 05] reveals significant local variations in accessibility (modeled on length of commute by road and specific indicators based on the number of each type of business that accessible within an 'acceptable' time interval) and their effect on single-family home prices through hedonic modeling. They have thereby controlled for the effect of cofactors. This 'acceptable' time interval is established by fuzzy logic according to mobility behavior observed during the OD study in order to reflect, through a series of indicators, specific accessibility (to grocery stores, restaurants, shopping centers, etc.), as perceived by consumers. In Québec City, a center highly affected by urban sprawl [VIL 02], the accessibility thus measured has an effect that is two to three times more significant on price variation (real estate rent) than the centrality measured using a gravitational index expressing the access potential of the agglomeration center.

These studies having revealed an interest in modeling mobility choices with GIS, we used this approach to establish market areas and model the commuting behavior of consumers. This methodology was developed during a study sponsored by the Québec Metropolitan Area (QMA) and was intended to generate an overall picture of retail business trends for its regional development and land-use planning.

3.3. Modeling market areas and consumer destination choices

As with most North American urban regions, the business structure of the QMA is marked by the rapid expansion of big box stores and power centers. During the period 2001–2006, the surface area of big box stores rose by 80%, while shopping centers decreased by 0.4% [MER 06][4]. In order to understand the current dynamic, we combined the use of GIS to spatialize outlets and itineraries with statistical analysis to model the spatial choice of consumers. Initially, GIS served to model the path taken for each commute, considering the means of access linked with road infrastructures. Then, the spatial competition among businesses – and, more specifically, the impact of the emergence of big box stores on commercial streets and shopping centers – is synthesized by the demarcation of market areas for each business group. Lastly, identifying a discrete choice model allows us to model consumer behavior with regard to destination choices.

3.3.1. *Spatial distribution of retail supply and definition of retail structures*

Identifying businesses is done based on information from a telephone directory of businesses in the region (ZipCom, version 2000). This directory has been geocoded [DES 03]. Each business was localized in the appropriate building by matching civic addresses with assessment units of the various municipalities in the region (according to an *address-matching* process). In 2001 the region's retail composition comprised some 6,571 retail businesses. Approximately 53% of these businesses were small firms scattered throughout the QMA region. The remaining businesses (47%) were located along the main commercial streets (22.4%), various types of shopping centers (23.9%), big box stores and power centers (0.6%).

To analyze the spatial and economic competition that characterizes supply outlets, it is necessary to identify the various types of businesses. However, the typological classification of retail businesses is not always accurately standardized in research dealing with urban retail structure. Most often these studies are based on discriminating variables such as:

– size of the store;

---

4 The number of shopping centers dropped from 83 to 80 during the 2001–2006 period; big box stores rose from 36 to 71; and the number of power centers rose from two to five [MER 06, THE 04b].

– product and service specialty/diversity;

– physical location;

– type of property;

– architectural style;

– etc.

For this study we combined and adapted the classifications established by the International Council of Shopping Centers and the Urban Land Institute to the region's context. The variables considered are: the building's commercial area; the number of stores in the building or the linear density of retail outlets on the street; the types (and variety) of products and services available (based on the Standard Industrial Classification or SIC codes); and the potential trade area.

Table 3.1 presents the various forms of shopping communities identified according to these criteria. The localization of the shopping clusters is done with a GIS by constructing a polygon around each one (see Figure 3.1). In addition to the typological differentiation criteria in Table 3.1, the construction of polygons also considered other specificities, particularly the characteristics of their location and the agglomeration economies that may result from each commercial structure:

– *Commercial street polygons* (72 segments) include a buffer zone 40 m wide on either side of the right-of-way[5]. The average length of a segment is 1.38 km (varies from 0.32 3.7 km), and shop density per linear km varies from 10 (minimum threshold) to more than 90 stores (with an average of 32 shops per linear km).

– *Shopping center polygons*[6] include a buffer zone built around their buildings, which vary from 20 m for small centers (neighborhood and community), up to 150 m for large centers (regional and super-regional) in order to reflect the scope of commercial centers. Thus, their polygons include not only parking zones but also those stand-alone businesses and services located in the immediate proximity and that benefit from agglomeration economies.

– *Big box store polygons* include the building and parking zone for each store or big box store. There are 36 autonomous superstores, owned by regional and/or national chain retailers such as: Wal-Mart, Réno-Depôt, Rona-l'Entrepôt, Maxi, Bureau en Gros, Canac Marquis, Canadian Tire, etc.

---

5 The width of the buffer zone allows businesses located on each side of the street to be included in the commercial street polygons. It is established according to our empirical studies of the region's commercial streets.

6 In 2001 there were 54 neighborhood shopping centers (which comprise 5.1% of all businesses in the region); 23 community malls (6.0%); three regional centers and three super-regional centers (12.1%).

Retail structure		Floor surface (m²)	Number of stores	Specific characteristics of the supply	Trade area
Commercial street		Varies according to store	Density ≥10 businesses/per km (Average = 32 businesses/km)	Variety of businesses and services; variety of transportation means to shop	Neighborhood and visiting clientele – large percentage of foot traffic
Shopping centers	Neighborhood (or locale)	<14,000	5–14 (with a large grocery store or pharmacy)	Daily-use products and services	Neighborhood clientele – commutes on foot or by car
	Community (or neighborhood)	14,000–50,000	15–99 (at least 1 large, national chain store)	Regular daily needs: clothing, shoes, etc.	Neighborhood clientele or from surrounding areas – served by public transportation (PT)
	Regional	50,000–80,000	100–199 (at least 2 large, national chain stores)	Large diversity of products	Regional clientele – accessible by car and PT
	Super-regional	>80,000	>200 (several large, national chain stores)	Very large diversity of products and services	Clientele exceeding the metropolitan area limits
Department stores	Big box store	>4,600	Main store that may rent custom boutique space	Wide range of mixed or specialized merchandise	Local and regional clientele (according to the size and specialty of stores) – primarily accessible by car
	Power center	>25,000	Several neighboring big box stores	Conglomeration of big box stores with some smaller, similar-style businesses	Regional and possibly super-regional – accessible by highway

Table 3.1. *Typology of retail agglomeration*

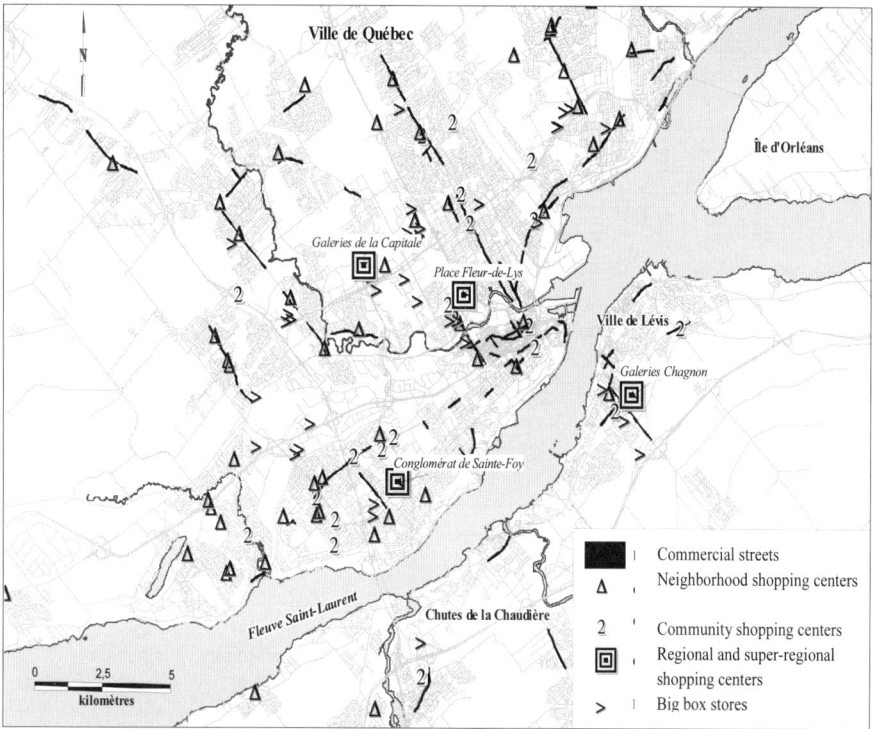

Figure 3.1. *Location of retail outlets in the QMA*

In addition to mapping the spatial distribution of stores, these polygons also permit the identification of the flow of customers generated by each shopping center. This identification is possible thanks to the use of the 2001 OD study, which details the daily commutes made by all members of the households surveyed (more than 8% of the regional population) and their main sociodemographic characteristics, reasons for traveling (work, school, shopping, leisure, etc.) and means of transportation used (car, public transportation, on foot, etc.).

For the purpose of this study, only shopping commutes were considered. A total of 24,519 commutes are distributed among four trip-purpose categories: groceries, sundry goods, foodservice and leisure. They were made by 17,943 persons from 13,074 households. After applying expansion factors to account for the local sampling fraction, we get a total of 258,304 shopping trips made during a typical weekday (Monday to Friday) by the region's population.

In 2001, commercial agglomerations siphoned off 63.6% of the total volume of daily shopping trips in the region (the remaining 36.4% were made to individual outlets distributed over the whole region). Flows to the various retail types were distributed as follows: regional and super-regional shopping centers 18%; commercial streets 16%; community shopping centers 12%; neighborhood shopping centers 9%; and big box stores 8%.

3.3.2. *Market area delineation: analytical approach*

The delineation of market areas is based on commute times by car. When the destination of each shopping trip has been identified, the trip's departure point is used to determine the market area of retail outlets. Using the departure point rather than that of the consumer's residence allows us to better evaluate the spatial influence of each business. It takes into consideration the multipurpose trips made by consumers and provides a better understanding of the effect of spatial organization on competition among retail outlets.

The market area of each shopping center is represented by a convex hull that includes consumers' trip departure point to the center (see Figure 3.2). However, two types of market areas were considered in order to reflect the demand curve and intensity of spatial competition among businesses: the primary market area, which represents 50% of the closest clients; and the secondary market, which adds another 30% share[7].

The procedure was repeated independently for each of the shopping centers, which allowed us to compare their trade areas (particularly in terms of spatial range; market share; competitor types and intensity of competition among businesses). This enabled us to study the impact resulting from the overlapping of market areas between retail agglomerations.

The characteristics of market areas (see Table 3.2) allow us to note that regional and super-regional shopping centers have the largest customer base and strong control over their potential market. However, this retail structure seems to face high competition from big box stores, especially in the renovation (e.g. Rona, Rona l'Entrepôt, Réno Dépôt) and general merchandise sectors (e.g. Wal-Mart), which

---

7 These 50% and 80% customer thresholds are established through empirical observation of the spatial distribution of consumers that frequent shopping centers. By keeping the fiftieth percentile of commute time as a limit for the primary market area, we only consider the geographical area that may include regular consumers for each commercial agglomeration. The secondary market area (eightieth cumulative percentile of commute time) thus includes a portion of less regular customers than for the primary area, but is an interesting segment of the market for business development. Lastly, consumers with commute lengths greater than the eightieth percentile can be considered as rather sporadic customers.

have market shares that allow them to cover the whole region. Commercial streets and community malls primarily serve a neighborhood clientele located at an average distance of 1–3 km from the outlet.

In addition to calculating indexes similar to those in Table 3.2, the delineation of market areas also allows us to identify the level of service offering available to consumers in the various urban zones (which indicates the intensity of spatial competition among retail structures). Thus, for the Québec City region, we see that the central neighborhoods are very well served by commercial streets (the consumer has the possibility of shopping on more than 10–15 streets, without spending a lot on transportation) and shopping centers.

Big box stores have a maximum retail service threshold in peripheral zones where the offering by regional and super-regional shopping centers begins to decrease.

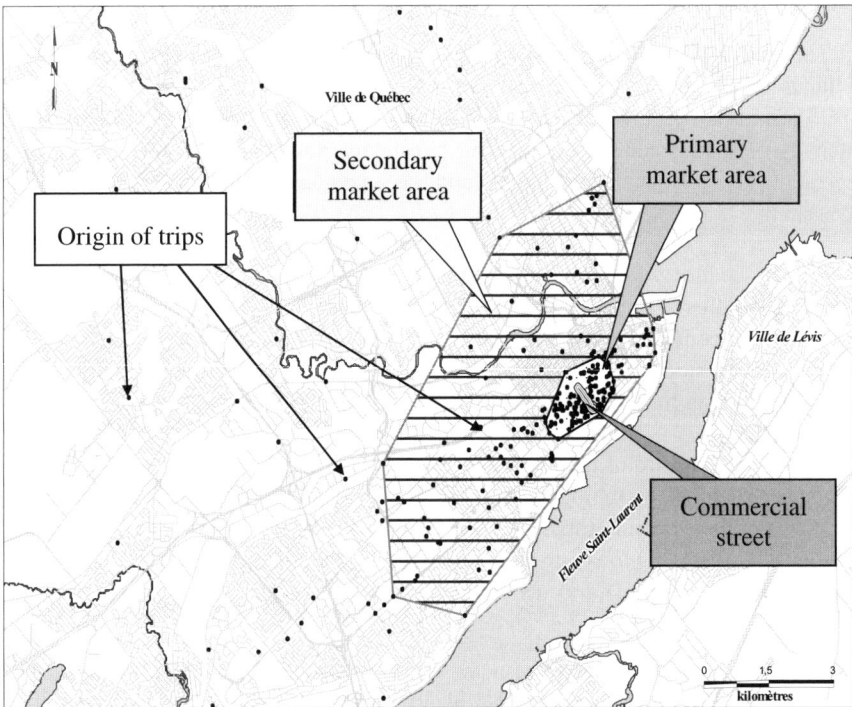

Figure 3.2. *Example of determining market areas for a commercial street*

Retail structure	Primary market area (50% of clientele)					Secondary market area (80% of clientele)					Market total (100% of clientele)	
	Average area (km²)	Potential clientele* (number)	Market coverage** (%)	Trip length (km)	Trip duration (minutes)	Average area (km²)	Potential clientele (number)	Market coverage (%)	Trip length (km)	Trip duration (minutes)	Trip length (km)	Trip duration (Minutes)
Regional and super-regional centers	48 (35)†	48,916	14.2	3.3 (2.1)	3.7 (2.1)	297 (263)	131,003	9.9	5.3 (3.3)	5.4 (3.1)	7.9 (6.8)†	7.3 (5.4)
Community shopping malls	8 (7)	10,452	11.9	1.7 (1.1)	2.4 (1.5)	54 (52)	39,375	9.5	2.8 (2.1)	3.5 (2.3)	5.0 (4.1)	5.5 (3.8)
Big box stores	26 (25)	21,443	3.3	3.3 (2.1)	3.8 (2.2)	72 (65)	54,328	1.8	4.8 (3.1)	5.2 (2.9)	7.1 (6.6)	7.1 (5.2)
Food sector	19 (15)	18,816	3.8	3.1 (1.8)	3.6 (2.0)	83 (59)	65,121	1.6	4.7 (2.9)	5.1 (3.0)	7.2 (6.8)	7.1 (5.5)
Renovation sector	66 (104)	45,852	1.3	4.4 (2.2)	4.9 (2.3)	118 (118)	83,405	0.9	6.3 (3.6)	6.6 (3.1)	8.4 (6.4)	8.3 (5.2)
Automotive sector	11 (4)	14,286	2.0	2.0 (1.2)	2.7 (1.4)	40 (17)	34,885	1.0	3.4 (2.3)	3.9 (2.3)	6.5 (5.5)	6.3 (4.6)
General merchandise	34 (29)	18,135	6.6	4.2 (2.1)	4.5 (2.0)	56 (39)	32,052	5.7	4.9 (2.1)	5.2 (2.1)	6.6 (5.5)	6.5 (4.1)
Other	10 (10)	12,614	2.9	2.7 (2.1)	3.3 (2.4)	48 (36)	38,769	0.8	4.6 (3.3)	4.8 (3.2)	7.3 (5.9)	7.0 (4.7)
Commercial streets and neighborhood centers	5 (4)	8,171	10.8	1.1 (0.8)	1.7 (1.2)	25 (21)	27,570	7.5	2.3 (1.9)	2.9 (2.2)	4.7 (5.2)	5.0 (4.7)
Streets with center	6 (4)	7,666	12.3	1.1 (0.8)	1.6 (1.2)	30 (19)	29,857	6.8	2.3 (1.8)	3.0 (2.1)	4.7 (5.3)	5.0 (4.6)
Streets without center	5 (4)	8,638	9.4	1.2 (1.0)	1.7 (1.4)	21 (19)	25,460	8.3	2.2 (1.7)	2.8 (2.2)	4.6 (5.2)	4.9 (4.7)

† standard deviation
* potential clientele = number of shopping trips made from a specific market area to a given shopping center (Source: 2001 OD study, after expansion)
** market coverage = trips to a given shopping center/total number of shopping trips made from a specific market area (%)

Table 3.2. *Characteristics of specific market areas for each type of retail structure (Source: 2001 OD study, before expansion)*

3.3.3. *Modeling consumer behavior*

Modeling consumer behavior choices enables us to estimate the probability of a consumer shopping at one place rather than another. By considering the individual characteristics and constraints for each consumer (and their household), this approach provides additional data on competition among retailers. In the following example, it is the consumer's propensity to frequent shopping centers (commercial street, community shopping center, regional and super-regional shopping centers, big box stores) that is modeled using the multinomial logistic regression model.[8] The consumer's choice is modeled according to approximately 10 explanatory variables (only those that revealed significant effects during model calibration are considered), grouped under the following aspects:

– travel purpose;

– trip attributes;

– socioeconomic profile of consumer and household; and

– urbanization period for the residential zone (representing the type of urban fabric).

The model's findings are listed in Table 3.3. The model's predictions with regard to consumer choice behavior are often surprising and enable us to identify the roles various factors play in the competition process among the different commercial structures. It also allows us to identify a certain number of plausible scenarios of the evolution of the commercial structure and its impact on the transportation system and land-use management.

Commercial streets are highly competitive in the leisure and foodservice sectors. They are more attractive to women, the elderly, people who shop on foot or by bus, households without children or a car, and those who live in older neighborhoods. Community shopping centers face direct competition from big box stores for their main product offering, which is grocery. They are more attractive than big box stores for trips made by bus or on foot, for women, the elderly and non-motorized households. Trip commutes by their clients show that this commercial structure has rather limited appeal in the space that corresponds to their primary market area (less than 2.5 minutes).

---

8 Given that randomly-located individual retail outlets constitute a rather heterogeneous commercial structure in relation to commercial centers (in terms of location, specialization and store size), they are not included in this model. The mathematical formulation of the model was presented in section 3.2.2 of this chapter.

Factors/alternatives		Commercial street	Community mall	Regional and super-regional center
Travel purpose (Ref. Consumer products)	*Groceries*	1.5***	--	0.1***
	Restaurants	27.3***	6.6***	2.9***
	Leisure	11.8***	4.6***	0.8*
Trip origin (Ref. Residence)	*Work/school*	--	--	1.4**
	Other	0.8**	0.7***	0.6***
Weekday (Ref. Monday, Tuesday and Wednesday)	*Thursday and Friday*	--	1.2**	1.6***
Departure time (Ref. 9am–6pm)	*After 6pm*	1.2***	--	0.8**
Transportation mode (Ref. Car)	*Bus*	5.9***	4.0***	6.0**
	On foot	4.5***	2.2***	--
	Other	2.2**	--	--
Trip duration (Ref. 2.5 to 4.9 minutes)	*<2.4 min*	2.3***	1.6***	1.6***
	5–15 min	0.5***	0.6***	1.2**
	>15 min	0.5***	0.4***	1.3**
Gender (Ref. Men)	*Women*	1.4***	1.5***	2.1***
Age (Ref. 25–64)	*<25 years*	--	0.8**	1.6***
	>65 years	1.3***	1.9***	2.2***
Type of household (Ref. Other types)	*Family with children*	0.8**	0.7***	--
Household motorization (Ref. With car)	*No car*	1.7**	1.6*	1.6**
Urbanization period of place of residence (Ref. Before 1961)	*1961–1978*	0.9*	--	--
	1978–2000	0.8**	--	--
	Scattered-site housing	--	--	1.2**

Reference category is "Big box store and power center"[9]
Model parameters: N = 14,627; 2LL = - 16,004; p <0.0001; Pseudo R^2 = 0.21
Statistical significance levels: -- insignificant (p >0.10); * p ≤0.10; ** p ≤0.05; *** p ≤0.01

Table 3.3. *Socioeconomic and spatial competition among commercial centers (odds ratios)*

---

9 Big box stores and power centers are pooled together. Their designation as a "reference category" serves to properly identify the effects they have on other forms of commercial structures and on consumer choice behavior.

Regional and super-regional shopping centers are more attractive than big box stores for trips made from work and for those who use public transportation. Women, individuals under 25 years, the elderly and non-motorized households are more likely to frequent this type of commerce. Odds ratios for trip durations show that the market areas for regional and super-regional centers are larger than for big box stores.

Big box stores are very competitive for groceries and sundries. They primarily attract men aged 25 to 64, large families (with children), and households with personal vehicles. With regard to consumers' commuting habits, this type of commerce promotes a sequence of commutes (departure point other than from home or work).

These results show that community shopping centers are most negatively-affected by competition from big box stores. Regional and super-regional centers also face direct competition from this new shopping structure, but they manage to meet the competition. Commercial streets appear less threatened by them, thanks notably to their diversified offering of products and services (foodservice, leisure) and their relatively specific customer base (seniors, women, small households). However, market area and consumer behavior modeling analyses indicate that the recent changes to urban commercial structures display effects that exceed regular competition among various retail outlets. The specific placement of new outlets and the expansion of their market share have a direct influence on consumers' behavior (in particular, an increase in the use of personal vehicles and, consequently, a reduction in the use of public transportation). They also have an effect on the spatial organization of economic activities at the metropolitan scale. Local and regional development planners and stakeholders must estimate the effects of these changes not only on employment and finances, but also on the increase in demand for new transportation infrastructures, land use and the environment.

3.4. Conclusion

The increase in consumer and retail mobility has had a profound effect on redefining concepts traditionally linked to the spatial analysis of commercial activity, such as those involving proximity, accessibility and competition. Reacting more to the duration of a commute rather than to the actual distance traveled, the consumer seeks to satisfy his or her needs while optimizing the scheduling of activities. To accomplish this, the consumer combines different activities (work–shopping–leisure) and locations (residence–work–other) that allow him or her to gain access to a large number of alternatives that may be located outside the residential neighborhood. The development of internet shopping also affords the consumer another accessible alternative, with no need to commute, since the distance becomes virtual. These changes in context and in commuting practices for consumers imply that, when choosing a location, the retailer must not only consider the "spatial" constraints, but also the "temporal" constraints of its potential clientele.

This chapter illustrates the analytical potential offered through the combination of GIS, spatial analysis and individual mobility analyses to study the destination choice process made by consumers and increase understanding of current commercial dynamics. Enriching mobility databases by means of spatiotemporal simulations of trip durations allows us to identify trade areas and their superposition (diversity of destination choices from a given location) while affording an analytical opportunity for studying spatial and economic competition among businesses. The statistical modeling of spatialized individual data leads to a better understanding of customer profiles for retail outlets and to predict behavioral changes brought about by a given action (e.g. construction of a new business power center in a given location). This is made possible through building a detailed model of the social, geographical and economic factors influencing the probability that a given type of consumer will choose a specific destination in a given situation.

The approach developed in this chapter may be further improved upon, both in terms of measuring the spatial influence of retail outlets by considering not only the speed of cars. It must also measure other means of transportation (bus, walking) and expand data on consumer mobility, by including each day of the week and, if possible, the size of each shopping basket for each trip. In conclusion, taking into account temporal data series on the development of commercial structures and consumer mobility behavior will enable us to test the effects of various phenomena discussed here (e.g. trip-chaining, complementarity of commercial structures, transportation demand, etc.). It will also enable us to revamp the spatial and economic portrait of urban commercial structures.

3.5. Acknowledgements

The authors would like to thank their colleagues at the University of Laval's *Center de Recherche en Aménagement et Développement* (CRAD), particularly Marius Thériault and François Des Rosiers for their enthusiastic collaboration in the field of spatial analysis of urban phenomena. They would also like to thank the *Fonds Québécois de la Recherche sur la Société et la Culture* (FQRSC) and the Social Sciences and Humanities Research Council (SSHRC) for their financial support.

3.6. Bibliography

[BAK 00] BAKER R.G.V., "Towards a dynamic aggregate shopping model and its application to retail trading hour and market area analysis", *Papers in Regional Science*, vol. 79, no. 4, pp. 413-434, 2000.

[BEN 85] BEN-AKIVA M., LERMAN S.R., *Discrete Choice Analysis: Theory and Application to Travel Demand*, Cambridge, MIT Press, 1985.

[BER 58] BERRY B.J.L., GARRISON W., "The functional bases of the central place hierarchy", *Economic Geography*, vol. 34, no. 2, pp. 145-154, 1958.

[BIB 06] BIBA G., DES ROSIERS F., THÉRIAULT M., VILLENEUVE P., "Big boxes versus traditional shopping centers: Looking at households shopping trip patterns – a Canadian case study", *Journal of Real Estate Literature, v*ol.14, no. 2, pp. 175-202, 2006.

[BON 00] BONDUE J.P., "Le commerce dans la géographie humaine", *Annales de Géographie*, vol. 611, pp. 94-102, 2000.

[BRO 92] BROWN S., *Retail location: A Micro-scale Perspective*, Aldershot, Avebury, 1992.

[BUL 04] BULIUNG R.N., KANAROGLOU P.S., "On design and implementation of an object-relational spatial database for activity/travel behaviour research", *Journal of Geographical Systems*, vol. 6, no. 3, pp. 237-262, 2004.

[CHR 33] CHRISTALLER W., *Central Place in Southern Germany*, (Translated by C.W. Baskin, 1966), New York, Englewood Cliffs, 1933.

[CLA 98] CLARKE G., "Changing methods of location planning for retail companies", *GeoJournal*, vol. 45, no. 4, pp. 289-298. 1998.

[DES 01] DESSE R.P., *Le nouveau commerce Urbain: Dynamiques Spatiales et Stratégies des Acteurs*, Rennes, Rennes University Press, 2001.

[DES 03] DES ROSIERS F., THÉRIAULT M., "Assessing retail trade areas, local economic potential and spatial competition: How origin-destination surveys may help", *GeoSpatial Solutions*, vol. 13, no. 11, pp. 46-51, 2003.

[DES 05] DES ROSIERS F., THÉRIAULT M., MENETRIER L., "Spatial versus non-spatial determinants of shopping center rents: Modeling location and neighbourhood-related factors", *Journal of Real Estate Research*, vol.27, no. 3, pp. 293-319, 2005.

[EPP 94] EPPLI M.J., BENJAMIN J.D., "The evolution of shopping center research: A review and analysis", *Journal of Real Estate Research*, vol. 9, no. 1, pp. 5-32, 1994.

[FET 24] FETTER F.A., "The economic law of market areas", *The Quarterly Journal of Economics*, vol. 38, no. 3, pp. 520-529, 1924.

[FOT 88] FOTHERINGHAM A.S., "Market share analysis techniques: A review and illustration of current US practice", in: N. WRIGLEY (ed.), *Store Choice, Store location and Market Analysis*, Wiley, London, pp. 120-159, 1988.

[GOT 03] GOTHER C., NIJKAMP P., KLAMER P., "The attraction force of out-of-town shopping malls: A case study on run-fun shopping in the Netherlands", *Tijdschrift voor Economische en Sociale Geografie*, vol. 94, no. 2, pp. 219-229, 2003.

[HÄG 53] HÄGERSTRAND R.T., *Innovation Diffusion as a Spatial Process*, Chicago, University Press, 1953.

[HAN 04] HANSEN T., SOLGAARD H., *New Perspectives on Retailing and Store Patronage Behavior*, Boston, Kluwer Academic Publishers, 2004.

[HAU 84] HAUSMAN J., MCFADDEN D., "A specification test for the multinomial logit model", *Econometrica*, vol.52, no. 5, pp. 1219-1240, 1984.

[HER 99] HERNANDEZ T., SCHOLTEN H.J., BENNISON D., BIASIOTTO M., CORNELIUS S., VAN DER BEEK M., "Explaining retail GIS: The adoption, use and development of GIS by retail organisations in the Netherlands, the UK and Canada", *Netherlands Geographical Studies* n° 258, Amsterdam, Amsterdam University, 1999.

[HOL 05] HOLM K., HOLME E., LINDGREN U., KALLE M., "The SEVERIGE spatial microsimulation model", *International Microsimulation Conference on Population, Ageing and Health: Modelling Our Future*, University of Canberra, Australia, 2005.

[HUF 64] HUFF D.L., "Defining and estimating a trading area", *Journal of Marketing*, vol. 28, no. 3, pp. 34-38, 1964.

[HUR 94] HURIOT J.M., *Von Thünen: Économie et Espace*, Paris, Economica, 1994.

[HYS 50] HYSON C.D., HYSON W.P., "The economic law of market areas", *The Quarterly Journal of Economics*, vol. 64, no. 2, pp. 319-327. 1950.

[JON 01] JONES K.G., DOUCET M.J., "The big box, the flagship, and beyond: impacts and trends in the Greater Toronto Area", *The Canadian Geographer*, vol. 45, no. 4, pp. 494-512, 2001.

[KES 04] KESTENS Y., THÉRIAULT M., DES ROSIERS F., "Impact of surrounding land use and vegetation on single family house prices", *Environment and Planning B*, vol. 31, no. 4, pp. 39-567, 2004.

[KIM 03] KIM H.M., KWAN M.P., "Space-time accessibility measures: A geocomputational algorithm with a focus on the feasible opportunity set and possible activity duration", *Journal of Geographical Systems*, vol. 5, no. 1, pp. 71-91. 2003.

[LEO 00] LÉO P.Y., PHILIPPE J., "Centers villes et périphéries commerciales: le point de vue des consommateurs", *Cahiers de géographie du Québec*, vol.44, no. 123, pp. 363-397, 2000.

[LON 05] LONGLEY P.A., GOODCHILD M.F., MAGUIRE D.J., RHIND D.W., *Geographical Information Systems and Science*, 2nd Edition, Hoboken (New Jersey), Wiley, 2005.

[LOS 54] LÖSCH A., *The Economics of Location*, (Translated by W.H. Woglom and W.F. Stolper), New Haven, Yale University Press, 1954.

[LOU 00] LOUVIERE J., HENSHER D., SWAIT J., ADAMOWICZ W., *Stated Choice Methods: Analysis and Applications,* Cambridge University Press, 2000.

[LUC 59] LUCE R., *Individual Choice Behavior*, Wiley, New York, 1959.

[MAA 05] MAAT K., VAN WEE B., STEAD D., "Land use and travel behaviour: Expected effects from the perspective of utility theory and activity-based theories", *Environment and Planning B*, vol.32, no. 1, pp. 33-46, 2005.

[MAN 77] MANSKI C.F., "The structure of random utility models", *Theory and Decision*, vol. 8, pp. 229–54, 1977.

[MCF 81] MCFADDEN D., "Econometric models of probabilistic choice", in: C.F. MANSKI, D. MCFADDEN (eds), *Structural Analysis of Discrete Data*, Cambridge, MA, MIT Press, pp. 199–272, 1981.

[MER 03] MÉRENNE-SCHOUMAKER B., *Géographie des Services et des Commerces*, Rennes, Rennes University Press, 2003.

[MER 06] MERCIER P., Évolution dans la Distribution des Commerces de Détail sur le Territoire de la Communauté Métropolitaine de Québec entre 2000 et 2005, Essai de maîtrise, ESAD, Laval University, 2006.

[NAK 74] NAKANISHI M., COOPER L.G., "Parameter estimation for a multiplicative competitive interaction model-least squares approach", *Journal of Marketing Research*, vol. 11, no. 3, pp. 303-311, 1974.

[PAR 95] PARR J.B., "Alternative approaches to market-area structure in the urban system", *Urban Studies*, vol. 32, no. 8, pp. 1317-1330, 1995.

[POR 04] PORTUGALI J., "Toward a cognitive approach to urban dynamics", *Environment and Planning B*, vol. 31, no. 4, p. 589-613, 2004..

[POT 82] POTTER R.B., *The Urban Retailing System: Location, Cognition and Behaviour*, Aldershot, Hants, England, 1982.

[REI 31] REILLY W.J., *The Law of Retail Gravitation*, New York, Pillsbury Publishers, 1931.

[ROY 01] ROY J.R., SMITH N.C., XU B., "Simultaneous modeling of multi-purpose/multi-stop activity patterns and quantities consumed", *Journal of Geographical Systems*, vol. 3, no. 4, pp. 303-324, 2001.

[SHE 80] SHEPHERD I., THOMAS C.J., "Urban consumer behaviour", in: J.A., DAWSON (ed.), *Retail Geography*, New York, John Wiley, pp. 18-86, 1980.

[SLI 02] SLIWINSKI A., "Spatial point pattern analysis for targeting prospective new customers: Bringing GIS functionality into direct marketing", *Journal of Geographic Information and Decision Analysis*, vol. 6, no. 1, pp. 31-48, 2002.

[THÉ 99] THÉRIAULT M., VANDERSMISSEN M.H., LEE-GOSSELIN M., LEROUX D., "Modelling commuter trip length and duration within GIS: Application to an O-D survey", *Journal for Geographic Information and Decision Analysis*, vol. 3, no. 1, pp. 41-55, 1999.

[THÉ 04a] THÉRIAULT M., DES ROSIERS F., "Modelling perceived accessibility to urban amenities using fuzzy logic, transportation GIS and origin-destination surveys", in F. TOPPEN, P. PRASTACOS (eds), *Proceedings of AGILE 2004 7ᵗʰ Conference on Geographic Information Science*, Crete University Press, Heraklion, Greece, pp. 475-485, 2004.

[THÉ 04b] THÉRIAULT M., DES ROSIERS F, BIBA G., LAVOIE C., *Le Commerce de Détail sur le Territoire de la Communauté Urbaine de Québec*, Québec, CRAD, Laval University, 2004.

[THÉ 05] THÉRIAULT M., DES ROSIERS F., JOERIN F., "The effects of accessibility on house values: Its links to households' daily mobility behavior", *Journal of Property Investment and Finance*, vol. 23, no. 1, pp. 22-54, 2005.

[THI 92] THILL J.C., TIMMERMANS H.J.P., "Analyse des décisions spatiales et du processus de choix des consommateurs: Théorie, méthodes et exemples d'applications", *L'Espace Géographique*, no. 2, pp. 143-166, 1992.

[THO 02] THOMAS C.J., BROMLEY R.D.F., "The changing competitive relationship between small town centers and out-of-town retailing: town revival in South Wales", *Urban Studies*, vol. 39, no. 4, pp. 791-817, 2002.

[THR 02] THRALL G.I., *Business Geography and New Real Estate Market Analysis*, New York, Oxford Press, 2002.

[TIM 82] TIMMERMANS H.J.P., "Consumer choice of shopping center: An information integration approach", *Regional Studies*, vol. 16, no. 3, pp. 171-182, 1982.

[VIL 02] VILLENEUVE P., VANDERSMISSEN M.H., "Accessibilité, mobilité et équité dans le Québec métropolitain", in: R. CÔTÉ (ed), *Québec 2003: Annuaire Politique, Social, Économique et Culturel*, Montréal, Fides, pp. 282-290, 2002.

[YEA 01] YEATES M., "La géomatique du commerce de détail: un domaine en développement", *Canadian Journal of Regional Science/Revue Canadienne des Sciences Régionales*, vol. 24, no. 3, pp. 387-398. 2001.

[ZEN 01] ZENG T. Q., ZHOU Q., "Optimal spatial decision making using GIS: A prototype of a real estate geographical information system (REGIS)", *International Journal of Geographical Information Science*, vol. 15, no. 4, pp. 307-321, 2001.

Chapter 4

Generation of Potential Fields and Route Simulation Based on the Household Travel Survey

4.1. Introduction

Urban sprawl, the scattered nature of living spaces, an increase in daily travel and new trends in mobility behavior following the introduction of the car are transformations that particularly influence people's daily mobility and lead us to reconsider city management methods. Improving our understanding of these urban dynamics is a tangible issue when it comes to promoting a sustainable city. Designing observation platforms allows us to plan out transportation supply according to the associated risks and nuisances (accidents, air and noise pollution). However, evaluating the impact of policies governing the mobility of goods and people or adapting urban rhythms to daily needs also requires proper analytical tools.

In this sense, the household travel survey is of critical importance when considering the cities of developed countries [CLA 00, TRE 01]. This standardized observation tool, which measures mobility on the urban scale, is in fact the main source of data currently available to analyze the mobility of every class of person and all modes of transportation available at the city level. The use of the information provided by the resulting dataset must, however, be adapted to the new issues

---

Chapter written by Arnaud BANOS and Thomas THÉVENIN.

arising from changes in urban mobility. Thus, for most specialists in this area, this data source, although quite prolific, is mostly underexploited. Current practice relies more often than not on a standard methodology, based on the construction of macroscopic aggregate indicators that are not adapted to reflect the growing variability of mobility behavior.

In that context, the microscopic analysis of individual mobility behavior is a promising avenue to explore because it generates qualitative and quantitative information that could shed new light on the understanding of daily urban mobility. Despite this, it often clashes with a widespread deficiency of the household-mobility survey: the lack of precision on the location of respondents and the trips they have made. This chapter aims to present a protocol that allows these individual trips and locations to be simulated, through the generation of a potential field for the spatial assignment of populations and activity places.

Three key steps are discussed in detail. First, we present a reconstruction of a virtual urban environment within a geographic information system (GIS). Then, within this environment, we calculate the potential fields required to implement disaggregated simulation models. One example allows us to demonstrate how to use these data to show the city in motion. Lastly, a second example/model allows us to demonstrate how the same potential field can be used to reconstruct individual trips through simulation.

4.2. Rebuilding the virtual city

A realistic reconstruction of a city must first be done in order to study its urban dynamics. The research conducted by Michael Batty's team at the CASA (Centre for Advanced Spatial Analysis) on the virtual representation of Greater London is the baseline reference in the area of digital city reconstruction. According to Batty, there is no unique definition to describe a virtual environment [BAT 98]. The most frequent approach taken is first to distinguish the digital environment, such as buildings, and then the individuals who live in this virtual space (e.g. users of public transportation). But, how do we connect these users to their environment?

4.2.1. *A systematically disaggregated model*

Environment-individual interaction can be modeled using three scale levels [BAT 98]. Frequently used in video games, the *immersive* model consists of reproducing the perspective of a single user, while the *semi-immersive* model considers a group of individuals. The *remote* model allows us to observe the phenomena that occur over an entire space. The semi-immersive and remote models seem better adapted to meet research needs in urban transportation modeling, and we use this combination to study the dynamic phenomena for a group of buildings or to reproduce all of the detectable movement within a population cluster.

This multilevel modeling requires an accurate description of reality in order to properly master the data-clustering process [FOT 00]. Although very accurate data on the urban environment are available, for privacy reasons, sociodemographic or mobility data are mostly disclosed using an administrative boundaries network that is not always conducive to observing urban dynamics [ROB 01]. Thus, there is a need for some systematic disaggregation of data that can be directly integrated into a GIS. This can lead to the "totally disaggregate approach" suggested by Chapleau [CHA 92], which differs from the classical approaches (e.g. four-step model), mainly given the precise referencing of the data it introduces.

The systematic disaggregated approach is seen here using three observation units [THE 02]:

– The *site unit* represents the space used for a human activity as accurately as possible. For example, the home, school and workplace are located separately by a geographical coordinates system. The population, generally considered at an aggregated level, can be assigned to these locations following the potential field method described below.

– The *time unit* is a reference day. The choice of a representative day to study the pace of activities performed by many people is open to criticism. Unfortunately, data-gathering constraints grow exponentially when we try to comprehend the exact weekly and monthly comings and goings of individuals [LEE 05, STO 07]. Thus, this work deals with the reconstruction of activity flow for a typical weekday. Two temporal views are considered – the exact time the trip is made (absolute temporal view; e.g. peak hours) and the actual trip duration (relative temporal view).

– The *transportation mode unit* is broken down into three major groups: soft modes, also known as active modes (walking, bicycle); individual motorized modes (car, motorcycle, etc.); and public motorized modes (bus, tram, metro). The schedules of public modes could be used to differentiate each vehicle in operation according to arrival time at service points (stops).

The integration of these observation units allows us to reconstruct the spatiotemporal dynamics of the city for a given reference period (e.g. a typical day).

4.2.2. *Structuring data through space and over time*

The disaggregated system used to model the virtual city requires that databases be appropriately structured [CLA 03] in order to reflect the localized supply and demand of urban services. To succeed, we must follow a formal procedure and specify, using a conceptual data model, the various relationships among the phenomena of interest. Used widely in computer science and in the field of geomatics, the so-called Unified Modeling Language (UML) fulfills this objective.[1]

Figure 4.1 provides a basic overview of the model's main data classes, grouped into three sub-models:

– The *Activity* sub-model groups together data on the population located at home, work (employees), as well as educational and university institutions. Businesses and leisure centers are also integrated. Opening and closing hours are added as an instance of a class based on a field survey.

– The *Land use* sub-model describes the geographical context of the city. It indicates the precise location of buildings, especially residential ones. Street addresses (*Address*) are used for the geocoding of businesses. Lastly, the *Parcel* class gathers precise data on buildings, especially land use and the number of dwellings.

– The *Network* sub-model includes classes for modeling private and public transportation modes; it is needed to simulate trips.

These three sub-models are needed for two distinct procedures: associations between the *Activity* and *Land use* sub-models; and associations between the *Network* and *Land use* sub-models.

First, the associations between the *Activity* and *Land use* sub-models are needed for locating activity places within the city. Four classes are thus modeled as both a work place and an activity place:

– *Business*: groups together consumption areas (e.g. department stores and retail businesses) that may also be work places;

– *Study*: associates students and children in school, as well as teaching and administrative staff;

– *Leisure*: groups together supervisory staff and members of clubs; and

– *Corporate*: combines employees of industrial and service firms (outside business sector).

These classes are related time-wise by an association with the *Opening Time* class (schedule of operation), and associated with the *Building* class using an

---

1 We used the prototype Perceptory software, developed by the team of Yvan Bédard at Laval University in Québec City [PRO 02]. This tool is particularly useful when considering both the nature of geographical objects and their changes over time.

address-matching procedure. The *Resident Population-Building* association is implemented using population redistribution rules detailed in the next section.

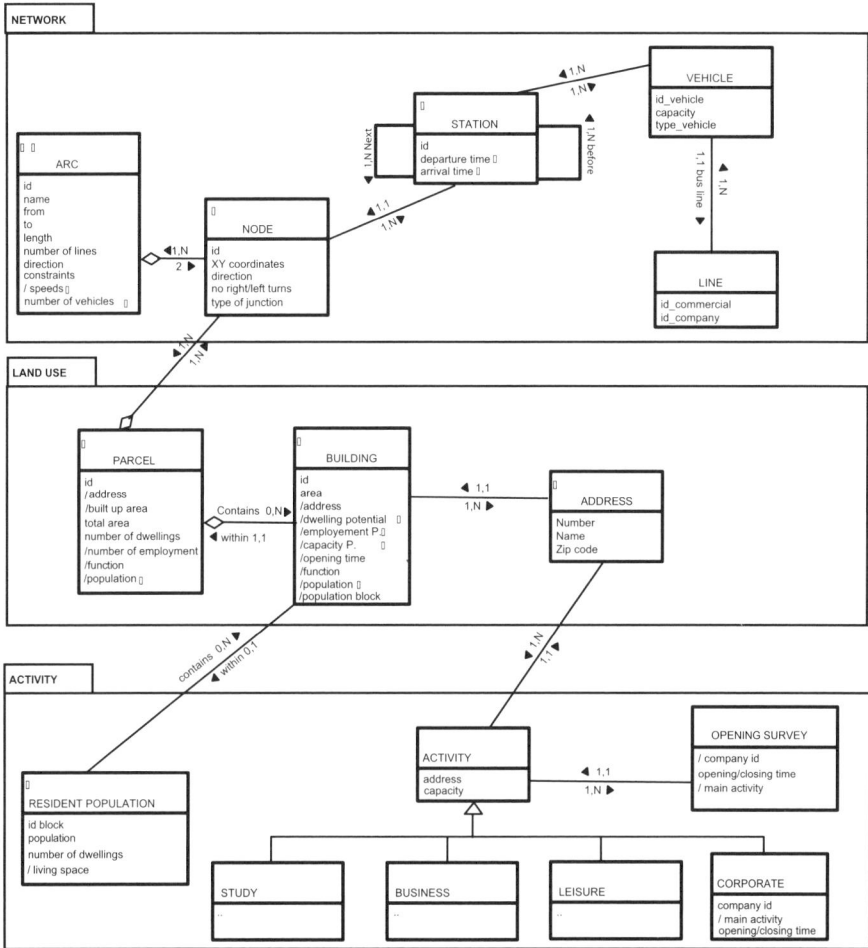

Figure 4.1. *A conceptual model of the virtual city*

Second, associations between the *Network* and *Land use* sub-models relate to the *Nodes* class. This, in turn, ties in all the transportation modes considered. The *Network* sub-model includes *Nodes* and *Arcs* classes, and topological relationships are modeled using *to* and *from* association links. Interconnection between the modes is ensured by the fact that each node may belong to both public and private transportation networks. The UML class diagram (see Figure 4.1) depicts how data on users and the transportation networks are integrated.

4.2.3. *Generating a potential field for spatial assignment of a population*

Understanding the daily movements in a city begins with a wide-scale study of trips. Origin-destination studies (OD) in Canada (see Chapters 2 and 3) or the *Micro-recensement-Transport Suisse* (Swiss Transportation Microsurvey) are very precise data-gathering protocols that enable us to recreate a true image of travel in a city for a typical weekday [BON 04]. The two preceding chapters provide examples of the use of Canadian OD surveys to compare and model men's and women's trips, and to study consumer destination choices.

France also has a daily mobility survey for larger cities (populations above 100,000), which is called the *Enquête Ménages-Déplacements* (Household Travel Survey or HTS) [CER 98a]. This survey initiative, coordinated by the *Ministère de l'Equipement et des Transports* via CERTU[2], collects data on a representative sample of the population for three levels of detail: trips, individuals and households (see Figure 4.2).

Integrating these data poses a basic problem, however, since the data stem from two types of spatial representation. Data gathered from the French HTS is collected from a zoning system, represented in the GIS by an area. The data pertaining to transportation are represented by a network, symbolized by arcs and nodes in the GIS. The difficulty lies in linking these two sets of spatial entities. To do so, we must enhance the data of each zone by deriving the point-based distribution of people surveyed within its boundaries, as accurately as possible.

Several methods have been proposed for making this transition between the area and a multitude of points in order to redistribute persons while considering land use [ROB 01, TOB 79]. We suggest using a distribution method based on the potential field concept and attempting to define the capacity of each location, and ideally, each building, to supply and attract individuals. Figure 4.3 presents the main steps of this process.

---

2 CERTU (*Centre d'Études sur les Réseaux, les Transports, l'Urbanisme et les constructions publiques*) is a governmental technical service of the *Ministère Français de l'Equipement, des Transports et du Tourisme*. It was inaugurated in February 1994 as a result of the merger of two services: the *Centre d'Études des Transports Urbains* and the *Service Technique de l'Urbanisme*.

Household

Number of individuals	Housing	Housing tenure	Number of cars	Number of Drivers licence		
3	Detached	Tenant	1	2		

Individual

Sex	Year of Birth	Status	Class	Drivers licence	Mode 1	Car park 1	Car park 2
Female	40	Education	Lower mid.	Yes	Car	Company	Free

Trip

Day	Departure number	Trip purpose Destination	Zone of origin	Hour of departure	Hour of arrival	Duration	Zone of destination	Mode
Tuesday	1	Work	1901	7:50	8:00	10	1702	Car
Tuesday	2	Chauffering	1702	12:00	12:05	5	1801	Car
Tuesday	3	Chauffering	1801	12:20	12:25	5	1901	Car
Tuesday	4	Home	1901	12:25	12:30	5	1901	Car

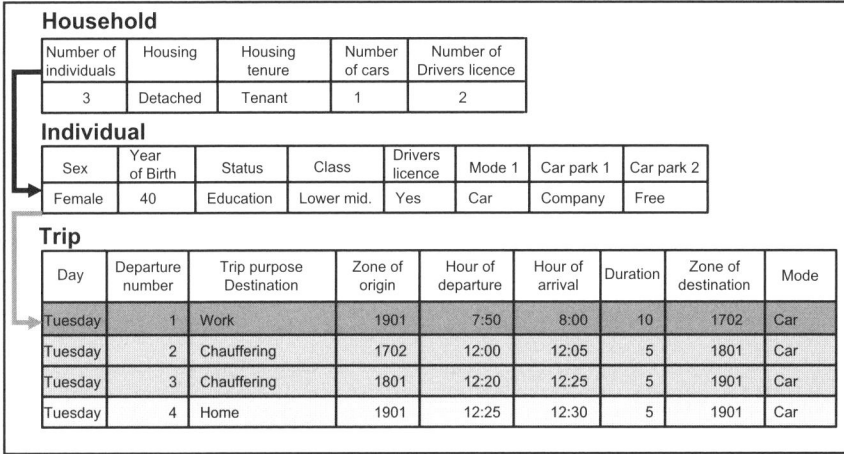

Figure 4.2. *The three levels of detail of a HTS*

Thus, the *Blocks*, *Parcels* and *Buildings* data levels are combined in order to estimate the resident population for each building belonging to each category. The population can be broken down using a proportional assignment rule taking into account the population per block and the number of housing units per parcel. Each building's potential (i.e. its resident population) is thus estimated using:

$$Potential\left(B_j\right) = Population\left(I_k\right) \times \frac{\#H.Unit\left(B_j\right)}{\#H.Unit\left(I_k\right)}, \forall B_j \subset I_k \qquad [4.1]$$

where B_j building j, I_k block k, and $\#H.Units$ the number of housing units in a given entity (building or block).

The effectiveness of this simple procedure lies mainly in the quality and abundance of geographic data. Generally, the more accurate data there are, the better the result will be. However, we can see that this simplicity is primarily linked to the fundamental simplicity of the estimator sought (building population) and that it holds for a more complicated estimator (for example, the population by age group or by social class per living area).

The potential of non-residential buildings (schools, businesses, companies, etc.) is assessed more directly using enumeration data compiled by various organizations: educational bodies, business registers, the National Statistics Service.

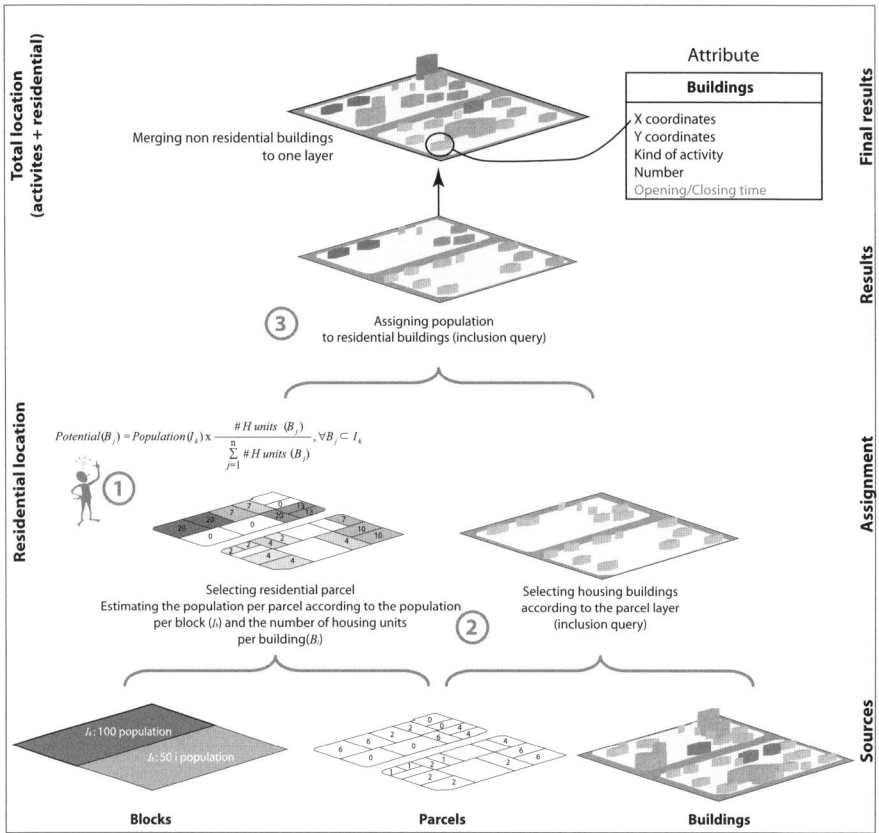

Figure 4.3. *Global overview of the method used to distribute people in buildings*

In some cases, it is not worth attempting to drop to the level of individual buildings, either because the database does not allow it, there are too many data, or the application purpose does not justify it. In this case, it is possible to select an alternative grid-based procedure, as shown in Figure 4.4. The focus is then to superimpose a regular grid on the residential building layer. Grid cells of the same shape and size become referential geographic entities. A simple aggregation function then allows us to estimate the living area for each cell. This value can be used directly in calculating the potential. Some ambiguity may, however, come from one cell possibly being linked to several survey zones. In this case, the intersection surface with the cell is used as an indicator to weight its share for the proportional allocation procedure.

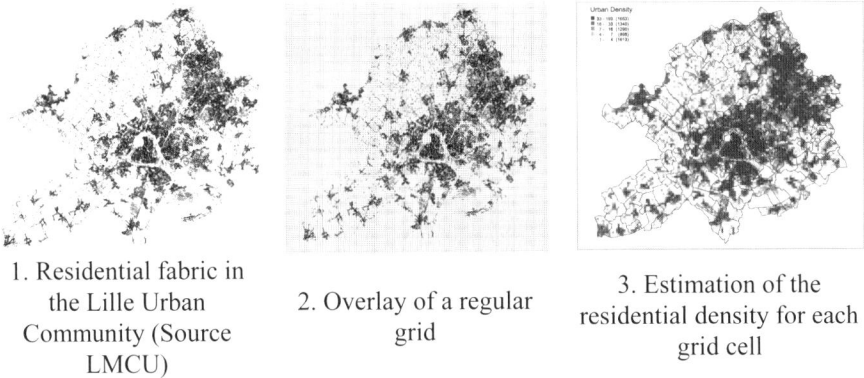

| 1. Residential fabric in the Lille Urban Community (Source LMCU) | 2. Overlay of a regular grid | 3. Estimation of the residential density for each grid cell |

Figure 4.4. *An alternative method for estimating the potential field*

Once defined, this potential field enables several applications, particularly in the spatial mobility sector. We show two examples: the first one looks at the city in motion by revealing its "pulse", while the second focuses on reproducing individual routes and trajectories.

4.3. From the city in motion to individual trajectories

It is difficult to understand and model the daily mobility within a city. No urban management policy today can avoid this challenge which, aside from its "transportation planning" aspect, falls back on a basic and essential function of the city: to permit and promote social interaction among its inhabitants. It is also a challenge, both conceptually and methodologically [MIL 07]. As a space-based social phenomenon, daily mobility of persons can be understood through the spatial prism of time geography [HÄG 70]. But how do we do justice to this incredible effervescence, this seething life that reveals itself to the observer's attentive gaze?

Coming into play are a multitude of individuals, who are autonomous within geometrically variable free spaces, driven by personal motivations, consuming and moving back and forth daily across a relative space made up of a specific combination of interconnected locations with their own individual and social values. Understood at this microscopic level, the buzzing beehive appears to be but an amalgamation of particular, specific individual behaviors. Yet, seemingly clear, lasting structures seem to emerge at times from this enormous diversity, if we agree to look from another vantage point: to temporarily stop focusing on the individual and concentrate on groups of people and the territory they visit. When we focus too much on space and its dynamics, however, it is easy to forget about the individuals.

Thus, the variation in geographical and observational levels must go hand-in-hand if we want to shed light on this truly complex "urban anthill".

4.3.1. *Revealing the city in motion*

The modeling process has to be reductionist, in that the focus is on reconstructing a phenomenon using only its basic components. It is also simplifying, in that the purpose of the model is to provide a representation of an often complex reality, which is simplified but also intelligible and comprehensive. This paradox is clearly behind the difficulties of the modeling process. One of the most classic examples of simplification in the area of mobility and transportation is given by the first two steps in Figure 4.5: 1) the division of space into a small number of supposedly homogeneous zones in terms of mobility; 2) the creation of OD matrices that estimate the number of trips made (emissions) and received by each of these zones, based on a segmentation logic of events. The classic first-generation models (four-step aggregated models) and, to a lesser extent, second-generation models (discrete choice models) basically fall under this approach [BON 04, CER 98b, KAN 83, MAN 79, MEY 84, ORT 94].

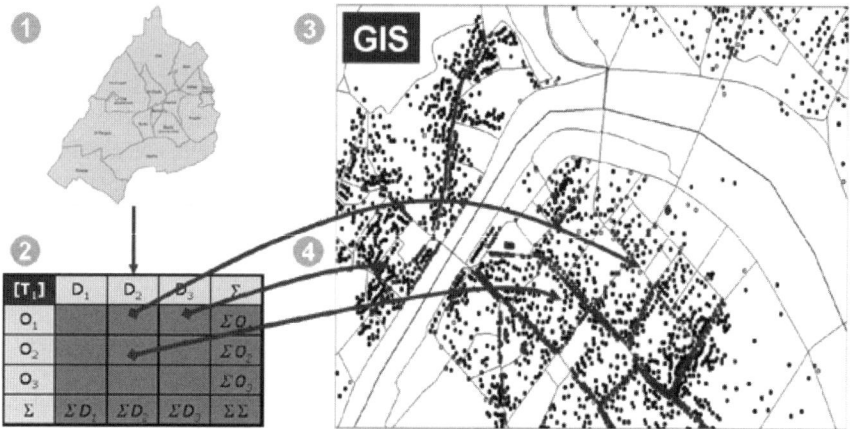

Figure 4.5. *Disaggregating mobility data with simulation*

Based on such aggregated data, how can we produce a finer image to depict realistic sets of individual trips compatible with the city-wide travel trends reported during one of these typical days (the OD matrix)? An interesting idea would be to assign the trips estimated by the matrix to the built-up areas. First and foremost, however, this operation implies admitting a segmented view of daily travel as mutually independent trips. Reproducing trip chains of individuals based on the data from this type of matrix would indeed be a seemingly impossible feat, given that the possible combinations of successive trips are infinite. In general, it is extremely difficult and risky to disaggregate aggregated data, especially without additional information that would help in reducing the number of combinations. This is a very general principle that we must take into account, even if the objective at this stage is not to reconstruct chains of activity. Indeed, in the absence of any information other than that described in Figure 4.5, it would be difficult to come up with anything better than a random model, assigning trips in the matrix randomly in time (over the course of a day) and in space (within each zone). Two types of complementary information have proven especially useful for getting over this hurdle and reducing any unknowns in the model, namely overall temporal flow distribution (Figure 4.6) and the potential at each built-up area.

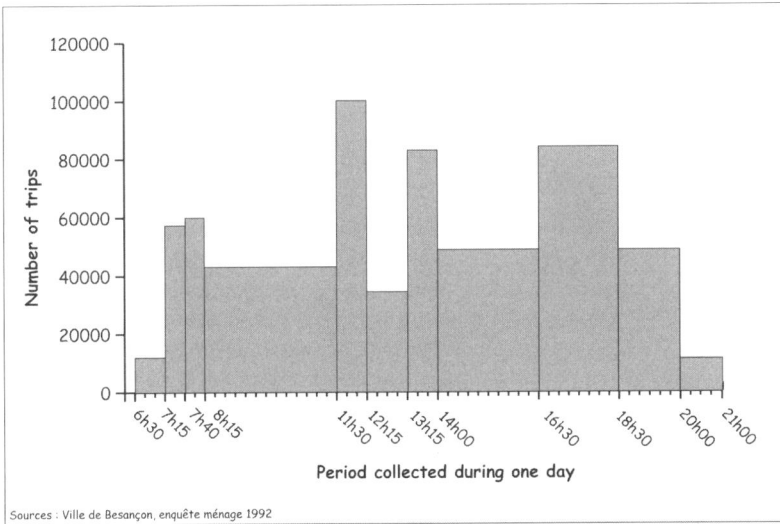

Sources : Ville de Besançon, enquête ménage 1992

Figure 4.6. *Estimated temporal distribution of trips in Besançon*
(Source: HTS, 1992)

Based on these assumptions, each of the 800,000 daily trips estimated by the OD matrix in 1992 for Besançon [BAN 05b] can be assigned to the built-up area level during a typical day[3]. For each period of time, t, this method can be used to modify the spatial assignment by a proportion, P_t (given in the histogram in Figure 4.6), of

---

3 The data presented herein represent workdays (not holidays) outside the school holiday period.

the trips in the matrix by randomly selecting an element j from the built-up zone in the corresponding zone i for each of these trips, based on a probability, P_{ij}, such as:

$$P_{ij} = \frac{Potential(B_j)}{\sum_{j=1}^{n} Potential(B_j)} \forall j \subset i \qquad [4.2]$$

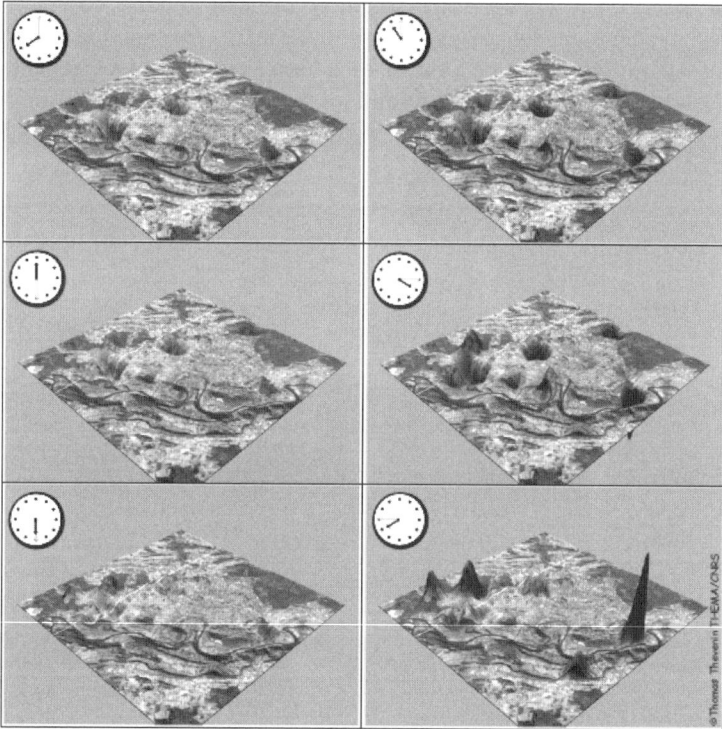

Figure 4.7. *Reconstruction by simulation of an area's "pulses" (Besançon in this case)[4]*

Since the 1970s, highly original mapping techniques have been developed to show dynamic phenomena, such as daily trips [JAN 84, JAN 88, MOE 76, TOB 73]. The example shown in Figure 4.7 is based on a spatial smoothing procedure, which portrays potential fluctuations in the urban space of Besançon. This mapping device shows that different sites constantly empty and fill up during the day depending on the movements of people in these areas. The valleys

4 A dynamic view is available at: http://arnaudbanos.perso.neuf.fr/geovisu/fourmiliere.html, accessed 22 September 2010.

correspond to sites that send out more movement signals than they receive, whereas the peaks indicate local accumulation of the population during a given period of time.

Revealing and identifying the "pulse" of population clusters in an urban area through the constant organization and reorganization of "empty" and "full" spaces over the course of the same day are undeniably major urban issues that deserve suitable methodology. By focusing like this on the space, however, it is very easy to forget about the individuals. The question that is raised then is: can we get past this initial level of analysis and keep track of these increasingly mobile individuals traveling the urban roads?

4.3.2. Rebuilding individual trajectories

By breaking down trips into a series of elementary events, we inevitably lose all data relating to the trip chains. This is a major limitation, pointed out at the end of the 1960s by partisans of the "space-time-activities" approach, largely inspired by prior studies by American sociologist Chapin [CHA 74] and by Swedish geographer Hägerstrand [HÄG 70] on daily activity programs [HEN 79, JON 79, KIT 88, KUR 97, RAU 83]. Working at the level of trip chains implies using disaggregate data, such as those provided by the household-mobility survey (see Figure 4.8).

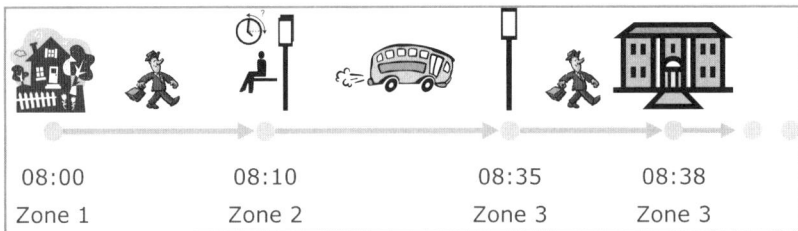

| 08:00 | 08:10 | 08:35 | 08:38 |
| Zone 1 | Zone 2 | Zone 3 | Zone 3 |

Figure 4.8. *Individual mobility data provided by household-mobility surveys in France*

What is striking here is the level of detail on the activities performed[5], the modes of transportation used, and the times (e.g. departure, arrival, duration). It must be mentioned, however, that the spatial dimension is more problematic: the exact location of individuals is unknown, given the location unit is the trip zone. Using this database and filling in the location gaps, notably with a potential field, is it

5 The French household survey includes all activities performed, and not only those involving a trip, as is the case in the Canadian OD surveys.

possible to reconstruct the actual trips made by the survey respondents? This task would imply some working hypotheses that compensate for missing data.

Thus, as an example, we assume that individuals always take the shortest route in time between a given origin and destination. Our motivation is both academic and methodological. From a theoretical viewpoint, to assume that individuals are capable of such a feat is to place them *ipso facto* in the conceptual framework of the omnipresent rational person, or *Homo oeconomicus*. In this context, the variability of individual behaviors can be summed up by an average synthetic behavior. This would be that of an average individual who is perfectly rational, with unlimited information, equipped with infinite calculation capabilities and constantly trying to maximize his utility by adopting an optimizing behavior. In short, a person who is always able to find the shortest route in a road network. This is a high and rather controversial assumption. Although some empirical research [BOR 86, LAU 74, SEN 86] shows that pedestrians and drivers try to reach their destination by the shortest route, other studies show that this is not the only criterion at work in the strategies chosen [GOL 97, SAI 86].

The second justification is more based on methodology: such a hypothesis refers to studies conducted over a lengthy period of time regarding operational research into the path of the graph. In fact, there are various families of algorithms with perfectly known properties and behaviors that enable us to identify, with impedance constraints, the shortest route between two points in a network.

The implementation of this type of model in a GIS allows us to reconstruct the possible trips and itineraries of individuals surveyed (i.e. nearly 13,000 people in Lille) [BAN 06a]. How then do we display the results obtained? There is an inherent cost to revealing complexity, linked in part to our ability (or lack thereof) to accomplish this. Figure 4.9 suggests a range of possible solutions. The road network is depicted with grey lines, moving people are represented by white dots, and immobile people at the time *t* by grey dots. Figure 4.9 (a) displays an overview of the urban anthill, while (b) focuses on the simulated trajectory of a selected individual. Lastly, Figure 4.9 (c) displays both moving individuals and the differentiated occupation of urban space, indicated as peaks whose height is proportional to the number of people present nearby. A very large number of charts can thus be produced using these disaggregated data, including traffic flow charts aggregated by road segments.

Beyond their apparent differences, these two approaches are based on the same statistical independence principle, which makes it much easier to implement them, but also considerably limits the potential for generalization. What understanding of the city in motion can we reach from these mutually independent trips and individual trajectories? According to Herbert Simon [SIM 62], it is now acknowledged that the whole is not only more than the sum of its parts, but it is eventually different, emergent phenomena often involving a qualitative leap. Given this, even if we are aware of all the factors coming into play in individual actions, we are still not able to predict the behavior of a group of individuals, precisely because of the many nonlinear interactions that come into play. This limitation led us, within the framework of the Miro project [BAN 06b], to favor a theoretical integration that we consider to be broad. This combines research regarding complex systems, the simulation of artificial societies and agent-based simulation, and targeting the better understanding of phenomena, not by deconstructing them but by reconstructing them.

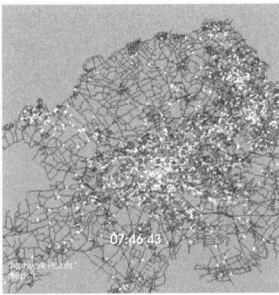

(a) An overview of the "urban anthill"   *(b) An individual trajectory*

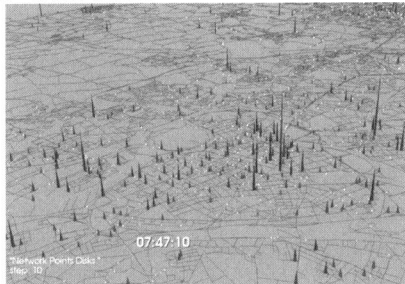

(c) Individual trajectories within a city in motion

Figure 4.9. *Rebuilding individual routes (taking the example of Lille)*[6]

4.4. Conclusion

Current issues in city management encourage the implementation of multilevel investigation strategies in order to take into account mobility and action spaces, but

---

6 Animations created by Julien Lesbegueries, and available at: http://arnaudbanos.perso.neuf. fr/geovisu/fourmiliere.html, accessed 22 September 2010.

also to reconstruct the spatial routes chosen by the individuals who make these trips. A task such as this is limited by the constraints of existing data. The disaggregation of aggregated data, notably through the generation of potential fields, is thus often a necessity that involves the production of spatial simulation protocols within a GIS.

While useful, these protocols do not allow us to truly track individual trajectories, unless the number of working hypotheses is unreasonably expanded. The use of more disaggregate data, collected from reduced samples of individuals, and *a priori* better adapted to the set objectives, meets with other difficulties. Such difficulties include ethical issues and generalization issues, among others.

A complementary course of action is to generate these individual routes from start to finish in a virtual, computerized setting that allows us to specify the behavioral rules and characteristics liable to result in the emergence of collective and structural behaviors that make sense.

4.5. Bibliography

[BAN 05a] BANOS A., CHARDONNEL S., LANG C., MARILLEAU N., THÉVENIN T., "Simulating the swarming city: A MAS approach", *Proceedings of the CUPUM'05 Conference on Computers in Urban Planning and Urban Management*, Paper 97, London, June 2005.

[BAN 05b] BANOS A., THÉVENIN T., "La carte animée pour révéler les rythmes urbains", *Revue Internationale de Géomatique*, vol 15, no. 1, pp. 11-31, 2005.

[BAN 06a] BANOS A., MARILLEAU N., THÉVENIN T., CHARDONNEL S., LANG C., BOFFET-MAS A., "Génération d'emplois du temps individuels pour une simulation multi-agents des mobilités urbaines quotidiennes", *Colloque SAGEO'06*, Strasbourg, September 2006.

[BAN 06b] BANOS A., "Geosimulating the swarming city: A bouquet of alternatives managing complexity", *Geoinformatics*, vol. 8, pp. 58-61, 2006.

[BAT 98] BATTY M., DODGE M., DOYLE S., SMITH A., Modelling Virtual Urban Environments, Working Paper, CASA, http://eprints.ucl.ac.uk/219/1/modelvue.pdf, 1998.

[BEN 84] BEN-AKIVA M., LERMAN S., *Discrete Choice Analysis: Theory and Application to Travel Demand*, MIT Press, Cambridge, 1984.

[BON 04] BONNEL P., *Prévoir la Demande de Transport*, Presse de l'ENPC, 2004.

[BOR 86] BORGERS A., TIMMERMANS H., "A model of pedestrian route choice and demand for retail facilities within inner-city shopping areas", *Geographical Analysis*, vol. 18, no. 2, pp. 115-128, 1986.

[CER 98a] CERTU, CETE de Lyon, CETE Nord-Picardie, Enquête Ménages Déplacements: Méthode Standard, CERTU, Lyon, 1998.

[CER 98b] CERTU, Comportements de Déplacement en Milieu Urbain: les Modèles de Choix Discrets: Vers une Approche Désagrégée et Multimodale, Dossier du CERTU no. 81, CERTU, Lyon, 1998.

[CHA 74] CHAPIN F.S., *Human Activity Patterns in the City: Things People do in Time and in Space*, John Wiley & Sons, New York, 1974.

[CHA 92] CHAPLEAU R., "La modélisation de la demande de transport urbain avec une approche totalement désagrégée", *World Conference on Transportation Research Proceedings*, Lyon, pp. 937-948, 1992.

[CLA 00] CLARAMUNT C., JIANG B., BARGIELA A., "A new framework for the integration, analysis and visualisation of urban traffic data within geographic information systems", *Transportation Research part C*, vol. 8, pp. 167-184, 2000.

[CLA 03] CLARAMUNT C., PARENT C., "Modelling concepts for the representation of evolution constraints", *Computers Environment and Urban Systems*, vol. 7, pp. 225-241, 2003.

[FOT 00] FOTHERINGHAM S., "GeoComputation analysis and modern spatial data", in: OPENSHAW S., ABRAHART R. (eds.), *GeoComputation*, Taylor & Francis, London, pp. 33-48, 2000.

[GOL 97] GOLLEDGE R., "Defining the criteria used in path selection", in: ETTEMA D.F., TIMMERMANS H.J.P. (eds.), *Activity-based Approaches to Travel Analysis*, pp. 151-169, 1997.

[HÄG 70] HÄGERSTRAND T., "What about people in regional science?", *Papers of the Regional Science Association*, vol. 24, pp. 7-21, 1970.

[HEN 79] HENSHER D., STOPHER P., *Behavioral Travel Modelling*, Groom Helm London, London, 1979.

[JAN 84] JANELLE D., GOODCHILD M., "The city around the clock: Space-time patterns of urban ecological structure", *Environment and Planning A*, vol. 10, pp. 179-189, 1984.

[JAN 88] JANELLE D., GOODCHILD M., KLINKENBERG B., "Space-time diaries and travel characteristics for different levels of respondent aggregation", *Environment and Planning A*, vol. 20, pp. 891-906, 1988.

[JON 79] JONES P., "New approaches to understanding travel behaviour: The human-activity approach", in: HENSHER D.A., STOPHER P.R., *Behavioral Travel Modelling*, Taylor and Francis, London, pp. 55-80, 1979.

[KAN 83] KANAFANI A., *Transportation Demand Analysis*, McGraw-Hill, New York, 1983.

[KIT 88] KITAMURA R., "An evaluation of activity-based travel analysis", *Transportation*, vol. 15, pp. 9-34, 1988.

[KUR 97] KURANI K., LEE-GOSSELIN M., "Synthesis of past activity analysis applications", *Activity-based Travel Forecasting Conference*, US Department of Transport, Washington DC, 1997.

[KWA 00] KWAN M.P., "Analysis of human spatial behavior in a GIS environment: Recent developments and future prospect", *Journal of Geographical Systems*, vol. 2, pp. 85-90, 2000.

[LAU 74] LAUSTO K., MUROLE P., "Study of pedestrian traffic in Helsinki: Methods and results", *Traffic Engineering and Control*, vol. 15-9, pp. 446-449, 1974.

[LEE 05] LEE-GOSSELIN M.E.H., "A data collection strategy for perceived and observed flexibility in the spatio-temporal organisation of household activities and associated travel", in TIMMERMANS H.J.P. (ed.), *Progress in Activity-Based Analysis*, Elsevier, pp. 355-371, 2005.

[MAN 79] MANHEIM M., *Fundamentals of Transportation Systems Analysis. Volume 1: Basic Concepts*, MIT Press, Cambridge, 1979.

[MCN 00] MCNALLY M.G., "The activity-based approach", in: HENSHER A., BUTTON J. (eds), *Handbook of Transport Modelling*, Elsevier Science Ltd, 2000.

[MEY 84] MEYER M., MILLER E., *Urban Transportation Planning, a Decision-oriented Approach*, McGraw-Hill, New York, 1984.

[MFA 79] MCFADDEN D., "Quantitative methods for analysing travel behaviour of individuals: Some recent developments", in: HENSHER D.A., STOPHER P.R., *Behavioral Travel Modelling*, Taylor and Francis, London, pp. 279-318, 1979.

[MIL 07] MILLER H, "Necessary space-time conditions for human interaction", *Environment and Planning B: Planning and Design*, vol. 32, pp. 381-401, 2007.

[MOE 76] MOELLERING H., "The potential uses of a computer animated film in the analysis of geographical patterns of traffic crashes", *Accident Analysis & Prevention,* vol. 8, pp. 215-227, 1976.

[ORT 94] ORTUZAR J., WILLUMSEN G., *Modelling Transport*, John Wiley & Sons, Chichester, 439 p., 1994.

[PRO 02] PROULX M.J., LARRIVÉE S., BÉDARD Y., "Représentation multiple et généralisation avec UML et l'outil Perceptory", in: RUAS A. (ed.), *Généralisation et Représentation Multiple*, Hermes, pp. 113-129, 2002.

[RAU 83] RAUX C., Modèles et prévision des comportements de mobilité quotidienne, Economic Sciences Thesis, University of Lyon II, 1983.

[ROB 01] ROBERT D., GRASLAND C., *La Pertinence des Zonages Géographiques pour l'Analyse des Transports*, Paris, Predit-Metl, 2001.

[SAI 86] SAISA J., SVENSON-GARLING A., GARLING T., LINDBERG E., "Intra-urban cognitive distance: The relationship between judgements of straight-lines distances, travel distances and travel times", *Geographical Analysis*, vol. 18, pp. 167-174, 1986.

[SEN 86] SENEVIRANTE P., MORALL J., "Analysis of factors affecting the choice of route pedestrians", *Transportation Planning and Technology*, vol. 10, pp. 147-159, 1986.

[SIM 62] SIMON H., "The architecture of complexity", *Proceedings of the American Philosophical Society*, vol. 106, pp. 467-482, 1962.

[STO 07] STOPHER P., GREAVES S., "Household travel survey, where are we going?" *Transportation Research, Part A*, vol. 41, pp. 367-381, 2007.

[THÉ 03] THÉVENIN T., JOSSELIN D., FAUVET M.C., "Modélisation spatio-temporelle d'un réseau de transport public: articulation intermodale d'un réseau de bus, d'un transport à la demande et d'un espace piétonnier", *Revue Internationale de Géomatique*, vol. 13, no. 2, pp. 157-180, 2003.

[TOB 73] TOBLER W, "Choropleth maps without interval", *Geographical Analysis*, vol. 5, pp. 262-265, 1973.

[TOB 79] TOBLER W., "Smooth pychnophylactic Interpolation for geographical regions", *Journal of the American Statistical Association*, vol. 74, no. 357, pp. 519-535, 1979.

[TRÉ 01] TRÉPANIER M., CHAPLEAU R., "Analyse orientée-objet et totalement désagrégée des données d'enquêtes ménages origine-destination", *Revue Canadienne de Génie Civil*, vol. 28, no. 1, pp. 48-58, 2001.

[YU 07] YU H., SHAW S.L., "Revisiting Hägerstrand's time-geographic framework for individual activities in the age of instant access", in: MILLER H. (ed.), *Societies and Cities in the Age of Instant Access*, Dordrecht, The Netherlands, Springer Science, pp. 103-118, 2007.

Chapter 5

Impacts of Road Networks on Urban Mobility

5.1. Introduction

A review of the literature on urban transportation shows that the road network is urban system since it defines its space of mobility. Road networks are taken for granted, an object that we readily modify by adding or reorienting lanes. Although engineering literature may focus on the technical aspects of streets and traffic (road cross-sections, pavement, traffic light sequencing, etc.), the network itself, the analysis of its shape and functioning, is an area that remains minimally explored. Form and function typologies of networks are barely defined in connection to possible side-effects of transportation practices and urban patterns. Questions arise as a result: are some networks more accident-prone than others, more vulnerable to traffic congestion, friendlier to this or that mode of transportation, more receptive to urban sprawl? Many of these issues remain unanswered, especially since we make little effort to subsume the form and function of road networks into several aggregated indexes [BEG 97]. This no doubt explains why work on the externalities of urban form, especially environmental, has up to now primarily focused on the built environment, using population density as a descriptive variable.

The work of Newman and Kenworthy [NEW 89] – which links density, energy consumption and automobile dependency on a very large scale – is just such an example. Although other studies are much more guarded regarding the link between high density and low energy consumption [BAN 97], they still follow the same

Chapter written by Jean-Christophe FOLTÊTE, Cyrille GENRE-GRANDPIERRE and Didier JOSSELIN.

logic, without taking into consideration the way mobility is organized through the conception and operation of road networks.

Some studies have revealed the existence of externalities linked to the particular structures of road networks. Thus, work resulting from space syntax, a scientific movement based on architecture [HIL 84, HIL 96], shows that flow distributions, whether pedestrian or vehicle, depend in large part on the configuration of the road network and its topology, more than on the origin-destination matrix [GEN 00]. Can some types of networks, especially those that have an intrinsic tendency to focus flow, be regarded as "naturally" more liable to congestion? Using space syntax, Shu and Hillier [HIL 00, SHU 00, SHU 03] have shown how the topological position of road segments is related to the location of crimes.

Connecting the shape of road networks to various forms of externalities is a challenging research topic that is still underexplored. In the field of urban planning, it is a valuable tool in that it allows us to study the constraints and possibilities offered by the various types of networks, especially in rebalancing modal shares. However, research must look beyond the simple structural analysis of networks and also focus on their operation. Just as the morphology and topology of a road network's sections are important criteria to connect to externalities, so are its quality of service and types of accessibility for the kind of vehicle used.

In this context, our study aims to identify the effects of form and function externalities of road networks on mobility practices and urban transportation services. We are not focusing on the advantages and disadvantages of a compact versus a sprawling city. We claim, however, that it is useful and possible to effectively combat urban sprawl in order to reduce car dependency and its inherent urban nuisances through the intelligent design of networks. Our discourse is globally constructed around this working hypothesis, which we assume, and brings into perspective the question of the sustainability of urban spaces.

Using three complementary examples based on observations and simulations, this chapter illustrates the depth of the issue under discussion. The first example concerns the influence of the form of road networks on the pattern of pedestrian flow and *in fine* on walking trends in the city. The second shows that the competitiveness of a public transportation service (in this case, a transport-on-demand system) is in large part determined by the structure of road networks. The third example concerns the metrics of the automobile system which, by shaping the relative positioning of urban components, exacerbates car dependency and promotes urban sprawl.

5.2. The urban road network: a major determinant of pedestrian flow

At a time when car-induced nuisances are seen as a danger to the welfare of citizens [NEW 89], the search for a better balance with collective means of transportation becomes an attainable objective, notably within the framework of reorganization of urban mobility (i.e. the French *Plans de Déplacements Urbains*). In this context, the revival of walking – an activity that has been declining for more than 30 years – is considered particularly strategic. The promotion of walking as a competitive means of travel, especially for short distances, is often just a question of setting up proper facilities to improve the safety and quality of sidewalks [ROD 04]. Although these steps are fully justified in offering pedestrians appropriate conditions, safety and comfort criteria are insufficient to create a context that promotes walking.

Upstream of the ability of each road section to facilitate pedestrian travel, the trip structure resulting from the road network as a whole plays a fundamental role. The way that road sections are interconnected determines the road network's ability to produce a given spatial distribution of pedestrian flow. This can incite or discourage people to walk, even before the local factors in favor of (or against) this practice come into play. Given this basic property, the network can be seen as a true instigating element of flows, and not just a simple, neutral support. The evaluation of a road section's user-friendliness to walking without considering its relationship to the whole network neglects the facts produced by the geometry and topology of the structure in its entirety. The road network's role on the distribution of pedestrian flow and the promotion of walking has rarely been considered in detail in analysis and in action plans.

Studies of space syntax using the effect of network configuration as a strong hypothesis are an exception to this tendency. The next section begins by presenting the contributions of this trend to understanding the role the urban framework plays on the practice of walking. We then introduce other, more isolated factors based on converging ideas that help enrich the "flow producing" network concept.

5.2.1. *The effect of the road network on the space syntax*

One of the key ideas of space syntax is the "natural movement" [HIL 93]. Natural movement expresses the fact that an arterial system "naturally" produces a given distribution of pedestrian frequency levels, basically due to the topological relationships between a city's streets. Based on theoretical elements drawn essentially from the psychology of space (spatial behavior), space syntax proposes original methods for the calculation of a number of structural indexes.

In the case of an arterial system, these indexes apply to a group of sections devoid of any functional attributes. An initial phase establishes the "axial map", where the segments of the arterial system that are adjacent and run in the same direction are aggregated to form axial lines (or sight-lines). These lines then form the nodes of a non-planar graph, where the intersections between the axial lines form the arcs. Based on this particular graph, several indexes can be calculated for each axial line; the one most frequently used, the integration index, quantifies the degree of separation (referred to as the "depth") of a line in relation to the other lines [JIA 99, JIA 00].

Given the placing of the axial map, the integration index is a type of accessibility measurement using a topological metric based on directional variance. A significant methodological development of space syntax, combined with the use of specially-developed computer programs, enables a large number of tests to verify the hypothesis of the link between the network's configuration and pedestrian flow. Numerous studies [CUT 99, DES 01, KAS 03, MAJ 00, PEP 97, REA 99] have shown that the levels of pedestrian traffic are well correlated to the integration index. More often than not, local integration – an index whose calculation is limited to a given topological proximity (depth) – revealed itself as the most relevant measurement.

In recent studies involving the city of Lille [FOL 07, PIO 06], space syntax was used and combined with landscape criteria that can have an influence on the choice of itineraries and thus on the resulting pedestrian flows.

5.2.2. *Applying space syntax to pedestrian flows as observed in Lille*

A travel survey was conducted in the town of Lille within the framework of a program funded by the French government (*Ministère Français de la Recherche*) [BAJ 06]. Precisely recorded pedestrian itineraries were used to illustrate the space syntax method. The study area was the central part of the town, clearly delineated by belt freeways and fortifications. In this area, a large number of pedestrian trips were recorded on a regular basis and enabled researchers to calculate the usage frequency of each street segment. The frequency of use was applied to a logarithm transformation in order to homogenize its statistical distribution (see Figure 5.1a). At the same time, the aggregation of these segments into axial lines resulted in topological calculations producing the integration index[1], whose values were then

---

1 To measure the degree of integration, we calculate the total number of intersections or "turns" we would encounter when traveling the shortest paths from a straight sight-line (axial line) [HIL 96] to all other axial lines in the network. In space syntax terminology, such a measure is called global integration in contrast to local integration, which limits the maximum number of turns traveled to reach other axial lines (specific depth).

disaggregated at the level of street segments. Several analyses have shown that the second depth of local integration was the most relevant level in order to estimate the flows observed (see Figure 5.1b). We tested this approach and applied a linear regression by seeking to "estimate" the frequency of pedestrian use applying the road segment integration index. In spite of a low overall explanatory power[2] ($r^2 = 0.23$), residuals were calculated and mapped (see Figure 5.1c).

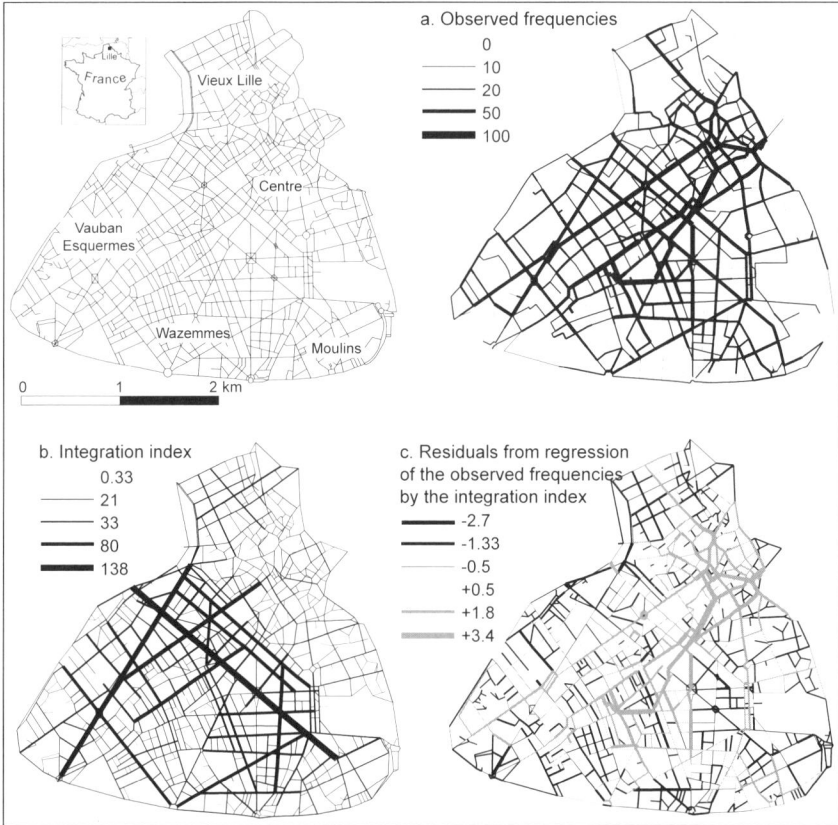

Figure 5.1. *Applying space syntax in Lille*

2 In our study, conditions are not optimal for using the integration index. This is because the observed frequencies are the result of the overlapping of a sample of trips, and not a true counting by the gate-to-gate method. Thus, we see a rather high effect of origin-destination trip locations of the sample coming into play, while the integration index considers each axial line as a potential origin-destination location.

With this method, frequency of pedestrian use of street segments becomes partially predictable using the local integration index. By construction, this index attributes a theoretical usage value to rectilinear streets that are closely connected to the entire network and clearly higher than usage values attributed to winding street sections or those with a weak connection to the network. Thus, integration underlines the strategic role of streets comprising large axes, which contributes to producing a strong hierarchy of spatial distribution of usage levels. According to the recorded pedestrian frequency map (see Figure 5.1a), this favored use of rectilinear streets is a reality. The eastern side of the city seems to be highly discordant, however, as we observe a high concentration of flows even though there is low integration, given that the roads are more winding. The residual map (see Figure 5.1c) clearly highlights this discordance by displaying a positive-value continuous zone in the East. Negative residuals also tend to be located in the zone's peripheral area.

Using the integration index, the space syntax provides an overview of the spatial distribution of pedestrian flows as linked to the structure of the road network. In comparison, we cannot reproduce the hierarchical motivation of pedestrian usage levels by using metric accessibility indicators. Despite this, the example presented does show that the relevance of integration varies according to the general shape of the network. Its relevance becomes higher when the road network follows a regular grid pattern. This finding opens the door to discussing the interest of using space syntax to measure the effect of the road network on pedestrian usage levels. An initial limitation stems from the very notion of the axial line, whose construction remains tricky [BAT 04b, TUR 05], although it is a determining factor of the topological calculations that result in the integration index. Moreover, this axial line concept introduces a perceptual factor, assuming that the choice of itineraries gives a preference to visual continuity axes, to the detriment of minimizing metric distance.

5.2.3. *Other models for assessing the role of a road network on pedestrian flow*

Studies of the role of networks in the distribution of flows for various trip modes have shown that the effect of spatial configuration linked to road network geometry could be determined by analyzing "random trips" [GEN 00]. By recording the shortest trips (in distance for walks, in time for car trips) linking a large number of pairs of points located randomly on the network, we built a theoretical usage map of the network. The usage frequency for each section can be seen as a quantification of its strategic dimension in the circulatory functioning of the network. We then see that this theoretical usage frequency is closely correlated to actual flow values. It looks as if, in the "urban anthill", it was the structure of the network that dictated the map of day-to-day flows rather than the location of traffic generators. This is the case even if we can imagine that in the long term, mobility behavior could trigger the creation of new infrastructures.

This method of generation of theoretical frequencies, based on a "passive" recording of simulated flows, is similar to other processes often conducted in limited spaces. This is the case for simulations carried out by Turner [TUR 02] to estimate the level of attendance for each room in the Tate Gallery in London, implemented by mapping the theoretical frequencies resulting from the simulation of a large number of partially random trips. In the field of animal ecology, Matthiopoulos [MAT 03] also used an individual movement simulator which, by accumulating trips, represented accessibility disparities produced only by the trips' spatial structure. Although quite distant from the urban setting, these examples have all used simulated trips to study the effect of travel space by using theoretical frequencies.

This type of approach differs from space syntax because of its stochastic aspect and the possibility of generating relatively realistic virtual trips, as a function of the initial hypotheses. Thus, the frequencies calculated by Genre-Grandpierre [GEN 00] are based on the hypothesis of a simple minimization of the range of trips, while in the Turner's studies [TUR 02] the rules for trips have integrated visual parameters, according to which virtual individuals seek to maximize the scope of their field of vision.

5.2.4. *Resorting to theoretical flows derived from the road network configuration*

If the configuration of the road network prefigures the distribution of pedestrian trip flows in the intra-urban setting, there is presently no integrated method to measure this effect. Calculating theoretical frequencies that represent this effect implies making assumptions on the way the trips are made. On one hand, the space syntax appears as a theoretical and methodological body separate from the classical analysis of transport mode and route choices [BAT 04a] by *a priori* retaining the local optimization of the scope of the visual field as the main criterion for the choice of an itinerary. On the other hand, more classical methods presuppose only a minimization of global impedance, like trip distance or duration.

Even if, in reality, flows observed cannot be deduced exclusively from theoretical flows defined by the networks' structure, the very fact of recognizing this principle and associating it with a measurement can be of great interest in various contexts. Being able to evaluate the strategic dimension of each segment of a network for a given trip can help achieve a better determination of priorities in urban planning that promote walking. The previously presented methods may thus serve to shape a land-use policy, by helping to reconcile the "natural" vocation of road segments and the type of use assigned to them. For example, if a segment lends itself "naturally" to pedestrian use, it may be counter-productive to put it to another use.

In the same context, steps to optimize the location of businesses can also depend on the distribution of potential trips resulting from the network. Hillier [HIL 93] underlined the existence of a "multiplier effect" by which the distribution of certain urban functions, like businesses, is often the consequence of a natural movement which in turn increases the strategic interest of certain parts of the network, through positive feedback. By acting on the trip structure itself, we can consider the modification of the network (e.g. its topology) as a possible lever to maximize co-presence and thus promote the tranquility and safety of urban neighborhoods.

In parallel to the network's role in flow distribution, we can also observe its ability to influence modal practices. If it is true that the choice of walking depends in the first place on the physical ability and willingness of people, it also depends on the structure of the road network and its inherent accessibility to pedestrians. The way roads are connected, the level of pedestrian access in a given neighborhood and the quantity of potential space available to the walker [GEN 03] are all factors that come into play regarding the "return on investment" of walking in a person's daily activity [VER 97]. Thus, the accessibility of a network is recognized as having a higher priority than the comfort or safety of pedestrians [ALF 05].

We have studied the structural influence of networks on pedestrian activity. However, this influence is not limited to this mode of travel. We will now look at its effects on the efficiency of a flexible public transportation service.

5.3. Influence of the road network on the efficiency of a transportation service

Studies of the performance of public transportation services generally do not include an analysis of the characteristics of the road network [BEN 91]. More often than not, this evaluation focuses on the inputs of the service (capital, work, etc.) with results being measured by various aggregated indexes (revenue/expenditures ratio, usage rate). However, we can imagine that the form and function of the network will have an impact on the quality of service and, *in fine*, on its performance, independently of the inputs mentioned earlier. To paraphrase Dupuy [DUP 91] [DUP 93], we can even say that the structure of the black box approach to modeling the effect of a network intrinsically integrates a good part of the service's economic performance, with regard to both cost and usefulness.

The second example falls into this setting. Based on simulations dealing with real cases of the operation of public transportation, we measure the impact of the road network's structure on the efficiency of a public transportation service, in this example, a flexible on-demand system. The analysis also allows us to compare the calculated efficiency of the public service with the hypothetical use of private cars, which gives us insight into the factors of modal competition that could be advantageous to public transportation. The key is to find effective levers to reduce personal vehicle use as much as possible and replace it with public transportation (e.g. a flexible transport-on-demand service).

After a quick overview of the transport-on-demand service *Evolis-gare* operating in Besançon, we describe the simulation method used to estimate the effect of six network configurations using a series of indexes describing the performances of this service compared to the personal vehicle in similar settings.

5.3.1. *Principle and operation of the optimized transport-on-demand service*

In order to assess the externalities of the road network in terms of the efficiency of a public transportation service, we need data on a linkage between the network graph structure and the supply of transportation, so that the services can be analyzed independently of the network structure. This would not be the case for traditional transportation services comprised of regular bus routes, as these are all too often directly modeled on the network's structure. Rather, a flexible transportation service would seem more relevant for this type of analysis, as its vehicles have the ability to operate over all roads and choose *ad hoc*, independent routes, while integrating the network as a constraint. Flexible transportation is characterized by a relative operational flexibility [BAI 01, JOS 05] and it provides individualized group transportation [THE 02], offering rule-based levels of flexibility in terms of schedule, stops and operational modes. This example considers a converging-diverging flexible transportation service based on a flow generator: Evolis-gare (for high speed trains: TGV) in Besançon in France.

Deployed in the city of Besançon[4] in 2001 and managed by the *Compagnie des Transports de Besançon*, the Evolis-gare service is a proven flexible transportation service (from 3,000–4,000 customers annually) between the TGV station and regular bus stops. It provides a service during early morning and late evening hours, when regular transportation is not running. It is an intermodal flexible transport service, running in conjunction with TGV arrival and departure times, and guarantees arrival 10 minutes before the departure of a TGV [BOL 06]. The service endeavors to promote ride sharing and minimize total travel distances, number of

---

4 117,836 inhabitants in 2007.

vehicles in circulation and related expenses. It makes use of optimization software specifically designed for this application (Resad2)[5].

Finding the best routes is done by calculating cost on the basis of the taxi rate as set by the department's prefecture. The cost includes a boarding fee and a per-kilometer rate. Without delving into details of the algorithm of optimization, the one important aspect of the method used concerns the use of time slots during which deviations from the fastest route are allowed within the general framework of the set time limits. Using this approach, passengers are grouped in vehicles using acceptable detours.

Overall, total trip time for the stops furthest from the station never exceed 45 minutes, including a five-minute walk to the stop and a 10-minute wait at the train station. Thus, the system makes optimal use of the relationship between the origin of demand and the road network structure to create routes. Taking into account the network structure and traffic constraints (often referred to as "impedance") assigned to the road segments ensures that the best routes are selected, and allows us to analyze the relationship between the structure of the network and the efficiency of the service.

5.3.2. Convergence-divergence simulation for transport-on-demand

In order to simulate the network structure's influence on the efficiency of the Evolis-gare service, we located 226 stops in the current database of the Ginko bus network of the *Compagnie des Transports de Besançon*, distributed throughout a portion of the *Communauté d'Agglomération du Grand Besançon* perimeter. Then, for simulation purposes (bearing in mind that this is not a prospective study), we randomly distributed 1,000 clients at these stops using a Gauss distribution equation (see Figure 5.2).

Based on these parameters, we were able to test the influence of six different configurations of the network. The first case corresponded to the actual network, where average speeds range from 15 to 110 km/h, depending on the road segment. For the second and third cases, the speed was set between 10 and 60 km/h, respectively, over all segments. The fourth network presented highly differentiated impedances: super highways, two-lane roads and segments longer than 500 m were limited by a commercial speed of 10 km/h, while other segments of the road maintained their true speed limits. The last two configurations modeled a restricted (downsized) network constructed on the basis of the original network, but with a

---

5 Software package marketed by Prorentsoft.

large number of non-radial segments removed to form a star-shaped network. This last configuration had a low connectivity aspect, but was highly polarized, and access to a stop was located at the end of a branch of the star required passing through the centre (see Figure 5.2). In one case the network maintained its actual differentiated speeds, and in the other the speed was set to 60 km/h.

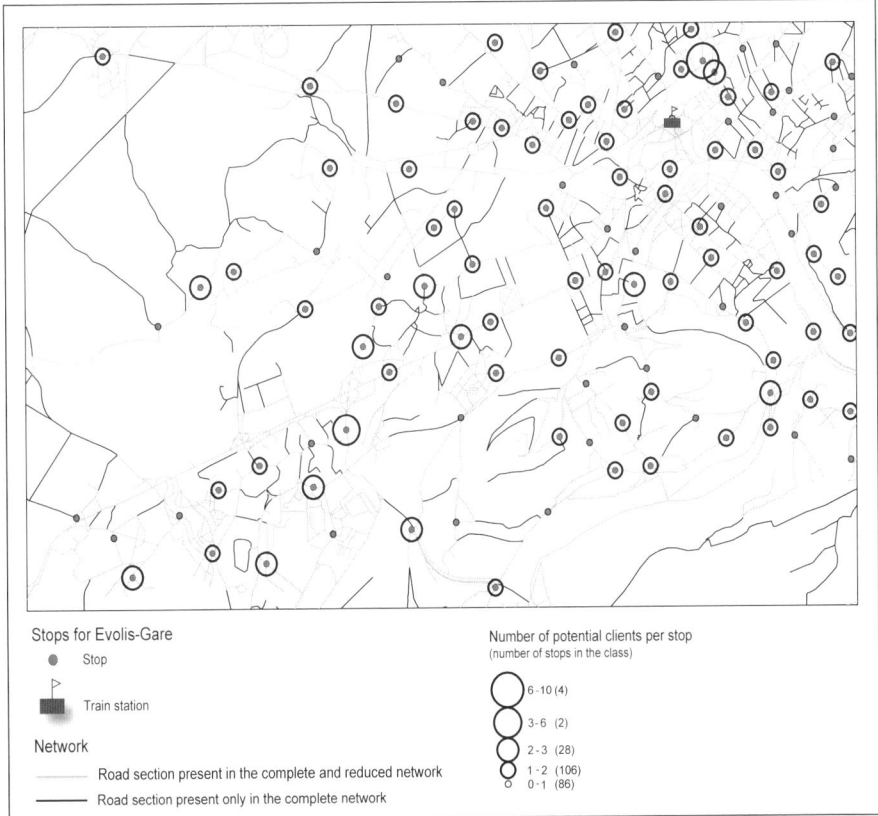

Figure 5.2. *Sample pattern of stops, clients and network used for simulations*

In order to evaluate the relationship between the network and efficiency of the service, we developed a software program to simulate passenger requests and train departures. In observing operations at Evolis-gare, we noticed that customer distribution on departing trains (based on station arrival times common to many clients) follows a gamma distribution, which balances out when the number of potential clients increases. In transportation geomarketing, dissymmetrical or hyperbolic distributions of this kind are frequently observed [JOE 01]. Thus, for low demand (in this case 1,000 passengers for 500 train departures), the frequency of trains attracting only one passenger to the service is high, while it is rare for trains to have more than three passengers (see Figure 5.3). Conversely, when demand increases, the middle class is represented the most.

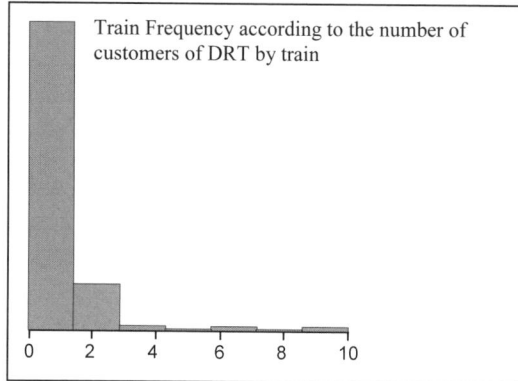

Train Frequency according to the number of customers of DRT by train

Figure 5.3. *Distribution of clients of the Demand Responsive Transport (DRT) Evolis-gare in trains (gamma distribution).There is a large over-representation of trains with less than three intermodal passengers*

The simulations were conducted under almost true conditions for each configuration. The simulator was calibrated using data from the client's database of the Evolis-gare service from 2001 to 2004. Its use to analyze the expansion of the service to all communes of the *Communauté d'Agglomération du Grand Besançon* revealed the quality of the forecast, with 10% maximum error on efficiency indexes. The simulation was carried out in several stages. First, using the gamma distribution, we randomly assigned customers to train departures. We then used the optimization kernel according to the demand and the distribution per train. Lastly, we conducted optimizations 1,000 times for each case and calculated the average efficiency indexes.

The selected performance indexes focused on the operation of the service, from an economic and environmental viewpoint, compared with the use of the personal vehicle:

– average occupation of vehicles (number of passengers per vehicle used);

– average rate of return (relationship between revenues and expenditures resulting from the service);

– reduction in the number of vehicles relative to personal cars (%);

– travel distance saved with regard to the use of the personal car (%).

5.3.3. *Functional and economic externalities of the road network*

Some indexes seemed relatively unaffected by the structure of the networks (see Figure 5.4). This was the case for the rate of return, for which the values are stabilized by the conditions of the simulation: relatively low average load of two passengers per departing train and low structural percentage of revenues in comparison with important costs linked to distances traveled. There would be greater variation if we were to simulate a much higher load rate. In spite of its weak elasticity, the rate of return remained an important factor to consider, since the smallest reduction involves significant expense for those territorial communities that introduced this type of grant-aided service.

For the other indexes, we observed that the network's structure greatly affected the efficiency of the service. An overall decrease in speed brought about a significant drop in the occupation of vehicles, with the time dimension of the service being an important optimization component across time slots (set at 15 minutes for our simulations). This effect was also apparent when moving from the total network to the downsized network (see Figure 5.4).

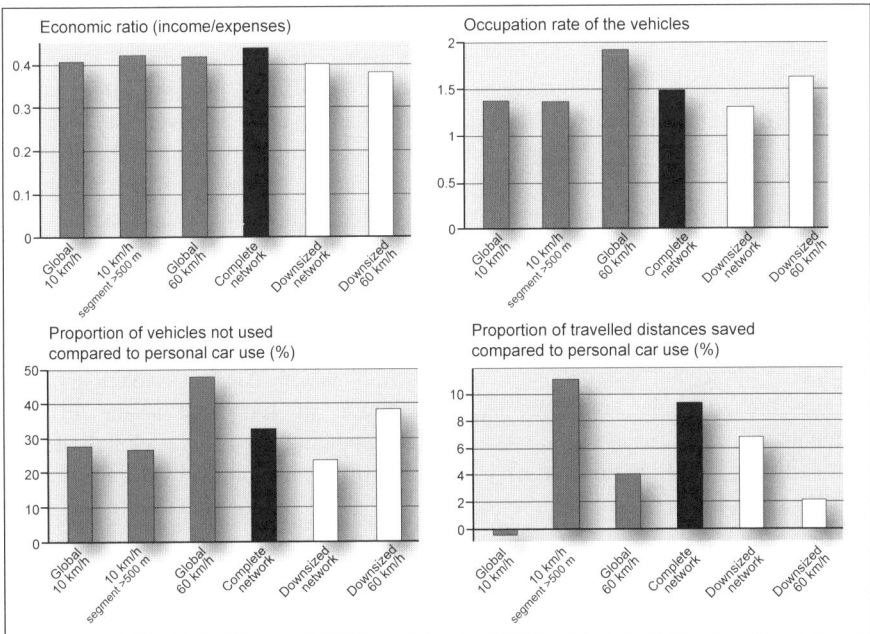

Figure 5.4. *Efficiency indexes of various network configurations*

It is also interesting to note that the service "saved" more vehicles when the network connectivity increased, thus ensuring that circuits followed the shortest routes and avoided detours. In addition, even on the downsized network, and regardless of speeds assigned to the road sections, there was a significant saving of vehicles for higher commercial speeds. Speed limits and the way they were organized on road segments appeared a good lever to control network structure externalities.

An analysis of the portion of distance reduction revealed several facts. For low-speed networks, research into passenger concentrations on the best routes and the reduction of the number of vehicles in motion did not compensate for detours caused by the relative slowness of the service and its access time. The service vehicle had to travel to pick up the client, creating a "deadhead" trip, while the personal vehicle remained close to key travel locations (household garage, parking lot near station). Given these conditions, the personal car became more economical in view of the distances traveled. The downsized network experienced a similar variation, albeit to a lesser extent. Its savings in distance traveled dropped slightly in relation to the entire network. Once again, network connectivity was of paramount importance to improve the efficiency of the flexible transportation service.

Lastly, reducing traffic fluidity on fast networks (highways and long road segments) by drastically reducing speeds created a major savings in distances traveled (case in point: speed limit of 10 km/h and segments over 500 m). This played in favor of shorter distances and faster speeds (over 10 km/h). This extreme configuration, aimed at slowing fast lanes, although politically difficult to justify, improved the ability of on-demand services to effectively compete against the use of private cars. The visible increase in the efficiency of the on-demand system was a result of the system's increased ability to speedily combine passengers. This was thanks to the privileged access of vehicles to short segments of the network that connected stops, segments that were densely distributed, highly connected and fast, in spite of the global slowdown in accessing the station. All that remained was to propose the sufficient and balanced distribution of speed limits on different segments to make this type of service more appealing than the personal car and politically acceptable in order to trigger an effective modal shift to the transport-on-demand service.

These results are consistent with the position we adopted, which aimed at improving access to dense urban facilities, rather than to facilities located further out which, with improved access, promote urban sprawl and car dependency. Thinking in this direction leads us to consider speeds, road segments and their topological organization as key elements in urban organization based on new metrics that promote research about modal competition during the initial design stage of networks.

5.4. Road network metrics, urban sprawl and car dependency

The focus of this last example was to demonstrate some of the induced effects of road network metrics[6] on urban morphology and car dependency. Here dependency corresponded to the growing necessity for people to use cars in order to perform their daily life activities [DUP 99]. It was a matter of better understanding how road networks operate in terms of accessibility, how accessibility could be modulated according to the vehicles that use them and the impact of accessibility on car dependency. This would better control this dependency by narrowing the performance gap between the various modes of transport early in the design of the networks.

Having described the metrics of current networks and explained how we might consider the phenomena of urban sprawl and car dependency as two cases of the networks' externalities, this section introduces the concept of "slow networks" [GEN 07]. Given their unique metrics, that intrinsically promote the efficiency of short trips over longer ones (as is the case for actual networks), we discuss how slow networks can offer a means for migrating towards a more sustainable mobility in urban areas. Initially this can be achieved by offering conditions so that modal competition may actually occur, mainly by promoting a reduction in the length of trips.

5.4.1. *Metrics of current networks that favor long trajectories*

The efficiency of trips is a simple but highly relevant measure to estimate the performance of a transportation system [GUT 98]. If we study the performance of car transportation using the "accelerating" metrics as a trip efficiency indicator, we see that the indicator increases sharply with the length of trips (e.g. straight-line distance between the origin and destination). Using the duration of the actual trip and the Euclidian origin-destination distance allows calculation of the nominal straight-line speed of the trip. This nominal "Euclidian" speed is an efficiency measurement that places the network metrics in relation to the Euclidian reality of physical distances. This is something not achieved by other indicators based on average trip speeds on the network, which may be higher without ensuring a good quality of service. This is notably the case with networks having many high-speed lanes that topologically and morphologically impose many detours. The graph representing the efficiency of car trips based on the length of trips for three test

---

6 In order to simplify the text, we introduce the *metrics* of a road network used by a given vehicle to designate the way the network makes two points seem closer or farther away due to time passed in traveling from one point to the other compared to their Euclidian distance. This is the case even if not all of the mathematical hypotheses that allow us to speak of metrics in the strict sense are verified.

zones (see Figure 5.5) shows a strong increase in efficiency, from less than 35 km/h for trips less than 5 km to more than 55 km/h for trips exceeding 35 km.

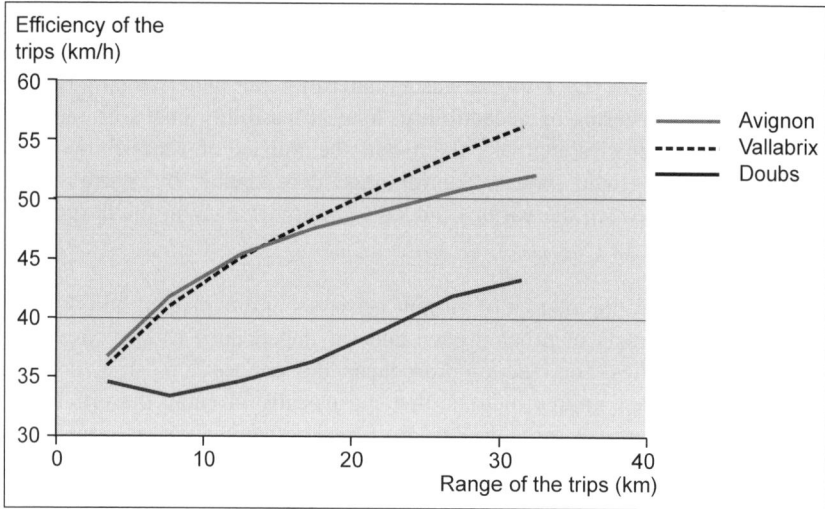

Figure 5.5. *Variation of trip efficiency as a function of trip length*

This "accelerating" metrics, which shows that car transportation efficiency increases with the length of the trip, depends on the functional hierarchical structure of road networks. To reach a destination, a driver seeks to leave the slow roads as soon as possible and travel along the fastest infrastructures, so as to reduce total travel time. Increasing the length brings about an increase in the portion of the trip made on fast networks, with an increase of its overall efficiency.

Although the observed increase in car travel efficiency and length of trips may seem trivial (and rarely measured), its side-effects seem very important, both in terms of modal competition and urban geography.

5.4.2. Network metrics, urban sprawl and modal competition

Given that efficiency increases with the length of trips, accessibility is not proportional to the duration of trips. In view of the relationship between the length of travel time and number of opportunities, each additional minute spent on a network with a car brings with it higher added value than the preceding minute. It is as if network metrics incited individuals to travel a bit further, since the marginal cost declines. A person living in a densely-populated area and looking to acquire a house will be naturally inclined to look increasingly towards the outskirts, since the supply of houses expands proportionately faster than travel time. Additionally, prices decrease globally from the urban center towards the town periphery (see Chapters 7, 10 and 11). This metric, which holds for all types of trips, and not only from the center towards the periphery, can be seen as a driving force for urban sprawl, as an incentive for a person not to organize his or ehr activity locations according to physical proximity criteria. The reasoning behind seeking a house also holds for the selection of locations for businesses, recreation areas or employment places. As the increase in distances traveled is the main contributor to the increase in traffic [WIE 02], we can clearly measure the problem at hand in terms of sustainability by the "accelerating" metric of efficiency, which fuels the dispersion of life spaces.

Focusing more specifically on transportation in urban networks, the "accelerating" metric of accessibility is especially favorable for car travel, while for non-motorized travel modes (bicycle, walking), accessibility is at best proportional to the trip time, as their speeds are relatively constant regardless of the route taken. As for buses and streetcars, the "accelerating" effect in the metric is present when the service is properly organized and when there are dedicated routes connecting the main urban poles with very few intermediate stops. Even in these particular cases, and *a fortiori* for classical urban public transportation networks (where vehicles provide proximity service with frequent stops), the "accelerating" metric is lowered by their need to make regular stops. These differences in characteristics make it difficult to balance service levels to promote balanced modal sharing. Thus, the current road network metric once again promotes car use or, to say the least, ensures its clear superiority in terms of efficiency with regard to long trips.

All current road networks facilitate the expansion of life space by maintaining a relatively stable time budget. This factor, along with the better access provided by individual cars, leads us to question whether it is at all possible to introduce alternate forms of transportation networks with different characteristics of accessibility. To provide such accessibility, different metrics would have to be designed to ease the promotion of public transportation and non-motorized modes. This would include ensuring balanced accessibility levels for the various modes, provided by the structure of the networks, including service for long trips.

5.4.3. *Principle and materialization of "slow metrics"*

Since speed is the basis of the hierarchy of roads which, in turn, generates the "accelerating" metric, we can assume that this phenomenon can be eradicated simply by homogenizing the speeds. Unfortunately, if this operation allows a reduction in the differential of given efficiencies proportionally to the length of trips, and with all other factors remaining unchanged, it is still greater for long trips. This can be explained by the fact that with longer trips, detours resulting from the morphological and topological configuration of the network have a lesser influence. A purely rectilinear network with a homogeneous speed helps to keep a constant efficiency with an increase in the length of trips, but it cannot inverse the efficiency ratios. As it is unrealistic to modify actual networks to set up rectilinear networks, the question of the existence of alternative metric networks remains open.

The work of physicians on fractal geometry shows the existence of structures for which trips follow a logic of efficiency diametrically opposed to the logic of existing networks. On a surface for which there exists a hyperbolic (fractal) relationship between the size of the gaps (impenetrable zones that must be by-passed) and their number, the probability of having to make a large detour increases with the length of trips, and the relative importance of the detours increases with increasing length of trips. More precisely, for a fractal structure where the size and number of gaps are related through a hyperbolic relationship, the actual length $l(L)$ of the shortest route for the Euclidian length L follows the equation [GOU 92, STA 86]:

$$ l(\lambda L) = \lambda^{d\,min} l(L) \quad \text{with } \lambda = \frac{1}{L}, \quad \text{we have } l \approx L^{d\,min} \qquad [5.1] $$

The relationship [5.1] expresses the fact that for a fractal structure, the increase in network distances occurs faster than for corresponding Euclidian distances. Short trips are thus "economically" preferred to long trips, since they are more efficient.

If the principle of this metric, which we refer to as "slow", where efficiency decreases with the length of trips, is relatively simple, its concrete expression remains complex. In urban space, it appears to be difficult to maintain a hierarchy of impenetrable gaps whose number and size are linked by a fractal relationship. In order to obtain "slow" metrics for real network examples, we started from the assertion that a hyperbolic distribution of gaps produced a "slow" metric. We considered as gaps the mandatory stops (traffic lights), distributed randomly over the network, while making sure that their number and duration were connected according to the statistical principle evoked earlier for gaps. We carried out two simulations: for the first, the duration of mandatory stops varies between 30 and 380 seconds, and for the second simulation from 15 to 232 seconds. In both cases the number of mandatory stops ranges from 750 for the shortest to 50 for the longest.

Figure 5.6 shows a clear inversion of trip efficiency relationships when compared to the classical case. In other words, this is a case of a "slow" metric where the probability of having to experience a long stop increases with the length of the trip, regardless of the trip's origin and destination. It is the inversion of the metric logic that is original here, rather than the simple decrease in absolute efficiency values (obtained by a simple reduction in speed). Slow metrics actually introduce positive side-effects in terms of sustainability. With slow metrics, short trips have a higher level of performance than long trips in terms of accessibility. Ultimately, by increasing traffic speeds while maintaining the hyperbolic distribution of stops, efficiency curbs would be shifted upwards, but the logic would remain the same: the efficiency of proximity trips would remain higher than for distant trips. We would then have "slow rapid networks," meaning that the performance level of car travel would remain high overall, but especially so for short trips!

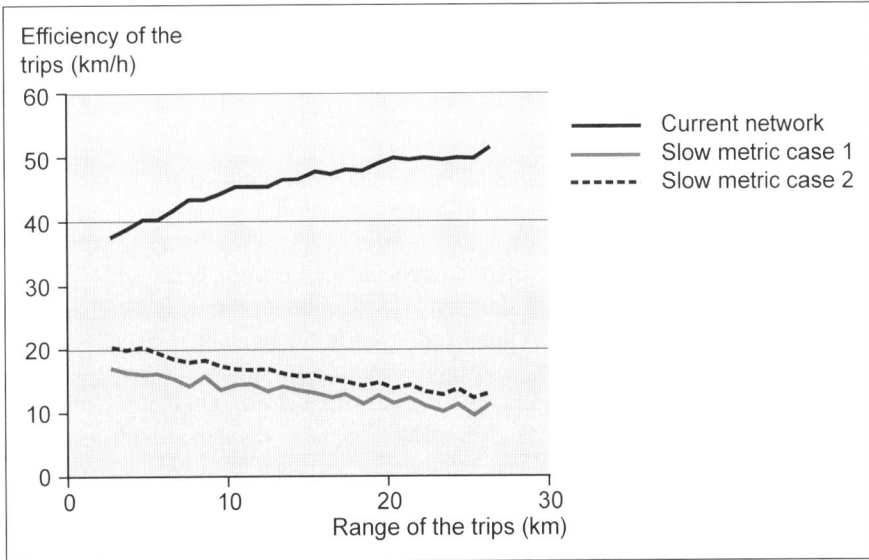

Figure 5.6. *Classical metrics and slow metrics around the town of Carpentras*

In practice, the mandatory stops could be made with programmable traffic lights, whose schedule would vary from one day to another, while maintaining the duration and number relationship as defined by a given rule over the whole area. This would avert a learning process on the part of drivers who, having waited at a long light, would try to avoid it the following day, pushing the network back into a classic metric operation.

Although these simulations show that it is possible to have a geographical zone function according to a slow metric, we still have to master the minimum number of parameters allowing us to set and calibrate this metric: how many stops, how long each stop, for which location and for what desired trip efficiency? It is also necessary to assess, for example by using multiagent simulations, the effect of the slow metric on the overall fluidity of traffic.

5.4.4. *What can be expected from slow metric?*

The "accelerating" metric of current networks tends to favor both urban sprawl and car dependency. A slow metric would fundamentally invert the "economic" logic of relationships by favoring short trips. Thus, we can assume that it would limit the pace of urban sprawl and consequently the increase in the length of trips, both of which go hand-in-hand with the prevalence of the car. The purpose is lowering differences in accessibility between the car and other transport modes through the design and operation of road networks. With an increasing commuting distance, slow networks can be seen as a primary means to enhance the modal competition. Balancing the performances of the various modes seems to be a necessary but insufficient condition to modal transfer; other aspects, such as comfort, flexibility, etc., also intervene in the modal choice.

This convergence of accessibility levels for the various modes can be difficult to reach by simply improving the non-motorized service, even with important investments. This is why we must also seek ways to reduce car transportation performance or, at least, not to favor it through the network structures [DUP 99, WIE 03]. It is in this context that slow networks can be used to balance the quality of the various service modes. On one hand, even in the unrealistic hypothesis of a drastic, overall reduction of traffic speeds, car accessibility would remain higher through the benefit of the "accelerating" metrics of networks. On the other hand, if we consider the introduction of a slow metric for cars, combined with a specific metric for public transportation that would lean as much as possible towards an "accelerating" metric, it would provide an advantage to public transportation within the very structure of the networks (the inverse of the actual situation). This would give it the means to compete, including for long trips.

In addition to this, slow networks can be seen as a lever to attain a "short distance" city, thus reducing negative environmental impact, especially that caused by motorized transportation. Nowadays, easy access to the "accelerating" car metric makes the choice of suburban areas a "winning" option. This is because the additional expense in time and money imposed by peripheral locations are mostly compensated for by advantageous local taxes and real estate prices. A suburban choice is even more interesting at the individual level, as costs are mostly borne by the community (infrastructure, pollution, etc.), while the advantages remain personal (access to property, privacy, sense of security, etc.).

This centrifugal push of living spaces (residence, work, leisure), which was made possible by the "accelerating" metrics of networks that ensure a relative constancy of individual travel time budgets, creates a number of negative environmental externalities [MAT 96]. A "short distance" city that is dense but not necessarily compact, where amenities are better distributed and accessible at close range, is the alternative most often thought of when speaking of sustainability (For an explanation, see the *New Urbanism* movement, www.newurbanism.org, or urban planning based on fractal models [CAV 04].). Here again, the slow metric, according to which we lose time in proportion to travel distance, can be obtained with a reasonable effort, by working with existing road networks through the regulation of speeds and by programming traffic lights. "Slow" metric emerges as a possible factor in triggering the reconfiguration of zones to make them more sustainable. The objective is to reintroduce the constraints of physical proximity in the choice of individual living spaces, with the penalty of seeing our time and travel budget increase substantially with longer distances. Putting strict constraints on long trips is a means to "economically" justify, notably in terms of time and transportation budgets, the layout of living spaces based on proximity, in accordance with the *rational locator* hypothesis [LEV 94]. With a "slow" metric, it is still possible to spread out the living spaces for questions of individual preference, but by paying the price in time. The current rise in the price of fuel and introduction of car user fee policies that make users pay the true cost of the automobile [LIT 98] combine to help this movement, which favors physical proximity of living areas.

While promising, the concept of "slow" networks and its application is not yet fully mastered. We must be able to calibrate the metric according to the demand of land developers wishing to use slow networks as a time-planning tool [SME 05]. The issue is how to rate car transportation in terms of overall efficiency. This brings us back to the number and duration of stops to introduce. There is also the issue of the necessary measures in terms of economic policy and lodging (what type of housing is needed for those who have been pushed to the outskirts because of slow metrics [DAC 05])? In this regard we must be careful not to undermine the fluidity of the job market by overly restricting individuals' mobility.

Alternatives to the car must be promoted to take its place, as car performance is degraded. The overall objective is to reduce car mobility, not mobility itself, with long trips being assumed by public transportation such as the train. The fact remains that through this example we become aware of the importance of better understanding of how the externalities of road networks operate, notably in terms of resulting accessibility, and the decisive role played by side-effects on the operation of urban areas.

5.5. Conclusion

The three examples presented all confirm that it is necessary to consider how networks are organized, including their topological structure and allowed travel speeds, in order to expand our understanding of the transportation supply and mobility practices and, more generally, to set up urban development policies. The impact of network structures on pedestrian traffic and on the efficiency of public transportation services has been clearly identified. Using metrics that have the capacity to help design new mobility spaces also delivers new perspectives for developing more sustainable cities in the long term. The deployment of homogeneous networks that are connective and impose a relatively low speed may be compatible, at the built-up city level, with an increased level of public transportation service and easier access to active modes (bicycle, walking). All this combines to lower the use of the personal car in urban zones.

In this context, we are seeing new and promising research avenues. Network economies require aggregate indicators that correctly sum up the form and function of networks, and simulations to evaluate the effects on cost function and service performance level of the various structural forms. Form indicators resulting from graphs as well as fractal and space syntax theories are already available, but need more investigation to measure the ability of networks to intrinsically generate/control the formation/distribution of flows. One of the objectives of this research is to define the ideal network, associating the optimal structure of lanes and the proper distribution of speeds in trip duration and space.

Work that maps appropriate transportation modes (and their combinations) to adequate mobility scales helps us imagine sustainable solutions as alternatives to the personal car, in particular to solve potential problems caused by the application of "slow" metrics over long distances. More generally, all network externalities must be carefully considered from the initial design stage of urban networks in order to anticipate their long-term impact on the development of cities.

5.6. Acknowledgements

The authors would like to thank the various organizations and projects that helped this work to become a reality: the National Centre for Scientific Research (CNRS), *Programme national de recherche et d'innovation dans les transports terrestres* (PREDIT), the Ministère de la Recherche (Action *Concertée Incitative Ville*), Avignon and Pays de Vaucluse University, Franche-Comté University, Communauté d'Agglomération du Grand Besançon, Communauté d'Agglomération de Lille, Syndicat mixte du Pays du Doubs Central, Compagnie des Transports de Besançon (CTB) and réseau Ginko, and Prorentsoft (Resad2 software editor).

5.7. Bibliography

[ALF 05] ALFONZO M.A., "To walk or not to walk? The Hierarchy of walking needs", *Environment and Behaviour*, vol. 37, pp. 808-836, 2005.

[BAI 01] BAILLY J.P., HEURGON E., *Nouveaux Rythmes Urbains: Quels Transports?* La Tour d'Aigues:l'Aube, Société et territoire, 2001.

[BAJ 06] BAJOLET E., MATTEI M.F., RENNES J.M., *Quatre ans de Recherche Urbaine 2001-2004*, François Rabelais University Press, MSH Villes et Territoires, Tours, 2006.

[BAN 97] BANISTER D., WATSON S., WOOD C., "Sustainable cities: Transport, energy and urban form", *Environment and Planning B: Planning and Design*, vol. 24, pp. 125-145, 1997.

[BAT 04a] BATTY M., A New Theory of Space Syntax, CASA working papers series, no. 75, University College London, 2004.

[BAT 04b] BATTY M., RANA S., "The automatic definition and generation of axial lines and axial maps", *Environment and Planning B: Planning and Design*, vol. 31, pp. 615-640, 2004.

[BEG 97] BÉGUIN H., THOMAS I., "Morphologie du réseau de communication et localisations optimales d'activités. Quelle mesure pour exprimer la forme d'un réseau", *Cybergeo: European Journal of Geography*, no. 26, April 9 1997 .

[BEN 91] BENZONI L., *Fonction de Coûts*, Preparatory text for *Journée Économie des Réseaux*, Paris, GDR and EDF, 1991.

[BOL 06] BOLOT J., Le transport à la Demande, une piste pour le développement durable. Approche géographique et mise en œuvre d'un système opérationnel à Besançon. Geography thesis, University of Franche-Comté, 2006.

[CAV 04] CAVAILHÈS J., FRANKHAUSER P., PEETERS D., THOMAS I., "Where Alonso meets Sierpinski: An urban economic model of a fractal metropolitan area", *Environment and Planning A*, vol. 36, p. 1471-1478, 2004.

[CUT 99] CUTINI V., "Centrality and land use: three case studies on the configurational hypothesis", *Cybergeo: European Journal of Geography*, no. 188, 1999.

[DAC 05] DA CUNHA A., "Régime d'urbanisation, écologie urbaine et développement durable", in: DA CUNHA *et al.* (eds), *Enjeux du Développement Urbain Durable*, Lausanne, Romandes Polytechnic and University Press, pp. 13-37, 2005.

[DES 01] DESYLLAS J., DUXBURY E., "Axial maps and visibility analysis", *Proceedings of the 3rd International Space Syntax Symposium*, Atlanta, no. 27, 2001.

[DUP 91] DUPUY G., *L'urbanisme des Réseaux, Théories et Méthodes*, Armand Colin, Paris, 1991.

[DUP 93] DUPUY G., "Géographie et économie des réseaux", *L'Espace Géographique*, vol. 3, pp. 193-209, 1993.

[DUP 99] DUPUY G., *La Dépendance Automobile*, Economica, Paris, 1999.

[FOL 07] FOLTÊTE J.C., PIOMBINI A., "Urban layout, landscape features and pedestrian usage", *Landscape and Urban Planning*, no. 81, p. 225-234, 2007.

[GEN 00] GENRE-GRANDPIERRE C., Forme et fonctionnement des réseaux de transport: approche fractale et réflexion sur l'aménagement des villes, Doctorate thesis, Franche-Comté University, 2000.

[GEN 03] GENRE-GRANDPIERRE C., FOLTETE J.C., "Morphologie urbaine et mobilité en marche à pied", *Cybergeo*, no. 248, 2003.

[GEN 07] GENRE-GRANDPIERRE C., "Des réseaux lents contre la dépendance automobile? Concept et implications en milieu urbain", *L'Espace Géographique*, vol. 1, pp. 27-39, 2007.

[GOU 92] GOUYET J.F., *Physique et Structures Fractales*, Paris, Masson, 1992.

[GUT 98] GUTIÉRREZ J., MONZÒN A., PINERO J.M., "Accessibility, network efficiency and transport infrastructure planning", *Environment and Planning Part A,* vol. 30, pp. 1337-1350, 1998.

[HIL 84] HILLIER B., HANSON J., *The Social Logic of Space*, Cambridge, Cambridge University Press, 1984.

[HIL 93] HILLIER B., PENN A., HANSON J., GRAJEWSKI T., XU J., "Natural movement: or, configuration and attraction in urban pedestrian movement", *Environment and Planning B: Planning and Design*, vol. 20, pp. 29-66, 1993.

[HIL 96] HILLIER B., *Space is the Machine: A Configurational Theory of Architecture*, Cambridge, Cambridge University Press, 1996.

[HIL 00] HILLIER B., SHU S., "Crime and urban layout: The need for evidence", in: BALINTYNE S., PEASE K., MCLAREN V. (eds.), *Key Issues in Crime Prevention, Crime Reduction and Community Safety*, London, Institute for Public Policy Research, 2000.

[JIA 99] JIANG B., CLARAMUNT C., BATTY M., "Geometric accessibility and geographic information: Extending desktop GIS to space syntax", *Computers, Environment and Urban Systems*, vol. 23, pp. 127-146, 1999.

[JIA 00] JIANG B., CLARAMUNT C., KLARQVIST B., "An integration of space syntax into GIS for modelling urban spaces", *International Journal of Applied Earth Observation and Geoinformation*, vol. 2, no. 3-4, p. 161-171, 2000.

[JOE 01] JOERIN F., THERIAULT M., VILLENEUVE P., BEGIN F., "Une procédure multicritère pour évaluer l'accessibilité aux lieux d'activité", *Revue Internationale de Géomatique*, vol. 11, no. 1, pp. 69-104, 2001.

[JOS 05] JOSSELIN D., GENRE-GRANDPIERRE C., "Des transports à la demande pour répondre aux nouvelles formes de mobilité. Le concept de Modulobus", *Mobilités et Temporalités*, University of Saint Louis Publishing, Brussels, pp. 151-164, 2005.

[KAS 03] KASEMSOOK A., "Spatial and functional differentiation: A symbiotic and systematic relationship", *Proceedings of the 4th International Space Syntax Symposium*, no. 11, London, 2003.

[LEV 94] LEVINSON D., KUMAR A., "The rational locator: Why travel times have remained stable", *Journal of the American Planning Association*, vol. 60, no. 3, pp. 319-332, 1994.

[LIT 98] Litman T., "Driving out subsidies: How better pricing of transportation options would help protect our environment and benefit consumers", *Alternatives Journal*, vol. 24, no. 1, pp. 36-42, 1998.

[MAJ 00] Major M.D., "The use of "Space Syntax" as an interactive design tool in urban development", *Planning Forum*, vol. 6, pp. 40-56, 2000.

[MAT 96] Mathis P., "Consommations d'énergie et pollutions liées à l'étalement des densités", in: Gaudemar J.P. (ed.), *Environnement et Aménagement du Territoire*, Paris, La Documentation française/DATAR (Research), pp. 95-106, 1996.

[MAT 03] Matthiopoulos J., "The use of space by animals as a function of accessibility and preference", *Ecological Modelling*, vol. 159, pp. 239-268, 2003.

[NEW 89] Newman P., Kenworthy, J., *Cities and Automobile Dependence, an International Sourcebook*, Gower, Aldershot, 1989.

[PEP 97] Peponis J., Ross C., Rashid M., "The structure of urban space, movement and co-presence: the case of Atlanta", *Geoforum*, vol. 28, no. 3-4, pp. 341-358, 1997.

[PIO 06] Piombini A., Modélisation des choix d'itinéraires pédestres en milieu urbain. Analyse géographique et paysagère, Doctorate thesis, Besançon, Franche-Comté University, 2006.

[REA 99] Read S., "Space syntax and the Dutch city", *Environment and Planning B: Planning and Design*, vol. 26, pp. 251-264, 1999.

[ROD 04] Rodriguez D.A., Joo, J., "The relationship between non-motorized mode choice and the local physical environment", *Transportation Research Part D*, vol. 9, no. 2, pp. 151-173, 2004.

[SHU 00] Shu S., "Housing layout and crime vulnerability", *Urban Design International*, vol. 5, no. 3-4, pp. 177-188, 2000.

[SHU 03] Shu S., Huang J., "Spatial configuration and vulnerability of residential burglary: A case study of a city in Taiwan", *Proceedings of the 4th International Space Syntax Symposium*, London, no. 46, 2003.

[SME 05] SMESSDRG (Syndicat Mixte pour l'élaboration et le suivi du Schéma directeur de la région grenobloise), Pour un "Chrono-aménagement" du Territoire. Vers des Autoroutes Apaisées, Les Dossiers Déplacement, Grenoble, Agence d'Urbanisme de la Région Grenobloise, 2005.

[STA 86] Stanley H.E., Ostrowsky N., *On Growth and Form: Fractal and Non Fractal Patterns in Physics*, Boston, Martinus Nijhoff Publishers, 1986.

[THE 02] Thevenin T., Quand la géographie se met au service des transports publics urbains: une approche spatio-temporelle appliquée à Besançon, Geography thesis, Besançon, Franche-Comté University, 2002.

[TUR 02] Turner A., Penn A., "Encoding natural movement as an agent-based system: An investigation into human pedestrian behaviour in the built environment", *Environment and Planning B: Planning and Design*, vol. 29, pp. 473-490, 2002.

[TUR 05] TURNER A., PENN A., HILLIER B., "An algorithmic definition of the axial map", *Environment and Planning B: Planning and Design*, vol. 32, pp. 425-444, 2005.

[VER 97] VERNEZ MOUDON A., HESS P., SNYDER M.C., STANILOV K., Effects of Site Design on Pedestrian Travel in Mixed-use, Medium-density Environments, Washington, Report for the Washington State Transportation Center, Federal Highway Administration, 1997.

[WIE 02] WIEL M., *Ville et Automobile*, Paris, Descartes et Cie, 2002.

[WIE 03] WIEL M., "Quelle place donner à la maîtrise de la vitesse automobile en ville?", *39th Colloque de l'ASRDLF*, Lyon, September 2003.

Chapter 6

Daily Mobility and Urban Form: Constancy in Visited and Represented Places as Indicators of Environmental Values

6.1. Introduction

Environmental values raise two interdependent methodological problems. Indeed, we are as much confronted with difficulties in identifying them as with their objectification. More concretely, it implies pinpointing information that respondents are not aware of. On the other hand, the current tools are linked to the stated preferences regarding urban forms that are not necessarily ecologically valid [MAT 88] from a behavioral point of view. Moreover, they pool the environmental values stemming from spatial practices with those from spatial representations. Thus, they introduce inaccuracies in the analysis of individuals' relationships with their living space.

Indeed, methodological tools produced to identify environmental values primarily fall under the survey method, whether the semi-structured interview technique is used [APP 69, HAR 72], or the more common questionnaire for evaluating the various environments presented as photographs or slides [HER 76, HER 03]. These researches, which are strongly oriented by environmental psychology, are based on the concept of preference. This concept was modeled according to a cognitive approach and includes four dimensions: the coherence, complexity, mystery and legibility of the environment [KAP 87]. Thus, the

Chapter written by Thierry RAMADIER, Chryssanthi PETROPOULOU, Hélène HANIOTOU, Anne-Christine BRONNER and Christophe ENAUX.

theoretical models and methodological tools are not directly related to the behavior of the individuals surveyed. Modeled on the problem of social psychology attitudes, these studies have the disadvantage of being sensitive to social desirability[1], which the survey situation inevitably generates. Moreover, as in studies on attitudes, the results obtained do not allow us to identify the relationships between this type of environmental value and the spatial behaviors of respondents.

A first "indirect" methodological approach (insomuch as it is not based on verbally-declared preferences) involves simulated negotiation. Although this survey method was initially developed in the field of training, as early as 1969 Raser [RAS 69] showed how negotiation simulation games can be used to enrich theoretical research on the urban environment. Thus, Hoinville [HOI 71] developed the priority evaluator to collect environmental preferences from different social groups to plan the layout of a university campus. In this type of application, preferences come from a set of methodologically interdependent environmental attributes, since the respondent is asked to make choices based on a situation where it is impossible to have all of his or her needs filled. Indeed, the respondent makes limited choices by assigning to an environmental attribute a number of points that will then be lacking in another attribute, given that the respondent only has a limited number of points at his or her disposal. Although it is possible to identify an individual's environmental values, these come from the cost-benefit process. This type of survey is therefore based on the concept of usefulness, which does not guarantee the absence of any discrepancy between the "indirectly declared answer" through negotiation and behavioral response. Lastly, the survey method, regardless of the technique used, anchors environmental values to the field of representations by referring to the concept of preference.

Taking inspiration from econometrics, the hedonic approach, whose conceptual foundations are presented in chapters 11 and 12, is based on economic behavior and allows one to leave the field of representations. Analysis using hedonic modeling of economic values for a geographical object, such as a lot or a dwelling, helps determine the weighting of environmental values, such as vegetation or, more generally, the urban landscape, on the price. In particular, this method was used to model the price of land in cities [KES 01]. However, these environmental values, given that they are measured using a single indicator that is part of a quantitative ratio scale, are based on the premise that all individuals use the same environmental

---

1 The effect of social desirability corresponds to the generation, whether conscious or unconscious, of answers for the survey perceived to be valued or desired by the investigator or society (showing your best face or a compliant image of yourself, not clashing with the investigator, etc.).

evaluation scale[2]. This is due to the fact that it is not the geographic space but rather a geographic object that is being evaluated in an "economic space".

Observing spatial behavior thus remains a method that allows us to identify environmental values outside the field of representations and declared responses, while offering the possibility of differentiating between social groups. In this case, the behavior does not reveal a preference, but rather a commitment of the individual in a sociospatial context of which he/she may be unaware. The commitment that the individual is subject to daily life thus implies searching for recurrences, even environmental constancy, in which he/she bases his/her activities. It is these recurrences that allow us to identify relevant environmental values without their having to be verbalized.

The method that we have developed to provide partial explanations for the geographic structure of daily mobility tends to deviate from the declared answers, and take into account both the representational and behavioral aspects of the environmental values. It is based on a methodological triangulation that combines the morphological analysis of the urban space, the observation of spatial behaviors and the cognitive representations of space. Thus, the information sought needs to take into account both the daily mobility of individuals and their relationship with space, structure of the urban space, environmental attributes and, lastly, the position of the individual in the social structure. In other words, psychological, geographical and sociological aspects converge when the point is to look for the processes in relation to visiting urban spaces based on their form. This is all the more so, when the processes are sought from the perspective of environmental values. In addition to the theoretical problems it produces, this disciplinary convergence poses many methodological issues. However, these problems cannot be dissociated from the first ones, since there is no theory-neutral method. We will therefore start by specifying the theoretical positions at the basis of this methodological work.

6.2. From landscape to eco-landscape

The method developed to identify the environmental values of individuals is based on the concept of the urban landscape. First of all, this concept refers, in the strict physical sense, to the morphological characteristics of the space. It also refers to the relationship of individuals to the physical space, i.e. to the representations that they make of the space and the environmental values underlying these mental images. Lastly, insomuch as the concept of landscape implies that the space is not isotropic but rather made up of specific forms depending on the geographic location,

2 Several devices and improvements to the method – among which the use of interactive terms in the regression – have been developed that allow market heterogeneity to be accounted for in estimating model parameters.

it refers to the concept of place, i.e. a categorized and spatialized expanse. Thus, this union between materiality, cognitive position and geographic position that comes out of the concept of landscape appears important to us, since it brings together the main supports that allow the individual to develop environmental values. The social positions of the individuals remain to be taken into account in order to complete the analysis, since each value only has meaning when considered in the context of social distinction and the homologous relationships they establish [BOU 79].

6.2.1. *Landscapes and environmental values*

Two basic components define the concept of landscape and have a large consensus. On one hand, landscape corresponds to what we can see. On the other hand, and correlatively, landscape combines the world of things with that of human subjectivity. In other words, nature and society, setting and sight are interacting [BER 95]. It is in this sense that the concept of landscape and, in particular, urban landscape, remains relevant to the understanding of environmental values. It is also this combination of materiality and its perception that, methodologically speaking, requires the development of tools that would join these two dimensions in order to simultaneously analyze them.

Theoretically, most approaches start with environmental values in order to define landscapes. Therefore, the majority of methods developed are based on evaluation of the environment by individuals. However, these approaches, especially when they involve understanding the impact of the landscape on individuals, stray from materiality. They reflect a tautology in which the individual ends up determining his/her own self, since it is his/her environmental values that define his/her behaviors and *vice versa*. In other words, the material dimension of the landscape has disappeared from the analysis. This position is untenable, especially with regard to urban landscape – a setting in which materiality refers to a set of signs and codes that are both physical and socially constructed.

Starting off with the fact that landscape is the spatial transcription of a social organization, the materialist and morphological approach actually integrates a human dimension and a system of values. This is unlike the physicalistic approach, which tends to only look at physical qualities of the setting (relief, biotic and abiotic components). It thus seems wiser to start with this human-oriented basis, which retains the link with landscape materiality in order to clarify its effect on the individual, whether in strictly behavioral terms or in relation with the associated environmental values.

This theoretical position is based on the structuralist approach rather than the phenomenological approach, which focuses on the experience of being in a relationship with space. Consequently, the landscape is considered to be a source of information whose signifier (materiality) is the result of the material conditions as well as spatial and morphological projections of all the systems of values that act on the space in question. Jointly, the landscape is the result of introjection of the landscape signifiers (environmental knowledge) based on the social and cultural position of the individual. Put otherwise, landscape materiality is a social construction from both the material and individual perspective. It is the social distance between the elements of materiality and those interiorized by the individual, with the result that the same space does not constitute the same landscape from one individual to the next. We are therefore talking about the social legibility of space [RAM 98]. Thus, the structuralist approach has the advantage of identifying permanencies and developments in the relationship between the individual and the materiality of the landscape; it allows us to understand the environment value signifiers. It does not, however, allow us to look for social meanings (the signified) that are associated with landscape materiality.

Lastly, in the urban environment, inasmuch as landscape materiality is not limited to the physical dimension but rather to a combination of social dimensions, we believe it important that the structuralist approach be associated with a systemic approach, in order to account for the main facets that make up this materiality. Thus, in addition to the morphological dimension, there are functional dimensions (services and retail), historical dimensions (morphogenesis) and sociological dimensions (populations living in the areas). In methodological terms, the "user" of an urban landscape is momentarily removed, in order to construct a landscape recognition grid on the basis of information materially presented in the space and interpreted by a single observer on all the land being studied. It is only then that the "user" is introduced to identify the environmental values on the basis of both landscape recurrences and the opposition between landscape recurrences depending on the social groups.

6.2.2. *Methodological orientation*

This theoretical position leads us to put aside urban landscape analysis methods based on the evaluation by the respondent him/herself, whether using a single note [FIN 68] or noting several elements that are supposed to contribute to the value of the landscape being combined with morphological indicators (CSW method)[3]. The phenomenological approach, like that developed by Bailly [BAI 90] – which consists of identifying through exploration and drift, the relationship between the

---

3 CSW stands for Coventry, Solihull, Warwickshire. For a more detailed typology on the different landscape analysis methods, see ROUGERIE G. and BEROUTCHACHVILI N., *Géosystèmes et Paysages: Bilan et Méthodes*, Paris, Armand Colin, 1991.

observer and a landscape he/she does not know – is not appropriate for the theoretical and methodological objectives that we are working on. Indeed, in all these cases, the meanings of the environmental values are what define the landscape, whereas we are looking to define the materiality of the landscape to bring out the signifier of environmental values.

The methods for operational urban analysis are more suited to our objectives, since they are focused on the diagnosis of urban space while taking into account its history, morphology, built space, etc. This type of landscape analysis of the urban setting was studied from two perspectives: from cognitive representations of space or from land cover. Put otherwise, we come across the idea that the landscape is a cognitive and material construction.

In the first case, the work of Lynch [LYN 60] and related works are based on the five types of urban elements that allow us to qualify the legibility of space. Here the urban landscape is fragmented, since it refers to a type of element (roads, nodes, etc.) or the element in its singularity (such-and-such a building or such-and-such a street, etc.). For us it seems interesting to use this sensory evaluation technique of the landscape to more generally identify the landscape context in which this element lies, in order to identify the cognitive component of the signifier of the environmental values.

In the second case, it is the composition and organization of biotic and abiotic elements of the space itself that are analyzed to classify the landscapes. The space is a source of information that must be interpreted to obtain a land cover map [PET 06]. To this end, it is important to simultaneously consider the three components making up the landscape: abiotic, biotic and man-made components [BRO 84]. Subsequently, the satellite image becomes an interesting tool in that, on one hand, this type of document can cover larger land surfaces and, on the other hand, the sharpness of the resolution is now adequate while these images have been subject to statistical and thematic treatments for the past four decades. Together, these three benefits allow us to study the complexity of land cover and to identify landscape typologies that bring together the three landscape components. This aspect will be partially reviewed in Chapter 12.

To conclude, instead of determining urban landscape categories on the basis of environmental meanings produced by the respondents, we are looking to structure biotic, abiotic and human-made components (urban morphology, urban functions, etc.) of land cover. We then want to map the land being studied based on the system of landscape categories built in order to analyze which landscapes are both visited (those associated with an individual's daily activity) and represented (those associated with an *ad-hoc* item of the individual's spatial representation). Landscape recurrences that emerge at these two levels of analysis allow us to identify the environmental values of individuals, while limiting them to their signifier (categorized and spatialized materiality). Thus, the geographic approach becomes the point of convergence, or medium of the merging between sociological, psychological and geographical approaches to the relationship with urban landscapes.

6.2.3. *Landscape ecology and the concept of eco-landscape*

The most appropriate theoretical model for our methodological aims is landscape ecology, which is a holistic approach that focuses on understanding landscapes, taking into account their heterogeneity, especially since it is specifically this heterogeneity that characterizes urban landscapes. Moreover, landscape ecology is based on the systematic observation of landscapes using aerial photos, and then satellite images, as they became available. These two methodological elements are adapted to developing a thematic map of urban landscapes, which is an indispensable prerequisite tool for building a landscape analysis grid.

The main analysis unit of landscape ecology is the ecotope, a concept defined as the smallest holistic unit of land. It is characterized by the essential attributes on the corresponding surface of land. Inasmuch as we believe that the homogeneity of the space does not exist in itself, as Claval [CLA 95] proposed for the concept of landscape, the ecotope has coherence and structure, and "it owes these qualities far more to recurrence or opposition of themes than to the unity of composition." It is the internal coherence and diversity of the region versus the closest neighbors, which gives the appearance of homogeneity and a unity built on its specificity. Consequently, the typology of urban ecotopes is the result of the researcher's interpretation on the basis of knowledge acquired on the ecology, history, physical and social geography of the space concerned. The image interpreted from the ecotope is also an eco-landscape [PET 03]. The ecotope can be mapped using a scale of between 1:5,000 and 1:25,000 [NAV 84].

Two limits characterize this methodological process. First of all, the results are largely dependent on the interpretation of the researcher and his/her field knowledge (whether historical, political, sociological aspects, etc.). This limit remains satisfactory, however, if a single researcher carries out all the interpretative work (homogeneity of processing on all territory) and the emphasis is put on comparing the visited and represented landscapes as well as the social groups to look for environmental values. Indeed, these comparisons remain possible, since the same analysis tool is used each time. Lastly, the precise qualification of urban landscapes can be found by going back to the very high-resolution satellite images, especially in regions identified in the analysis. The second limit is based on the fact that it is a "bird's eye" landscape analysis, and thus not the daily perception of the environment. This process has the advantage of minimizing the interpretative bias by eliminating esthetic and subjective ambiance data the researcher would be confronted with if he/she should have to interpret an urban scene. In other words, the procedure allows us to conserve a general interpretation and avoid getting lost in the diversity that characterizes urban space. Thus, this last limit minimizes the first while retaining the central idea of the landscape concept; namely that it must refer to what we see, without it necessarily being what the "user" consciously sees.

6.2.4. *Method of analysis: eco-landscape cartography of urban spaces*

Our analysis of landscape features of urban space is based on the eco-landscape approach developed by Petropoulou [PET 03] in the urban environment. The eco-landscape maps are created based on a visual interpretation of colored compositions from SPOT satellite images and using Quick Bird images based on nested scales logic (trans-scale).

SPOT XS (multispectral) images provide, at a scale of 1:25,000, interesting information on vegetation and the built-up spaces as a whole (urban and peri-urban space), while their composition with SPOT P (panchromatic) images provide better distinction between urban fabrics. For their part, Quick Bird images provide information on urban morphology and help distinguish the different types of urban fabrics within built-up spaces, especially at a scale of 1:5,000.

Three types of media are thus necessary for the analysis:

– high and very-high spatial resolution satellite images:

   - 1998 SPOT multispectral (XS) images; 20 m spatial resolution,

   - 1992 SPOT panchromatic (P) images; 10 m spatial resolution,

   - 2000 Quick Bird Multispectral images; 2 m spatial resolution,

   - 2000 Quick Bird Panchromatic images; 1 m spatial resolution;

– topographical map of land cover (public buildings, services, cemeteries, etc.);

– digital map of the road network.

The satellite images are processed before being used. It is during essential operations and a critical presentation of results that we obtain the following:

– Multispectral analysis to obtain an XS colored composition and superposition of the hierarchized road network on the resulting image. This first procedure is used to choose samples from the supervised analyses and allow us to distinguish the major types of land cover and the imposing building zones versus other types of urban spaces.

– Calculation of the normalized difference vegetation index (NDVI) on SPOT XS images over the entire urban space. The analyses using NDVI showed strong contrasts between chlorophyllous vegetation zones (in white) – with or without trees, dense urban zones and water with almost no vegetation (in black). They allow us to choose samples from the various types of vegetation in order to interpret the eco-landscapes.

– Classification of the major types of land cover using a discriminant analysis and a supervised clustering on the SPOT XS image. This classification allows us to build an initial typology of land cover. At this step, we distinguish between the dense city center, "*grands ensembles*" suburbs and residential areas. The classification is not precise enough, however, given that only biophysical information was taken into account. The lack of morphological information and the strong proximity of spectral signatures do not allow us to improve the categorization of eco-landscapes.

– Merging of SPOT XS and P images. This is followed by digitization of the rail network and waterways on the SPOT XS image using the topographic map and to create classes of networks.

– First visual interpretation of the XS plus P image (1:25,000 – 1:50,000 – 1:100,000 scales). We started with an inventory to define the zones (and not the objects, as is done in urban morphology), and highlight the properties that distinguish them in order to establish categorization criteria. Then we did a first grouping by classifying the zones belonging to the same reading level of the urban environment to form the different types. The analysis started in the larger areas (1:100,000 scale view) and then continued to the zones at a scale of 1:25,000. At this analysis scale, the set of ecological characteristic classes is defined, but their geographical definition remains inaccurate due to the low spatial resolution of the images. Moreover, the zones occupied by large buildings cannot be distinguished from one another, and certain classes (cemeteries, dumps and other zones) cannot be defined given the lack of information on land cover. Lastly, the majority of zones within Type 1 "Urban Zones – Predominant Habitat" cannot be sufficiently differentiated.

– Merging of the Quick Bird MS and P images over 10 parts of urban space. This is followed by formation of a mosaic of Quick Bird images so as to distinguish the types of urban fabric within built-up spaces, the essential features of urban morphology and predominant building types in each zone. These images are useful for finer spatial definition of classes already defined on the SPOT images, especially the borders of the various zones and the smaller zones of two to five hectares. They are also useful for better interpretation of interstitial zones that are difficult to interpret on the SPOT images (dumps or interstitial vegetation-free surface zones, linear vegetation, limits between different types of urban zones in which the habitat predominates).

– Use of land-cover maps to distinguish public buildings, certain industrial zones that are difficult to distinguish on the image, certain cemetery and other zones.

– Digitization of the eco-landscape zones and land cover on the SPOT XS plus P image using the information from the mosaics of Quick Bird images and topographical maps. Visual interpretation of urban eco-landscapes based on the presence of vegetation and water, structure and organization of the urban fabric, density of the built-up areas, height and right-of-way of buildings and type of roof covering them. The interpretation is done at the second level of typology for types 1 to 4 in Table 6.1 (interpretation at scales of 1:25,000 to 1:50,000) and up to the first level for types 5 to 10 (interpretation at a scale of 1:50,000). At this step in the analysis it is impossible to distinguish the high buildings or zones of vegetation, water and vegetation-free zones of less than two hectares when the urban fabric is very dense. The definition of types is done using the concepts of "landscape ecology" adapted to urban spaces. Indeed, an industrial zone (e.g. type 2.1) may include (at this interpretation level) the vegetation-free and water zones, as well as a few dwellings and sports areas, which will be distinguished at the third level (2.1.1–2.1.5) using images with finer spatial resolution.

– Validation on the field investigation by randomly taking verification samples. Note that the classes concerning the different types of urban fabric were defined following familiarization with the landscape studied based on the work and on-site visits. Moreover, the majority of uncertain sectors were delimited on the map and then defined and classified based on the field investigation.

– Creation of the final typology (see Table 6.1).

– Verifying the eco-landscape zones and uses of soil on Quick Bird images (spatial resolution of 1–2 m) neighborhood by neighborhood. Overlay of the topographical database (TDB) on the image to check the results at the neighborhood level (scale of 1:5000). The image shows the relationship between urban morphology and urban eco-landscape.

– Comparison of eco-landscape maps with the local urban policy using the documentary sources available and local INSEE[4] data.

---

4 INSEE: *Institut National de la Statistique et des Etudes Economiques* (National Institute of Statistics and Economic Studies).

Types	Eco-landscape definition
1. Urban Zones – Predominant Habitat	
1.1 Historical center	Eco
1.2 Center with large buildings and mixed zones	Eco
1.3 Old centers integrated into the city and extensions	Eco
1.4 *Grands ensembles* (high buildings and open areas)	Eco
1.5 Residential with garden	Eco
1.6 Residential with large garden	Eco
1.7 Village and extensions	Eco
1.8 Diffuse (small buildings scattered around the city)	Eco
2. Industry, Public Buildings and Infrastructures (greater than 2 ha and/or outside of Type 1)	
2.1 Industrial zones and mixed zones (industry and residential)	TM and Eco
2.2 Zones of public buildings (universities, hospitals, etc.)	TM
2.3 Fire stations and other areas with large buildings	Eco
2.4 Train station and train station buildings	TM and Eco
2.5 Port and port industrial zone	TM and Eco
2.6 Airport	TM and Eco
2.7 Open areas of warehouses, garages, etc.	TM and Eco
3. Vegetation-free Surfaces, Quarries, Work Sites (greater than 2 ha and/or outside of Type 1)	
3.1 Interstitial vegetation-free surfaces	Eco
3.2 Very large traffic hubs of roads and work sites	TM and Eco
3.3 Quarries and extraction areas	TM and Eco
4. Green Spaces and Sports Areas (greater than 2 ha and/or outside of Type 1)	
4.1 Parks and small wooded areas	TM and Eco
4.2 Stadiums, golf courses and other sports areas	TM and Eco
4.3 Community garden	TM and Eco
4.4 Vegetation areas around streets	Eco
4.5 Vegetation areas around water	Eco
4.6 Cemeteries	TM and Eco
4.7 Other areas of vegetation	Eco
5. Water (greater than 2 ha and outside Type 1)	Eco
6. Agricultural Areas (greater than 2 ha and outside Type 1)	Eco
7. Forests (greater than 2 ha and outside Type 1)	Eco
8. Forest and Water (greater than 2 ha and outside Type 1)	Eco
9. Sand, Pebbles, Rocks (greater than 2 ha and outside Type 1)	Eco
10. Undefined (greater than 2 ha and outside Type 1)	Eco

Table 6.1. *Typology of urban eco-landscapes–urban community of Strasburg (from satellite images: Eco; from topographical maps: TM)*

6.2.5. *Building the landscape analysis grid*

At the end of this eco-landscape analysis, and once the map is produced, the objective is to build an analysis grid that will allow the introduction of survey and observation data from individuals, who, in the end, are the ones who will allow us to identify the signifier of environmental values.

The methodological requirements are based on the fact that:

– the eco-landscape data correspond to highly heterogeneous zones with regard to form and size;

– that the data from the representations are geo-referenced as points; and

– that the behavioral data can be identified either *ad hoc* (address) or within a network defined by the researcher.

In other words, each type of data tends to ignore the format of the others. Therefore, the methodological objective involves harmonizing the spatial unit of these three types of data so as to be able to qualify the landscape context of visited and represented places based on the same procedure. Establishing relationships between these data is then possible by using the network found in the detailed city-plan booklet for the urban space being studied (a commercial booklet for addresses in order to meet daily needs).

Choosing this grid provides many advantages. On one hand, it is easy to use at the time when the respondent codes the site he/she visited, which allows him/her to maintain a certain level of confidentiality regarding his/her private life and to easily code the same place. Indeed, the coding procedure is identical to the procedure for searching for an address in daily life. On the other hand, grid regularity offsets the irregularities in form and size of the landscape map. Lastly, the cells allow us to qualify the context of the places visited or represented on the basis of an expanse rather than on that of a punctual spatial correspondence between a represented or visited urban element and its landscape attributes. This choice is due to the fact that the concept of landscape implies qualifying a space on the basis of an area, since our focus is on the landscape context for the urban element recorded in the survey, and not only on the quality of the element insertion point. Therefore, the cell becomes the spatial unit for data analysis – an analysis unit that must initially be qualified from a landscape point of view.

The spatial intersection of the grid used with the urban eco-landscape map allowed us to qualify each cell, knowing that each cell is very often composed of several types of landscapes. A classification based on the proportion of landscape surfaces recorded in each cell allowed us to show the main landscape qualities making them up. Six types of cells were identified using the following criteria:

– *uniform cells*: cells with 95–100% of their surfaces covered by a single type of landscape;

– *predominance cells*: cells having landscapes that cover at least 50%, and at most 94.9%, of the surface, even if other landscapes are present;

– *minor predominance cells*: cells having one type of landscape covering 40–50% of the cell surface, and whose proportion of other landscapes is at least 10 points less than the predominant landscape;

– *concomitance cells*: cells having two types of landscape each covering 50% of the cell surface, with a maximum difference of 10 points more or less than 50% for each type of landscape;

– *minor concomitance cells*: cells having two or more types of landscape with a difference of more than 10 points between them;

– *other cells*: these are generally cells with a landscape surface of <10% of the total area of the cell, or that do not correspond to any of the other cell categories.

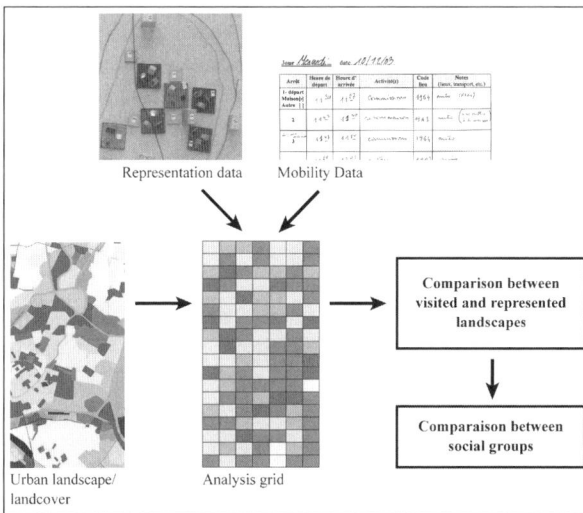

Figure 6.1. *General procedure for landscape analysis*

The diversity of possible combinations and the large initial number of urban landscapes do not allow us to group cells based on the type of landscape. Therefore, each cell is, in the end, qualified according to the landscape(s) that dominate(s) in its composition. Thus, it is possible to consider several types of eco-landscape for each visited or represented place, which corresponds to a geographical reality, particularly in dense urban areas where landscape diversity is very high.

Qualification of the overall grid on the basis of the landscape map is the tool that allows us to compare represented landscapes and visited landscapes, as well as social groups.

6.3. Behavioral and representational data collection

The survey method makes use of a semi-structured interview, a questionnaire and respondent self-observation. It requires an initial face-to-face meeting lasting at least 90 minutes, and then three telephone interviews, each of approximately 15 minutes. Searching for complementary information on the signified of the environmental values is possible; this implies adding a second semi-structured face-to-face interview to identify reflexive data based on response time used.

The face-to-face meeting consists of a questionnaire for initially identifying the sociodemographic characteristics, residential mobility of the individuals and the level of equipment in the household. The person is then invited to express his/her representation of the space using a simple model that records the spatial organization of environmental knowledge. The final stage in the interview involves training the respondent in the self-observation of his/her mobility behavior over seven consecutive days following the meeting. The telephone interviews take place during this self-observation phase.

6.3.1. *The spatial reconstruction set (JRS[5]): a cognitive spatial representation data collection technique*

The objective of this task involves recording the urban elements that make up the individual's spatial representation. Freehand drawing is the most common technique. It has been used for more than 40 years in various fields, such as geography, sociology, anthropology and psychology. It has the enormous advantage of being inexpensive in terms of the materials required (paper and pencil). Moreover, the procedures for carrying it out are simple and easily adapted to different survey situations (individually or in a group, on different sites, etc.). Lastly, the respondent can potentially refer to the scale of his/her choice in his/her response. It has many methodological biases, however, since drawing is a highly discriminating form of expression from the point of view of social groups [RAM 06a]. It implies having either previously integrated the graphic codes of mapping or developed a graphic language *in situ* that is accepted by both the investigator and the respondent. This situation limits the expression of spatial knowledge and tends to lead to a self-censuring of knowledge in certain groups.

---

5 *Jeu de reconstruction spatiale* (in French).

The JRS consists of a series of eight standardized pieces adapted to the externalization of cognitive representations of the urban space from neighborhoods to cities.

In the test protocol put in place for the intergroup comparison of JRS and freehand drawing, the following question was asked: *What knowledge do you have of downtown Strasburg?*

To answer this question, we are asking you to *reproduce downtown Strasburg* on this tray (show the tray) using a set of elements that I will show you:

Small houses to represent the buildings or houses that you know and that are important.

Wooden blocks to represent high or wide buildings, or buildings that cover a lot of land.

Plates with three houses on them to represent neighborhoods or blocks of houses.

Green plates to represent green spaces, parks, gardens, etc.

Blue plates to represent places or parking lots.

The plates can be juxtaposed (demonstrate) to obtain the desired size.

Red, black or blue wire to represent, respectively, roads (streets, routes, avenues, highways, etc.), railways, waterways or ponds, etc.

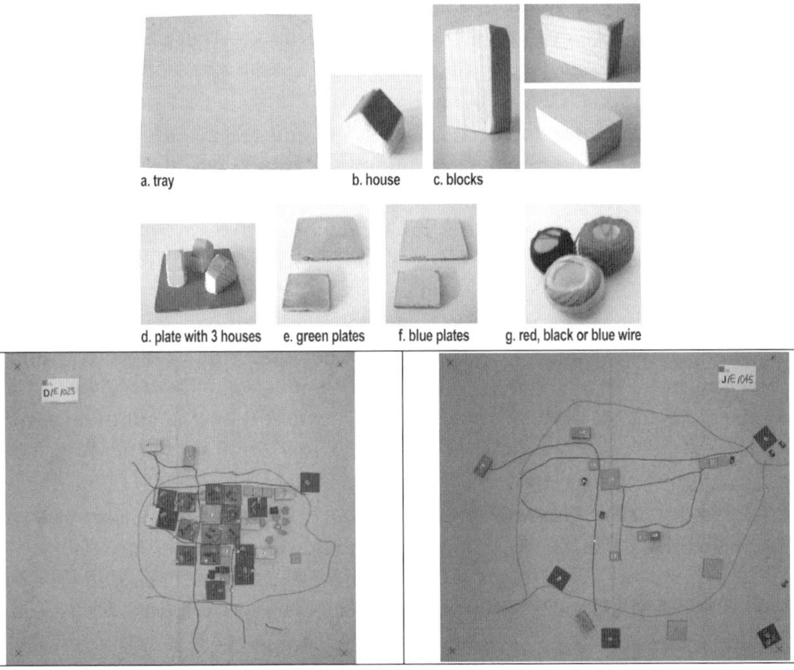

a. tray   b. house   c. blocks

d. plate with 3 houses   e. green plates   f. blue plates   g. red, black or blue wire

Figure 6.2. *Synthetic presentation of special reconstruction set and examples of results*

Furthermore, the fact that this type of expression uses "paper and pencil" strongly differentiates respondents, since the relationship to this type of task is highly dependent upon the individual's level of education. This is significant to the point that certain groups are more reluctant than others to carry out this exercise, which definitively amounts to putting them in the position of failing. The alternative involves using a modeling task, in this case the spatial reconstruction set (JRS), in order to improve expression of the cognitive representation of space and the comparisons between social groups [RAM 06a].

This game, made up of eight series of standardized pieces (see Figure 6.2), enables the individual to use a relatively flexible "language" to express the overall environmental knowledge making up his/her spatial representation. Consequently, the mental loading required for this exercise is lightened, which reduces the bias generally introduced in the paper-and-pencil task. Lastly, the limited number of pieces allows us to use the set with highly diverse populations in terms of age. Thus, this technique can be used from six years of age [RAM 06b] well into senior years [RAM 06c]. Moreover, the standardization of JRS pieces allows for comparisons that are often difficult to make using a drawing. It is therefore easier for the investigator to directly carry out quantitative analyses (number and types of elements represented, etc.) and qualitative analyses (comparison of spatial structures regardless of graphic style).

This technique, more appreciated than drawing [RAM 06a], also allows the respondent to adjust the position of the urban elements mentioned while moving the pieces. Each time that an element is put on the tray, it is numbered using a pre-printed label, and the investigator notes its identifier as formulated by the respondent.

The instructions are very open-ended and focused on the spatial knowledge of individuals. Their execution is limited to 15 minutes, since only the main urban elements are useful to the analysis. The spatial production is then photographed and archived.

6.3.2. Collecting travel behavior data

Collecting behaviors, which is done over a period of one week while dispensing with an observation method, is always a risky venture. This is because behaviors, when disengaged from an ecological situation (as in experimentation), are cut off from the issues that hinge spatial, social and cognitive dimensions and that define the commitment of the individual to the situation (ecological validity). Moreover, identifying declared behaviors means running the risk of only identifying those whose social interaction stemming from the survey allows it. Consequently, the declared responses emphasize the behaviors that reflect common sense and social representations, or that are strongly filtered by the social desirability effect, unbeknownst to the two people involved in the survey. Lastly, the disassociation between verbal behavior and spatial behavior accentuates both the omissions and selections taken from memory [AUR 96]. Therefore, doubting the truthfulness of the response is not what is important; rather it is getting beyond biases from the investigator and respondent that stem from the social situation that the survey inevitably implies.

Direct behavior observation is also too expensive, if not impossible, to carry out without technological input (e.g. GPS) when the observation place is not stable and unique. We believe that the trip log technique is the most appropriate solution for identifying spatial behaviors, especially because we had chosen to identify all mobility behavior over a full week. This tool is based on an indirect observation method in which the respondent observes him/herself. It is only when he/she recounts his/her self-observations that the survey situation comes into play with its methodological biases.

The benefit of this tool is that it does not separate verbal behavior from spatial behavior. Therefore, it is easier for the person taking the survey to refuse this type of method or to stop the procedure, often on the pretext that the process is too cumbersome, than to consciously or unconsciously get involved in a social game. For the investigator, it is thus easier to see the difficulties that individuals encounter in recounting their behaviors. The fact that the respondent observes him/herself without the control of the investigator is another considerable benefit. Yet another asset is the length of the behavior observation period. While an interview only allows us to look at the behaviors of the day before the meeting, self-observation allows us to easily piece together those of an entire week. The final advantage of this method is the possibility of collecting trips whose reasons are insignificant for the respondent (mailing a letter, talking to a neighbor in the street, etc.) and that are difficult to detect with GPS (stop time often too short). This implies a simple instruction for the respondent (e.g. note all your trips outside the home).

The main disadvantage of this method is the tediousness of the procedure, especially when the self-observation period exceeds two consecutive days. It generally results in simplifying the filling in of the log, repeating the routine activities periods in both time and space (e.g. work) and, gradually, leaving out small "insignificant" trips.

The tool produced and given to the respondents is a small booklet, the log (in A5 format) containing eight pages (see Figure 6.3). The first page lists the instructions for the log, with the seven other pages corresponding to each of the seven days following the presentation of the tool and constituting the observation period. Each of these seven pages is dated when it is presented. The respondent is instructed to use this log as a reference. He/she is encouraged to keep it with him/her during travel. The document can be folded, erased and scribbled on. The respondent must note the hours that he/she leaves and arrives at an activity outside his/her home. The activity is defined by a stop, even if it is brief or impromptu (stopping to talk to a friend on the sidewalk, for example).

Day _Mercredi_   date _10/12/03_

Stop	Departure time	Arrival time	Activities	Place code	Comments (places, locomotion...)
1-Departure Home [x] Other [ ]	11 24	11 27	Commissions (shopping)	19 B 4	auto (Atac). (by car)
2	11 42	11 45	retour maison	19 A 3	auto (sac oublié à la maison)
départ maison 3	11 47	11 50	commissions	19 B 4	auto
4	11 58	12 01	retour	19 A 3	auto
départ maison 5	14 02	14 04	boîte aux lettres	19 A 3	auto
6	14 05	14 20	permanence association	29 B 2	8 Av. F.N auto . parking rue pla..
7	18 49	19 24	retour maison	19 A 3	auto .
8					
9					
10					
11					
12					

Laboratoire Image et Ville, ULP / CNRS ; 3, rue de l'Argonne, 67000 Strasbourg

Figure 6.3. *Example of a page in the log*

The respondent is then asked to code, every evening, all of places where each activity took place using the booklet with detailed maps of the city being studied, which is offered to him/her at the end of the observation period. Coding involves noting the page number and then the reference in the cell in which the activity took place; this reference generally comprises a letter and number (e.g. 19A1 for page 19 and cell A1). This coding is taught when the log is handed out. Every 48 hours, the investigator reaches the respondent at the time of the telephone meeting set up previously, to find out for each activity the times, place-code, transportation mode, people who were with the person taking the survey during the travel and activity. The telephone interview enables us to prevent the respondent from growing tired of using the tool and also to check for any possible "holes" along the way in time-use taken and declared.

6.4. Behavioral and representational data processing

We must keep in mind that the purpose of the analysis is to understand the eco-landscape characteristics of places visited and represented in order to find the signifier of the environmental values involved in people's daily mobility.

6.4.1. *The processing of visited places*

From the time of the interview, insofar as the spatial reference is the cell, it is possible to directly geo-reference the places visited the on the basis of the geographical unit of analysis described above. This spatial coding allows us to collect two types of information:

– information on the spatial structure of the places visited, which corresponds to their spatial distribution (focused around the home, spread over a sector of the urban area, etc.);

– information on landscape qualities associated with the places visited.

These qualities allow us to obtain a landscape profile for each group:

– on all places visited during the week without taking frequency into account;

– on all places visited by weighing the landscape quality by visit frequency;

– or on places visited for each type of activity.

Activity coding can be broken down into seven categories:

– consumer activities (purchase of goods, shopping);

– service activities (use of public or private services: hairdresser, mechanic, dyer, photo agency, etc.);

– visits (any activity to a private address in one's social network);

– association activities (clubs, courses);

– religious activities (denominations, cemetery);

– recreational activities (sports, artistic activity, restaurants, cafés); and

– support activities (going with or picking up a member from the social network).

6.4.2. *The processing of space cognitive representations*

The first processing step involves distinguishing the *points* (specific places on the reduced scale and with identifiable contours, such as a park, a building), *lines* (thoroughfares) and *neighborhoods* in order to analyze the landscape quality for points alone. This choice is essentially guided by the current possibilities of using the lines and polygons with this type of landscape analysis grid. Indeed, thoroughfares, just like neighborhoods, are not urban elements whose cognitive reality corresponds to an administrative reality. Therefore, their limits cannot be specifically located[6]. Moreover, a large number of thoroughfares generally cannot be identified at all, because they are either unnamed or are based on their daily function (the street to go to the supermarket). However, we are seeking to develop methods that will offset these problems.

The first product of this processing is a pattern of dots (in the form of a GIS layer), which intersected with the landscape analysis grid allows us to associate a cell in the grid with each of them. The format of this set of dots is the same as that for data from places visited. On one hand, the landscape analysis is carried out using the same protocol, i.e. using the quality of the cell as the analysis unit, and on the other hand the places represented and places visited can be compared.

Complementary processing of cognitive representations is possible. Indeed, we can qualify the spatial structure of the representation by using the general form of the representation based on:

---

6 For thoroughfares, particularly when they are long, only a portion of the road or boulevard has a psychological reality for the individual. For neighborhoods, should we base ourselves on administrative limits (which generally have no meaning for the respondent) or morphological limits, at the risk of over-interpreting the elements present in the cognitive representation, etc.?

– the order that elements appear (did the respondent start his representation with the neighborhoods, thoroughfares or specific elements?);

– the type of identification of elements (common name, functional identification); and

– the proportion of thoroughfares, neighborhoods and specific elements.

This processing does not provide any direct information on environmental values. In particular, it allows us to complete the analysis of individuals' relationships with the urban space as a whole by looking further at the relationship between representation and spatial behaviors.

6.5. An application example: the Cronenbourg district pensioners' mobility

An exploratory analysis of the relationship between daily mobility and urban eco-landscapes was carried out with 19 retirees, over 60 years of age, all living in a residential suburb near Strasburg: the Saint Antoine sector, built in the 1950s to 1960s in the Cronenbourg neighborhood.

The eco-landscape throughout the urban community of Strasburg enabled us to produce the typology presented in Table 6.1. A total of 769 trips were identified. The spatial structure of mobility led to distinguishing four groups of individuals, as illustrated in Figure 6.4:

– those who visited places that are primarily *concentrated* around the home (*concentrated distribution*);

– those who visited places that form an *axis* between the home and downtown (*axial distribution*);

– those who visited places that are limited to a *sector* of the city and suburbs (*sectorial distribution*);

– those who visited places that are spread over several sectors of the city and suburbs (*multisectorial distribution*).

By searching for landscape constants related to visiting and representing places, regardless of the group of individuals, four types of urban eco-landscapes are highly visited by elderly people (at least twice a week): high-density housing areas (*grands ensembles*); industrial zones and/or mixed commercial-residential zones; former nuclei integrated into the city; and the historical center of the built-up area. However, those that are represented are limited to heritage and institutional landscapes, generally in the downtown area: historical center, the downtown area with large buildings, large-scale public buildings (hospital, university, etc.) and parks.

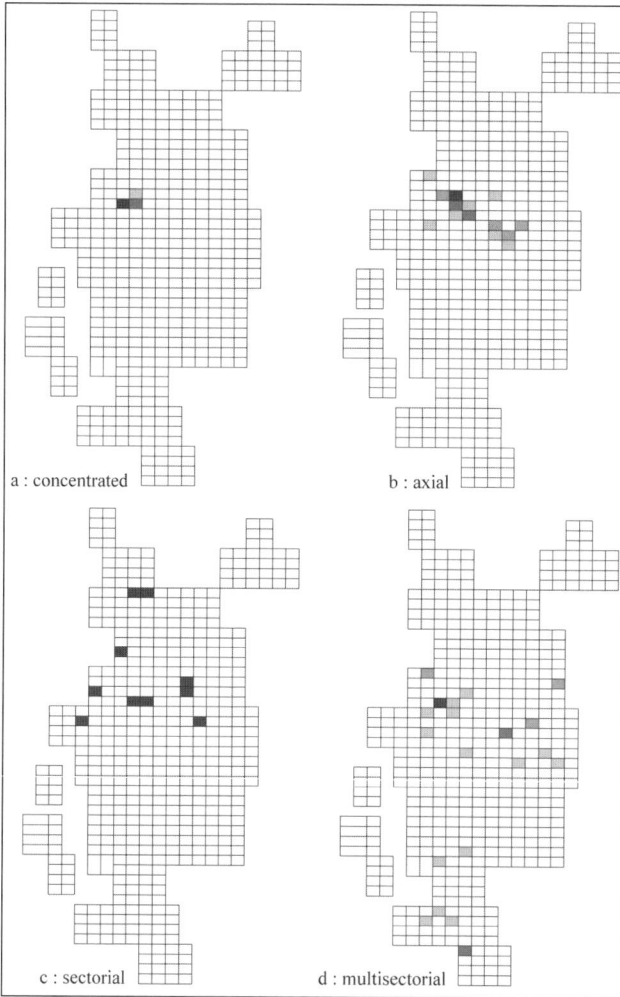

Figure 6.4. *Examples of the spatial structure of mobility – four remarkable distributions*

However, major differences appear based on the groups that were created from the spatial structure of daily mobility. Note that the average occurrence of an urban eco-landscape visited is calculated based on all trips of the individual during the week. Consequently, the average frequencies for the landscapes visited are generally greater than those represented.

When distribution is axial, the landscapes represented seem to conflict with the landscapes visited (see Figure 6.5). Indeed, when the heritage landscapes (historical center, downtown made up of large buildings, old centers integrated into the current urban fabric) are highly visited, it is the sectors where public buildings predominate that are represented. Moreover, seven types of landscapes are either only represented or only visited. Two types of environmental values seem to appear. First, heritage environmental values based on both the representation and visiting of places. Second, functional environmental values in the suburbs from both the landscape point of view (*grands ensembles*, mixed residential-industrial zone) and the property point of view (essentially behavioral base).

Unlike the first group, when trip distribution is multisectorial, the urban heritage landscapes become more apparent in the representation than they are actually visited (see Figure 6.6). Isolated villages are visited without, however, being mentioned in the representation of the city. Also, urban forests, and especially landscapes made up of houses with large gardens, community gardens, *grands ensembles* and sports centers in the surrounding areas do not appear in the representation, even though they are visited. Here the environmental values are in opposition with the heritage city in terms of nature and greenery, the first being identitary, with the second having a strong behavioral basis.

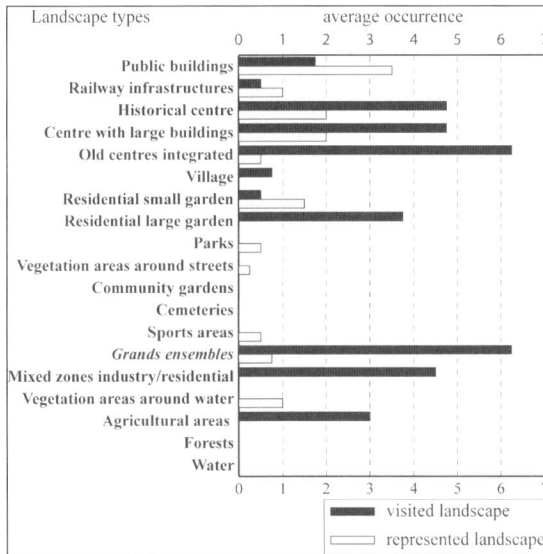

Figure 6.5. *Comparison of average occurrences between landscapes represented and landscapes visited for axial distributions*

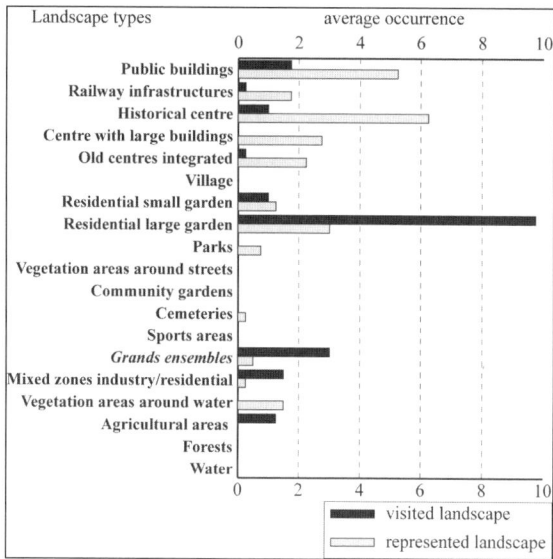

Figure 6.6. *Comparison of average occurrences between landscapes represented and landscapes visited for multisectorial distributions*

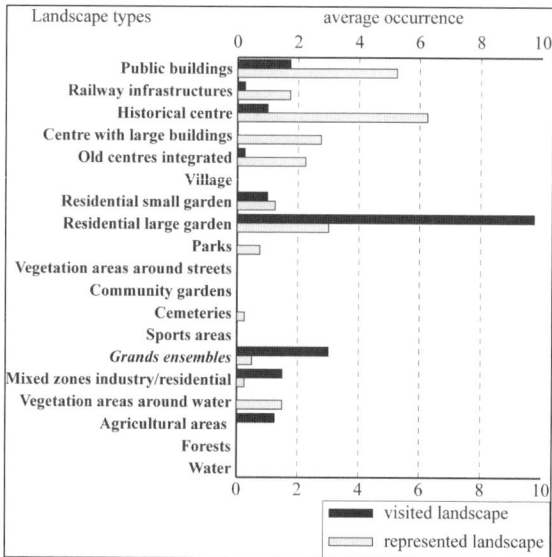

Figure 6.7. *Comparison of average occurrences between landscapes represented and landscapes visited for concentrated distributions*

When the distribution of weekly trips is concentrated (see Figure 6.7), as for the previous group, the representation is essentially made up of *ad hoc* elements from all urban heritage landscapes, while the landscapes visited are strongly limited to residential areas with large gardens. Here, the environmental values are essentially based on urban vegetation and residential areas, along with a strong urban identity.

Lastly, when the spatial distribution of trips is sectorial (see Figure 6.8), the landscapes visited are concentrated in the *grands ensembles* and the mixed industrial-residential zones, whereas the landscapes represented are primarily urban historical landscapes found downtown. Here the environmental values are essentially urban, with a symbolic heritage component on one hand and a functional suburb component on the other.

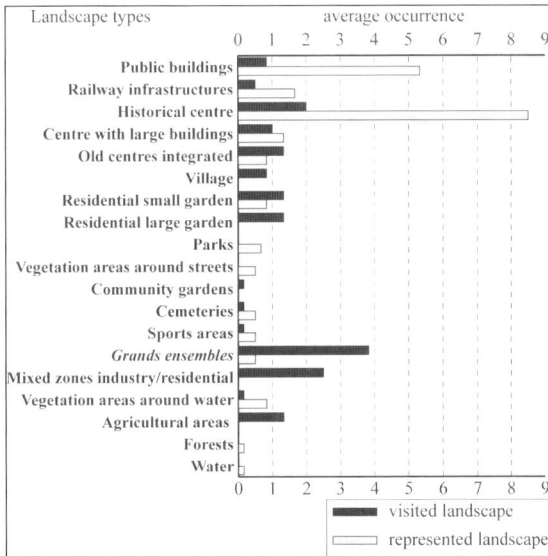

Figure 6.8. *Comparison of average occurrences between landscapes represented and landscapes visited for sectorial distributions*

This example, which comes out of an exploratory study on a few individuals that are specific because of their age, shows major differences in the morphology of the places visited, although the entire sample lives in the same neighborhood and in the same type of housing. In order for the environmental values that confirm these differences to be more explicit, more advanced statistical analyses (comparisons of averages, multivariate analysis of variance, factorial analysis, etc.) should be carried out in order to better understand the structure of data collected, which this sample did not allow us to do. Adding an interview at the end of the self-observation phase would also help identify certain dimensions of the signified environmental values, in particular using the spatiotemporal grid for the level of spontaneity of trips [RAM 05].

This example shows that the signifiers for the environmental values can be described by simultaneously taking into account their cognitive or behavioral dimensions.

The sociological dimension should not be ignored. Indeed, the level of education of the respondents is the variable that appears to best explain the four empirically formed groups of citizens. An analysis of a larger sample would allow us to statistically ascertain this and would open up other avenues of exploration regarding the link between daily trips and sociospatial segregation. Moreover, a larger sample would allow us to carry out more in-depth analyses by studying each trip mode separately, by comparing them or even comparing the results obtained based on the classes of activities behind the trip.

6.6. Conclusion

Understanding urban forms that are associated with daily trips is a problem to which several studies have attempted to find answers. Indeed, the questions associated with this scientific challenge are at the heart of those on urban planning. They primarily concern issues of equal access by city-dwellers to urban resources (businesses, public and private services, urban amenities, etc.) from their home, to itineraries taken during trips, but also to the pollution that intra-urban daily trips entail. All these urban development issues pose the problem of the relationship between urban morphology and spatial behaviors. However, environmental values internalized by the individual are crucial for analyzing this relationship.

Responses allow us to understand the various constraints and room individuals have to maneuver, based on their social group, caused by the related fragmentation of living spaces and temporal tensions. It is thus the differentiated accessibility to places and social segregations in space that produce the environmental values that are discussed in this chapter. They also allow us to understand what the contributions of each lifestyle may be to air and sound pollution. Studies on spatial media of the areas of activity are partly focused on motorized trips, and partly on studies of the distribution of chemical or sound emissions produced by two- and four-wheel vehicles. In this field the methodological advances will most probably have non-negligible repercussions since, in France, transportation produces 26.5% of greenhouse gases, and it is in precisely this area that its increase was the strongest in 1990–2004 (22.7%).

The investigation of environmental values associated with daily trips definitely allows us to search for solutions to the main problems produced by dominant urban practices. These studies are thus based on a definition of the urban space that is formulated around movement: "urbanization is defined here as the process in which mobility organizes daily life" [REM 92]; "urban is movement" [BAS 01]. Lastly, the convergence of psychological, geographical and sociological aspects implies that we no longer separate the effects of places, social groups and representation in the analysis of intra-urban daily movement.

6.7. Acknowledgements

The methodological developments presented in this chapter are from the research project *Morphologies de l'étalement urbain et exclusion par l'automobilité* (morphology of urban sprawl and exclusion by automobility, scientific coordinator: D. Pinson) for the Sustainable Urban Development Program at the CNRS. They are also taken from the research project *Approche éco-paysagique révélatrice des identités de déplacement: une contribution interdisciplinaire à l'éco-développement urbain* (eco-landscape approach reveals movement identities: an interdisciplinary contribution to urban eco-development, scientific coordinators: T. Ramadier and L. Wassenhoven) of EGIDE's Platon Program. The two projects were conducted at the *Laboratoire Image et Ville* (UMR 7011 CNRS/ULP), at Louis-Pasteur University in Strasburg.

6.8. Bibliography

[APP 69] APPLEYARD D., "Why buildings are known", *Environment and Behavior*, vol. 1, pp. 131-159, 1969.

[AUR 96] AURIAT, N., *Les Défaillances de la Mémoire Humaine: Aspects Cognitifs des Enquêtes Rétrospectives*, Paris, PUF, 1996.

[BAI 90] BAILLY A., "Paysages et representations", *Mappemonde*, vol. 3, pp. 10-13, 1990.

[BAS 01] BASSAND M., "Métropole et métropolisation", in: BASSAND M., KAUFMANN V., JOYE D. (eds), *Enjeux de la Sociologie Urbaine*, Lausanne, Romandes Polytechnic and University Press, 2001.

[BER 95] BERQUE A., *Les Raisons du Paysage de la Chine Antique aux Environnements de Synthèse*, Vanves, Editions Hazan, 1995.

[BOU 79] BOURDIEU P., *La Distinction*, Paris, Editions de Minuit, 1979.

[BRO 84] BROSSARD T., WIEBER J.C., "Le paysage: trois définitions, un modèle d'analyse et de cartographie", *L'Espace Géographique*, vol. 1, pp. 5-12, 1984.

[CLA 95] CLAVAL P., *La Géographie Culturelle*, Paris, Nathan, 1995.

[FIN 68] FINES K.D., "Landscape evaluation: A research project in East Sussex", *Regional Studies*, vol. 2, pp.41-55, 1968.

[HAR 72] HARRISON J.D., HOWARD W.A., "The role of meaning in the urban image", *Environment and Behavior*, vol. 4, pp. 389-411, 1972.

[HER 03] HERZOG T.R, LEVERICH O.L., "Searching for legibility", *Environment and Behavior*, vol. 4, pp. 459-477, 2003.

[HER 76] HERZOG T.R., KAPLAN S., KAPLAN R., "The prediction of preference for familiar places", *Environment and Behavior*, vol. 4, pp. 627-645, 1976.

[HOI 71] HOINVILLE G., "Evaluating community preferences", *Environment and Planning*, vol. 3, pp. 33-50, 1971.

[KAP 87] KAPLAN S., "Aesthetics, affect and cognition: Environmental preferences from an evolutionary perspective", *Environment and Behavior*, vol. 1, pp. 3-32, 1987.

[KES 01] KESTENS Y., THÉRIAULT M., DES ROSIERS F. "Nature de l'utilisation du sol et valeurs immobilières résidentielles: analyse par modélisation hédonique", *Les Cahiers du GRATICE*, vol. 21, pp. 111-141, 2001.

[LYN 60] LYNCH K., *The Image of the City*, Cambridge, Mass., MIT Press, 1960.

[MAT 88] MATALON B., *Décrire, Expliquer, Prévoir. Démarches Expérimentales et Terrain*, Paris, Armand Colin, 1988.

[PET 03] PETROPOULOU C., Étude comparée des changements périurbains. Les quartiers spontanés à Athènes et à Mexico, Human geography doctorat thesis, Louis Pasteur University, Strasbourg, 2003.

[PET 06] PETROPOULOU C., PANGAS N., "Χωροταξική προσέγγιση του περιαστικού χώρου με βάση την τοπιο-οικοσυστημική θεώρηση (Urban planning approach of peri-urban space based in landscape ecology theory)", Γεωγραφίες (*Geographies*), vol. 12, pp. 69-83, 2006.

[RAM 98] RAMADIER T., MOSER G., "Social legibility, the cognitive map and urban behavior", *Journal of Environmental Psychology,* vol. 3, pp. 307-319, 1998.

[RAM 05] RAMADIER T., LEE-GOSSELIN M., FRENETTE A., "Conceptual perspective for explaining spatio-temporal behaviour in urban areas", in LEE-GOSSELIN M.E.H., DOHERTY S.T (Eds) *Integrated land-use and Transportation Models: Behavioural Foundations*, Elsevier, Oxford, pp. 87-100, 2005.

[RAM 06a] RAMADIER T., BRONNER A.C., "Knowledge of the environment and spatial cognition: Jrs as a technique for improving comparisons between social groups", *Environment and Planning B: Planning and Design*, vol. 33, pp. 285-299, 2006.

[RAM 06b] RAMADIER T., DEPEAU S., "Approche méthodologique (JRS) et développementale de la représentation de l'espace quotidien de l'enfant", presented at *International Pluridisciplinaire Les Enfants et les Jeunes dans les Espaces Quotidiens*, Rennes, November 2006.

[RAM 06c] RAMADIER T., PETROPOULOU C., HANIOTOU H., BRONNER A.C., ENAUX C., Morphologie de l'Étalement Urbain et Exclusion par l'Automobilité: le Cas des Adolescents et des Personnes âgées de Cronenbourg, une Banlieue de l'Agglomération de Strasbourg, CNRS, program of durable urban development, Strasbourg, June 2006.

[RAS 69] RASER J.R., *Simulation and Society: an Exploration of Scientific Gaming*, Boston, Allyn & Bacon, 1969.

[REM 92] REMY J., VOYÉ L., *La Ville: Vers une Nouvelle Définition?*, Paris, L'Harmattan, 1992.

Chapter 7

Household Residential Choices upon Acquiring a Single-Family House

7.1. Introduction

This chapter investigates the reasons for moving and the selection criteria of neighborhood and residence, as revealed by new owners of single-family houses. Using data gathered in a telephone survey in Québec City for households that acquired a single-family house between 1993 and 2001, selection criteria are analyzed according to household attributes. The results show the relationships between lifecycle and residential choices. They shed light on location strategies, particularly with regard to the perception of neighborhoods and location in the city.

Since Rossi [ROS 55] published his seminal work on residential mobility in relation to lifecycles, several studies have attempted to resolve the complex question surrounding residential choices [DIE 02]. On one hand, studies on residential mobility deal primarily with people's propensity to move and the motives behind their decision. On the other hand, studies on residential choices focus more on the preferences, choices and satisfaction that can be studied with either declared (contingent valuation, conjoint analysis) or revealed methods (discrete choice models, hedonic models). These approaches have each have a drawback, however. In the first case, they depend on hypothetical facts (declared preferences) and in the second case they have potential sampling biases in the selection of data analyzed (revealed preferences).

Chapter written by Yan KESTENS, Marius THÉRIAULT and François DES ROSIERS.

Although many studies on residential choices have analyzed the influence of property attributes through declared or revealed preferences – or both [EAR 98] – very few of these analyzed the variability of selection criteria according to the socioeconomic profile of households. We have very little understanding of the *relative* role of a home's attributes and residential characteristics in the selection process of a residence [CLA 04]. In order to better document this process, a telephone survey was conducted in Québec City among 774 households who had purchased a single-family residence between 1993 and 2001. The data collected includes self-reported reasons for moving and criteria for choosing the neighborhood and property, as well as social and financial data on those households surveyed. The central hypothesis of this research is that these factors (reasons for moving and selection criteria) vary significantly according to household profiles, and that this variation can be modeled in reference to spatial cognition notions. More specifically, regarding the reasons for moving, we attempt to analyze: the balance between attachment to the neighborhood and the desire to be close to services; and access to landownership and improved living conditions. Special attention is given to the effects of revenue and lifecycle on housing selection criteria, compared to that for the residence's neighborhood.

7.2. Spatial cognition and perception of activity places

The specificity of the residential selection process involves much more than the simple acquisition of a material good, since the purchaser inherits, *de facto*, the neighboring characteristics. Consequently, we must consider the potential purchaser's or renter's prior understanding and perception of the area in order to model the selection process.

In a study inspired by Gibson [GIB 50] and Gärling [GÄR 93], Reginster and Edwards [REG 01] propose a conceptual model of spatial perception that integrates the concepts of location and activities. Their concept of perceptual regions is based on a combination of characteristics of the milieu and trips made, the feeling of belonging increasing with the frequency of visits to a given location. Environmental characteristics and activities, combined with the frequency with which they are called upon, are the focal point of their hierarchical schema for perceptual regions. They distinguish three nested levels:

– The *vista space* (place-perspective) is a portion of the surrounding area having characteristics seen as homogeneous and unique, without necessarily being limited by a single perspective. It corresponds to a sense of belonging linked to activities that take place there (my residence, my workplace, my school, etc.). It is located in immediate proximity to the home and possibly to other recurring activity sites.

– The *local displacement-reinforcement space* includes the normal communication routes taken by citizens to attend to business outside the home. Its representation is reinforced by the frequency of visits and trips. It is the normal commuting space.

– The last level, the *large displacement-reinforcement space* refers to the area that includes the various local displacement spaces. Sometimes reaching the city-wide level for more mobile individuals, it is essentially viewed as a network and includes numerous unknown locations. It is a "gap-filled" space.

Filion *et al.* [FIL 99] propose a conceptual model derived from geographical concepts and adapted to the location selection process, distinguishing *place*, *space and proximity*. Space can be evaluated in terms of potential access to activities in the relevant zone (for example, in a metropolitan region). Thus, the choice of *space* aims at increasing the temporal and economic *accessibility* to activities. Place refers to the immediate neighborhood of the residence [DUN 93], differing primarily to the physical attributes of the environment (a type of *vista space*). The *proximity* concept serves as a go-between and refers to the desire to get closer to activity sites that are frequented often, thus *reducing the size of the local displacement space.*

To this geographical vision of the perception of spaces the concept of *attachment* is added. This has been developed by environmental psychologists [ALT 92, FEL 96, FRI 82, GIU 91, TWI 96]. According to Sundstrom *et al.* [SUN 96] this research stream focuses on attachment to various spaces, in particular the home and neighborhood. It aims at shedding light on the development and nature of *affective links* between individuals and living places and the manner in which they contribute to the *identification of the area* [BON 99]. Breakwell [BRE 86, BRE 92] distinguishes four area identification elements: distinctiveness, continuity, self-esteem and self-efficacy. With the desire to preserve continuity of the self-concept, two distinct self-environment relationships are discussed in the literature. First is the place-referent continuity, whereby specific places that have emotional significance play the role of continuity markers between past and present and present and future. Second is the place-congruent continuity, referring to the generic features of places assuring continuity from one place to the next. In fact, the affective bonds between self and environment may transcend the relationship with a unique or specific place, and attachment may be developed throughout space(s) for types of places with similar characteristics [PRO 78, TWI 96]. Feldman [FEL 90] has extended this notion to the idea of settlement-identity.

Geographical and perceptual concepts are relevant to studying the location identification cognitive process and residential selection methods, as the latter involves a selection of locations that roughly meet expectations. Thus, according to Feldman's settlement identity principle, the sense of belonging is reinforced through frequency of use and activities, and partly inherited from previous place attachments. The transfer of a sense of belonging of one place to another (place-congruent continuity principle) provides an explanation as to how people can "feel at home" right from the initial visit to a new property they may acquire. A part of the place-identity linked to the newly-acquired property stems from former residential locations, and consequently influences selection criteria. Similarly, the sense of belonging is often influenced by the fact that the person is born or grew up in this or a similar milieu; all else being equal, childhood or adolescent memories probably count among the factors for choosing a residence. Thus, we put forward the hypothesis that hierarchical concepts of spatial cognition and psychological identity components must be considered jointly to study housing selection processes.

7.3. Residential mobility

In the continuity of Rossi's thesis [ROS 55], several studies have stressed the role of the lifecycle in the moving process [ART 78, HEN 83]. Most of these also stress the influence of the sociodemographic characteristics of the neighborhood and the household. In literature dealing with residential mobility, Dieleman [DIE 01] identifies three types of correlations between:

– the rate of residential mobility and lifecycle;

– mobility, residence size and dwelling status; and

– residential trajectory and other lifecycle aspects, such as education level, work and family trajectories [DIE 02, MUL 99, VAN 99].

In addition, mobility is studied specifically for given household types: young families [CLA 00], senior adults [MEG 99], divorced people [TIM 96] or ethnic groups [DEN 03, GAB 03]. In a multiple-attribute housing disequilibrium model of residential mobility, Onaka [ONA 83] shows that household and property attributes are jointly linked to the decision to move. Using a stated preference method, Kim *et al.* [KIM 05] have shown that characteristics of the residence and household attributes influence the probability of moving, and that the relocation choice is the result of a compromise between transportation costs and available resources.

According to Rossi [ROS 55], however, it is difficult to distinguish the inherent reasons for the decision to move, as a simply "why?" may result in a myriad of piecemeal answers. Often, respondents confuse the motives that led to the move and the selection criteria for location and the residence itself. Thus, it is a combination of the three aspects (motives for moving, choice of location and choice of property) of this complex decision that will be discussed in this chapter, considering its hierarchical spatial structure [LOU 90, PEL 02].

7.4. Residential choice and location

The *stated preferences*, used mainly to study the residential choice process, are based on hypothetical statements or a finite range of theoretical options. *Revealed preferences* are based on effective choices (made) or on the actual sales or rental price (transaction cost). They are used notably to analyze choice and satisfaction. The choice process is central to preference and satisfaction, as choices stem from preferences and satisfaction results from past choices.

Among the main *stated preference* methods and choice analysis methods are:

– the contingent valuation estimates of the willingness to pay, mostly applied to the valuation of environmental amenities [CUM 86];

– conjoint analysis methods that compare different goods and their characteristics [GOO 89]; and

– methods based on choices where respondents choose a combination of characteristics from a range of predetermined options [TIM 92, TIM 95].

Conjoint analyses of declared preferences are obtained by extension of the discrete choice method.

Discrete choice models were initially developed to study real choices (*revealed preferences* method) and are based on Thurstone's random utility theory [THU 27], later adapted by McFadden [MCF 78] to define the multinomial logit model (MNL), which is based on logistic regression.

As noted by Earnhart [EAR 98], several authors have adopted this model to conduct residential choice studies [FRI 81, LON 84, NEC 84, QUI 76, QUI 85]. The hedonic approach explores *revealed preferences*, based on the principle that goods traded draw their value from the marginal utility of their attributes.

Hedonic modeling allows us to estimate the monetary value of the attributes of properties, the neighborhood and local externalities [ROS 74]. Most hedonic models evaluate a single coefficient per measured characteristic. However, some spatial statistics techniques (spatially weighted regression, Casetti's expansion method, multilevel analysis, etc.) allow us to estimate the variation of the marginal value according to other dimensions. This is the case whether spatial localization [BEN 98, KES 04, ORF 99, THE 03, THE 05, WOL 00] or the sociodemographic characteristics of consumers are being studied [KES 06, DES 07].

It is difficult to measure heterogeneity in tastes using a random utility model, however, and this phenomenon has rarely been studied [ADA 01, BOX 02]. Molin and Timmermans [MOL 01, MOL 03] measured the relationships between the attributes of a household and the characteristics of residences or locations with a structural equation model. As expected, their results indicate positive ties between the purchaser's education level on one hand, and the size of the house and land tenure, as well as between the buyer's income level and housing costs, on the other. In this study we endeavor to go beyond this rather trivial assessment.

7.5. Mobility survey and residential choices in Québec City

In order to study the heterogeneity of residential choice, this chapter analyzes the reasons for moving and the criteria for choosing a neighborhood and residence, as given by 774 buyers of single-family properties in Québec City. The data come from a telephone survey conducted in 2001–2002 among homeowners who purchased their homes in Québec City between 1993 and 2001 (88% of purchases were made between 1993 and 1996). Among the 2,521 people contacted by phone, nearly half (45%) agreed to participate in the study, and 774 respondents completed the whole questionnaire. This sample is spatially stratified (13 municipalities in the urban community) and represents approximately 5% of single-family home sales transacted in Québec City over this period.

This study analyzes the responses to three questions regarding the latest residential choice made by the person surveyed:

1) What reasons prompted you to move?

2) What criteria did you use to choose your new neighborhood?

3) What criteria motivated the choice of your new property?

These open questions did not include any predefined response, and the number of responses was unlimited. The responses obtained (in open format) were classified into 21 categories for reasons for moving, 19 residential selection criteria, and 20 house choice criteria. Socioeconomic attributes describing the household were also gathered (see Table 7.1). Homes of the respondents were located in the appropriate building and positioned on the topographic map at 1:20,000 as paired with the Québec City municipal assessment role. This location allowed us to conduct relevant spatial analyses and linkage to actual transaction and valuation role data.

Socioeconomic profile	Sample proportion (%)	Average age (in years)	Average income ($)
Education			
University degree	55	36	77,482
University without degree	4	37	65,714
College degree	29	36	71,452
High school	11	38	55,960
Household type			
Single-parent family	7	38	44,035
Single person	6	42	49,286
Couple with no children	18	39	71,844
Two-parent family	67	34	73,480
Household revenue sources			
Working couple	70	34	75,595
Only one spouse working	30	40	54,830
Prior status (before purchasing current home)			
Owner	52	39	72,040
Renter	48	33	66,263
Respondent gender (telephone interview)			
Female	54	35	68,095
Male	46	37	70,649

Table 7.1. *Socioeconomic profile of respondents (buyers of single-family homes)*

The majority of respondents have a university diploma (55%) and nearly a third (29%) have a college diploma. Couples represent 85% of the survey sample, with 8 out of 10 of them living with children. Single-parent families and people living alone account for 7% and 6% of the sample, respectively. Approximately 70% of households have two salaries. Annual household income is $69,264, with a standard deviation of $22,705. This value is underestimated, however, as the highest salary level recorded during the survey was $100,000 or more. As expected, single-parent families have the lowest average salary, set at $44,000. The average age at the time of the transaction was 33 for former renters (48% of the sample) and 39 for former owners.

The sample includes a slight imbalance in favor of women (54%), who answered the telephone survey in greater number, especially in new neighborhoods and affluent areas where their proportion reached 58%. Moreover, the gender representation is very balanced in the downtown core and older neighborhoods, and in more low- and middle-income sectors. This slight selection bias may relate to the time the survey was conducted (early weekday evening and Saturday afternoon), because of low the availability of male respondents at these times in some population categories [ROU 89]. Given the low amplitude, there is little chance that this had a notable impact on the results.

The frequency of self-reported reasons for moving are listed in Figure 7.1, by distinguishing the reasons for which the respondent chose to freely indicate one (54%), two (32%) or more than two answers (14% of the sample). While, as expected, access to property, the desire to increase residence size and changes in the household size were primarily responsible for deciding to move, being close to the workplace and urban services (accessibility) were also important motivators to move. Another, less important motivation included seeking a location for its identity (peaceful, safe, retirement, moving back to childhood neighborhood, etc.). Moving closer to work (with or without a new job location) was cited by 28% of respondents. Seeking closer access to services was cited by 5% of respondents, compared with 5% who sought a more peaceful neighborhood. Four per cent wished to move closer to schools and 3% wanted to be closer to extended family. Of the remainder, 2% wanted a safer environment, 2% were returning to the place they grew up, 2% were seeking to move closer to the downtown core, 1% were seeking a livelier sector, and 1% were moving because of retirement. Decisions to move are basically motivated mostly by rational criteria of an economic or utility nature, with the concept of identity attachment coming far behind. We can, however, assume that the decision not to move (excluded by our respondent selection methodology) would comprise deep roots in the feeling of belonging. This question falls outside our research focus.

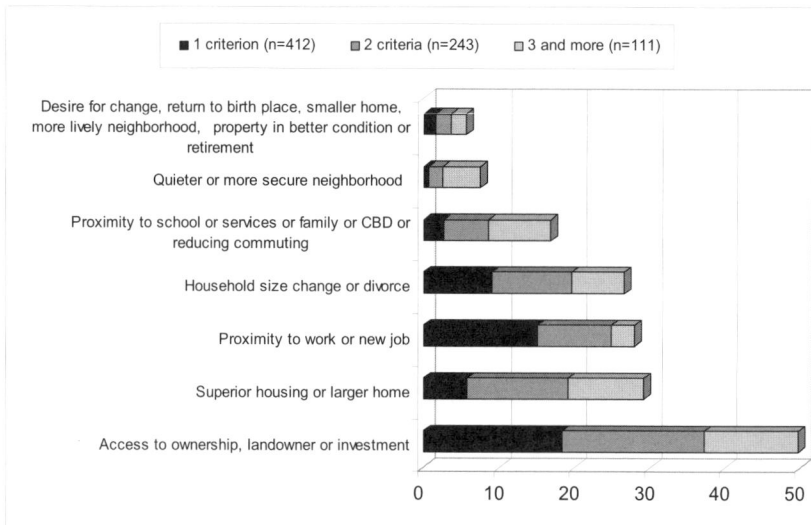

Figure 7.1. *Frequency of reasons for moving according to the number of responses provided*

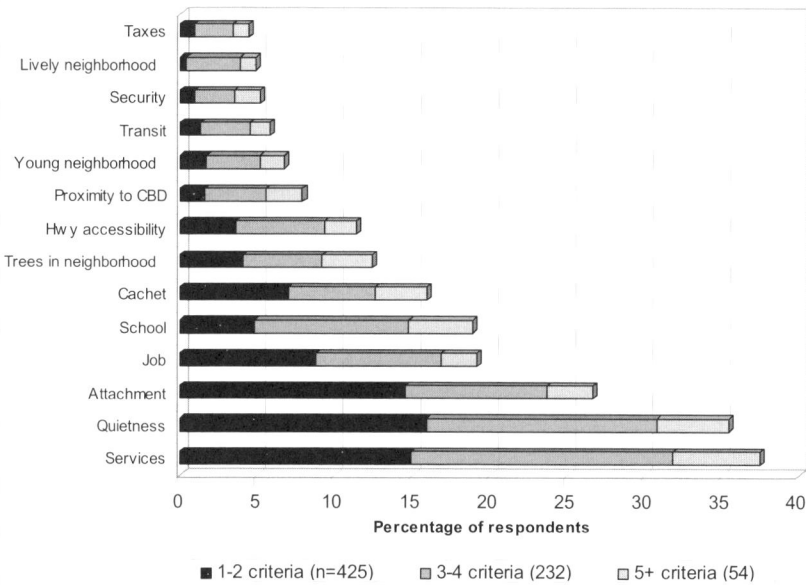

Figure 7.2. *Frequency of neighborhood selection criteria according to the number of responses*

On the contrary, neighborhood selection criteria are clearly more influenced by spatial cognition and identity perception of locations. Figure 7.2 lists the frequency of responses for selection criteria for the new residential neighborhood, considering the number of responses for each respondent. Proximity and accessibility factors linked to the concept of location choice and local displacement areas (services, employment, school, autoroutes, downtown core, public transportation services). These alternated with responses linked to the perception of the neighborhood in terms of attachment, identity and local continuity (tranquility, attachment, cachet, trees, security, and activities).

Aside from a few responses regarding property taxes, economic reasons are not a concern for respondents. Neither are criteria of socioeconomic continuity-conformity, which are only marginally mentioned. Young families, for example, report appreciating the young age structure of the neighborhood. Respondents spontaneously listed more criteria for the choice of neighborhood (2.2 on average) than reasons for the decision to move (1.6).

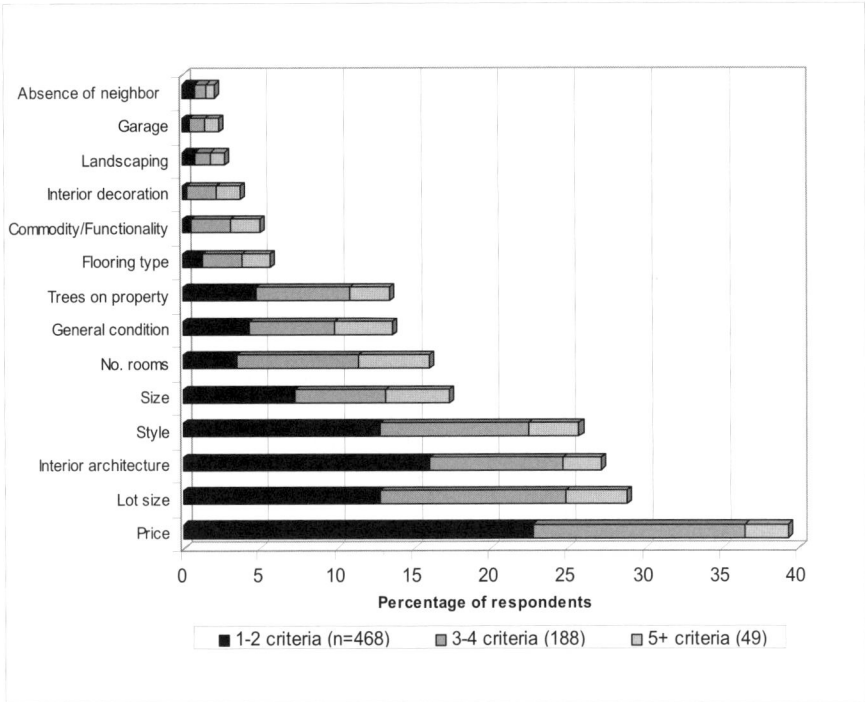

Figure 7.3. *Frequency of property selection criteria according to the number of responses*

Lastly, economic and utility criteria (sales price, lot and house size, depreciation rate, functionality and equipment) dominate the esthetic and identity criteria (architectural style, landscaping, interior decoration), relegating to second place the continuity of self-concept in the choice of residence itself. Figure 7.3 presents the details of responses provided (an average of 2.1 responses per respondent).

We can see the importance of architectural style and interior design listed among the major residence choice factors (more than 25% of responses). We can also see the importance of trees on the property and its landscaping, which respond to the presence of trees and the neighborhood cachet as a neighborhood selection criterion. This clearly illustrates the congruence among property and neighborhood selection criteria through vista space and local continuity concepts.

	Group	Group criteria list	Total (n=774)	Component 1		Component 2	
				City center and established neighborhoods (n=528)	New neighborhoods (n=246)	Low- and middle-income neighborhoods (n=358)	Affluent neighborhoods (n=416)
Choice of residence	Size	Lot size, house size, number of rooms	48%	46%	52%	**40%**	**54%** ***
	Interior	Architecture, quality of floors, functionality, decoration, garage	37%	36%	40%	35%	39%
	Building	Architectural style, overall condition	36%	36%	36%	**28%**	**42%** **
	Environment	Trees, landscaping	15%	14%	17%	9%	20%
Choice of neighborhood	Accessibility	Proximity to services, jobs, schools, autoroutes, downtown, public transportation	60%	**54%**	**73%** ***	**53%**	**66%** ***
	Ambiance	Tranquility, youth, safety, activities	43%	**47%**	**34%** **	47%	40%
	Attachment	Feeling of belonging	27%	26%	28%	22%	30%
	Esthetics	Cachet, trees	25%	26%	22%	19%	29%

The header "At least one criterion mentioned by …% of respondents" spans Component 1 and Component 2.

Table 7.2. *Synthesis of selection criteria for residence and neighborhood according to city location and socioeconomic status of neighborhood (importance of proportional difference: ** = 5%, *** = 1%)*

Table 7.2 presents a synthesis of the residential and neighborhood selection criteria by establishing groups that best represent the theoretical concepts presented in section 7.2. Thus, for the choice of a residence, the criteria from the "Size" category meet the desire to increase home size, which can also motivate the decision to move but then becomes a selection criterion for the property sought. The "Interior" category regards the residence's attributes of functionality and quality. The "Building" category concerns an appreciation of the architectural style and overall condition (physical depreciation) halfway between the economic rationality and subjective appreciation of its esthetics, which refers to the continuity of self-concept. Lastly, the "Environment" group describes the private land surrounding the residence, the interface between the property and its vista space. If we consider all of the 774 respondents, economic rationality[1] clearly dominates, with criteria pertaining to the building and the exterior environment only being mentioned by 36% and 15% of the respondents, respectively.

Some variations are notable if we distribute the responses according to the socioeconomic characteristics of neighborhoods. To do this, we used a Québec City neighborhood typology designed by Des Rosiers *et al.* [DES 00] based on census data (1996) for the Québec City region's population by enumeration area, which was subject to a primary component analysis. This chapter looks at the first two axes following a Varimax rotation of four factors. The first component (center-periphery gradient) illustrates the urbanization steps and expresses the contrast between the old historical center and neighborhoods developed before the 1970s and new neighborhoods developed through the extension of the regional highway network at the start of the 70s. The second component expresses a socioeconomic gradient (education, income, employment rate, etc.) that expresses the contrast between privileged and underprivileged areas.

Depending on its location, each locality receives a specific grade for each component. Enumeration areas were then divided among each distribution tercile, which were then agglomerated into two classes: centers and established neighborhoods (two terciles of component 1) versus new neighborhoods (last tercile); and low- and middle-income neighborhoods (two terciles of component 2) versus affluent neighborhoods (last tercile). As the transactions are not distributed equally over the area, we have an imbalance in the total number reported in Table 7.2, given that respondents were selected from among all single-family house transactions between 1993 and 1996.

Tests of the significance of proportional differences were conducted among the classes of each component in order to evaluate the probability that the deviations recorded were a coincidence of the sampling. Thus, it appears that exterior architecture styling and the immediate environment are more commonly favored by buyers who opt for affluent neighborhoods. This is the result of their ability to optimize the quality of their immediate environment, with financial constraints being visibly less salient in their decision-making framework even though they are also looking for large properties.

Local differences are more evident with regard to the choice of a neighborhood. Contrary to expectations, buyers who chose new neighborhoods (mostly in remote locations) speak more frequently than the others about criteria from the

---

1 Among the residence selection criteria, the sales price is cited by nearly 40% of respondents.

"Accessibility" group (proximity to services, schools, jobs, autoroutes, downtown core, public transportation). Is it possible that proximity to services is so evident for those who have chosen a more central location that it is not even relevant to mention it?[2] Would this be an effect brought about by the massive use of the car, which represented a modal share of more than 75% in 2001 in Québec City [BIB 08]? Could this be related to an economic rationality compensation aimed at offsetting a feeling of loss? This question exceeds the range of our current research, but deserves more in-depth study. As concerns the high score accorded to proximity by affluent neighborhoods, it is well justified with regard to their location to the immediate vicinity of the main regional services [DES 00].

Second most talked about was "Ambiance". This corresponds in good part to the continuity concept since it refers to social neighborhood characteristics sought after by buyers: tranquility, youth, safety, activities. The city center and older, established neighborhoods are clearly favored in this respect. The feeling of belonging ("Attachment" forming an emotional bond) is in third place, almost equally ranked with the "Esthetics" group linked to the cachet and presence of vegetation, which expresses the heritage or natural character of the environment. Here, too, we see that these aspects are more prevalent in affluent sectors, accrediting the hypothesis of a certain heterogeneity of the decisional criteria according to social groups, even though these differences only register rejection thresholds of 20% (attachment) and 15% (esthetics). Given the relative frequency of these responses, a survey that included more respondents would be required to confirm this trend.

Table 7.3 outlines the socioeconomic profile of buyers according to the choice of residential neighborhood made relating to center-periphery gradients (component 1) and neighborhood socioeconomics (component 2). Contrary to expectations, we see that the household types are distributed quite evenly over the two gradients; proportional differences are low and insignificant. Young buyers (below 40 years) are clearly more attracted to central neighborhoods than baby boomers (40 to 49 years of age), who contribute significantly to the urban sprawl process.

Family revenue allows more affluent households (earning over $80,000) not only to select affluent neighborhoods for their residence but also, alternatively, to choose properties located more on the periphery (new neighborhoods). Given the geography of Québec City, the two phenomena are seemingly more complementary than additive, which would justify a more in-depth behavioral study of this segment of the residential clientele. On the contrary, less affluent households (earning below $50,000) are significantly more confined (if not trapped) in low-income sectors.

---

2 Results obtained by Hoesli *et al.* [HOE 97] in the center of Bordeaux suggest that accessibility perception has no effect unless there is a high variation over the territory.

			Total	Component 1		Component 2	
				City center and established neighbor-hoods	New neighbor-hoods	Low- and middle-income neighbor-hoods	Affluent neighbor-hoods
	Variable	**Definition/categories**	**n=774**	n=528	n=246	n=358	n=416
Category-specific	Household type	Family (with children)	68%	69%	64%	66%	68%
		Couple with no children	18%	18%	18%	19%	18%
		Single-parent family	8%	7%	8%	8%	7%
		Single person	6%	6%	10%	7%	7%
	Age of respondent	Less than 30 years old	23%	**27%**	* **15%**	27%	19%
		30 to 39 years old	45%	47%	42%	46%	45%
		40 to 49 years old	26%	**21%**	** **35%**	22%	29%
		50 and over	6%	6%	8%	5%	7%
	Household income	More than $80,000	56%	**52%**	*** **65%**	47%	*** **63%**
		From $50,000 to $80,0000	28%	31%	22%	31%	26%
		Less than $50,000	16%	17%	15%	**22%**	* **11%**
	Education level of respondent	University degree	55%	**47%**	*** **72%**	39%	*** **69%**
		College degree	32%	**38%**	** **20%**	40%	** **25%**
		High school or below	13%	15%	8%	**21%**	** **6%**
Binary	Household structure	Couple (with or without children)	86%	**87%**	** **82%**	85%	86%
		Single person or single parent	14%	13%	18%	15%	14%
	Children	No children	25%	24%	28%	26%	25%
		Household with children	75%	76%	72%	74%	75%
	Attachment	No mention of attachment	73%	74%	72%	**78%**	** **70%**
		Mentioned attachment as a neighborhood choice criterion	27%	26%	28%	22%	30%
	Revenue source	Single-income household	30%	29%	34%	32%	30%
		Double-income household	70%	71%	66%	68%	70%
	Access to property	Already landowner	52%	52%	52%	**45%**	*** **58%**
		First property acquisition	48%	48%	48%	**55%**	** **42%**
	Gender	Male	46%	48%	41%	49%	43%
		Female	54%	52%	59%	51%	57%

Table 7.3. *Socioeconomic profile of respondents according to type of neighborhood chosen (significance of proportional differences: * = 10%, ** = 5%, *** = 1%; shaded areas indicate reference categories in logistic regressions)*

Among all the elements of buyers' socioeconomic profile, education is the most important in the choice of location. University graduates (highly correlated to income) primarily choose affluent central or peripheral locations; college graduates and households with less formal education seem more inclined to acquire older properties in the city center, in new suburbs and/or in poorer neighborhoods. While the former promote urban sprawl, the latter group recycle older housing stock and repopulate central neighborhoods. Is this trend compatible with a move toward gentrification? Are the new arrivals more or less fortunate than the populations they are replacing? With no details on sellers, a successive census survey of economic data would help shed light on this question. However, not surprisingly, we can already conclude the existence of a socioeconomic segregation founded on income and education in favor of new neighborhoods and affluent districts.

Quite surprisingly, the presence of children in the family does not appear to make much difference in the choice of type of neighborhood. It is an important determinant in the decision to move to a larger home but less important, no doubt, than the "price" factor, as suggested in Figure 7.3. The mention of attachment to a neighborhood is more frequent among respondents who chose a home in affluent neighborhoods. Prior status (renter versus owner) has no bearing with regard to moving to the city center or periphery, but is a significant factor (along with revenue) in access to more affluent neighborhoods (through the mobilization of capital earned from the former home). Slight proportional imbalances in the gender of respondents had no significant effect on the choice of location.

7.6. Conjoint modeling of household stated preferences

Results of the preceding section consider only bivariate relations. Conjoint preference models enable us to weigh the action of cofactors and to estimate the marginal probability of the mention of a given criterion according to the type of respondent and household attributes (age, revenue, double-income household, education, household composition, prior status, etc.). At the same time it allows us to control for gender and the total number of criteria mentioned by the respondent[3], possibly including effects of interaction (for example, age × income). In order to verify its marginal effect, the respondent's mention of attachment to the neighborhood serves as an additional factor to model the reasons for the move and property selection criteria.

The conjoint analysis of stated preferences uses logistic regression to identify the socioeconomic and conjunctural factors related to significant variations of the

---

3 This precaution is necessary to our study as the probability of citing a specific criterion naturally increases with the total number of responses provided. We must control this phenomenon because the number of responses varies among the respondents.

probability of mentioning one selection criterion among others. In logistic regression, the overall significance test of a model follows a chi-squared distribution (χ^2). According to Nagelkerke [NAG 91, p. 691], the Nagelkerke R^2 is an attempt to provide a logistic analogy to R^2 in classic OLS (ordinary least squares) regression and indicates that the logistic model is a good fit. The Wald statistic tests the significance of the coefficient associated with each variable or category. For the models presented below, only the significant relationships with a 5% threshold were considered. Lastly, the odds ratio (probability of an event occurring/probability of it not occurring) is derived from the regression coefficient (B). Tables 7.4, 7.5 and 7.6 present several conjoint analysis models developed for this research, odds ratios and associated significance levels.

Variable (reference category)	Category (measurement unit)	Model A Acquire landownership	Model B Increase size/ Improve residence	Model C Improve accessibility
Age * Income (<30 years old * income >80K)	30–39 * income <50K	0.228 ***		
	30–39 * income 50–80K	0.528 ***		
	40–49 * income <50K	0.081 ***		
Income	(per $10,000)	0.825 ***	1.13 ***	
Income source (one earner)	Double-income	2.27 ***		0.613 **
Household type * Income (couple with children * income >80K)	Single-parent family * income <50K	4.72 ***		
	Couple without children * income <50K	4.91 ***		
	Single person * income <50K	11.70 ***		
Prior status (owner)	Renter	22.2 ***	0.530 ***	0.279 ***
Attachment (not mentioned)	Attached to neighborhood	1.58 **		0.461 ***
Respondent (male)	Female			
Number of motives (>2)	1 motive	0.365 ***	0.057 ***	0.077 ***
	2 motives		0.324 ***	0.320 ***
Model adjustment	Chi-square	385.2	166.8	112.0
	Sig	0.000	0.000	0.000
	-2 log likelihood	645.5	729.1	492.8
	Nagelkerke R^2	0.534	0.284	0.188

Table 7.4. *Odds ratios and adjustment of conjoint models of reasons for moving (significance levels: ** = 5%, *** = 1%; Accessibility: proximity to family, services, schools, work and shorter commutes)*

7.6.1. *Conjoint analysis of moving motivations*

The primary motives for moving (see Figure 7.1) are:

– the desire to own property or for investment (43%);

– a person's desire to improve their living conditions, especially with regard to property size (26.9%);

– being closer to work (26.2%); or

– aspects related to the lifecycle and household composition (25.2%).

Accessibility to schools, services, family and the downtown core, or the desire to reduce commute times fall within the fifth most important priority (13.3%), while less than one buyer in 15 wished to live in a more peaceful or safe neighborhood. This very low concern for safety is typical for Québec City, which has the lowest crime rate among the 25 metropolitan regions in the country [STA 01].

Some criteria seem clearly more "exclusive" than others. For example, more than half of respondents who indicated a work-related motive (new job or move closer to the workplace) did not invoke any other element. In contrast, only one person in five among those who moved to improve their living conditions or acquire a larger property gave no other reason (80% gave at least one more reason). This is an additional reason to consider the total number of motives (or criteria) in the specification of conjoint models. With the reference category being defined as those who specified several motives, a very weak odds ratio (almost zero) for those who specified fewer criteria indicates that the modeled criterion is highly unlikely to appear alone or is relatively marginal in the scale of priorities. On the contrary, an odds ratio that tends to the unit (and especially the absence of significant relationship) indicates that this criterion is often cited alone or among a small number of motives.

Logically, the desire to *become a landowner* or to invest in a property is strongly linked to former occupancy status (cited especially by former renters, with an odds ratio of 22.2 to 1). This relationship with income is negative: the higher the revenue, the less this factor is mentioned (see Table 7.4, Model A) insofar as richer households are often already homeowners. Moreover, this objective is impeded for middle-aged households with more modest income (<$50,000 for the 40–49 age group and <$80,000 for those aged 30–39). It is exacerbated for double-income couples (2.27 to 1 for single-income households) and buyers who declare being attached to the neighborhood (1.59). Moreover, this desire to acquire property is a deciding motivator for single-parent families (4.72), couples with no children (4.91) and especially singles (11.7) with low income. Inversely, the desire to expand or improve home (Model B) increases as revenue increases (13% increased probability for each additional $10,000 wage bracket[4]), and is less frequent among first-time buyers (0.53 to 1) than among households that already owned their previous home.

The desire to *increase accessibility* to urban services, rather than to the workplace or family (Model C) is less evident among double-income couples (0.613), who must negotiate a location compromise, than among households with one income, which moreover are often faced with mobility issues (lower motorization rate, necessity to live within walking distance to schools). It diminishes with attachment to the neighborhood (0.461) and is not highly valued by renters (0.279) in search of their first property. Moreover, the last two motives (Models B and C) are rarely mentioned alone, indicating that they are essentially complementary objectives. Lastly, the predictive power (adjustment) of Model A largely outclasses that of Models B and C, which still remain very significant.

7.6.2. Conjoint analysis of residential locations

Four criteria groups become evident as much for the choice of home as for the choice of neighborhood (see Table 7.2). Reasons for choosing a neighborhood are listed as:

– accessibility (at least one criterion of this group mentioned by 60% of respondents);

– urban context (43%);

– attachment to the neighborhood (27%); and

– esthetics (25%).

For the choice of a home, size tops the list (48%), followed by interior layout (37%), architectural style (36%) and landscaping (15%). For property choice, interior layout is a more exclusive criterion than the size of the home or land. For the neighborhood, proximity to services, while often cited, is not an exclusive criterion, being more often matched with at least three other criteria. This exclusivity response analysis should be studied more closely, as it reveals the relatively single- or multicriteria character of the decisional process.

Table 7.5 presents five conjoint neighborhood selection models pertaining to access to public transportation (Model D), to schools (Model E) and the workplace (Model F), as well as the importance of neighborhood cachet (Model G) and the attachment to the neighborhood (Model H). Overall, the adjustment of neighborhood

---

4 This relationship reflects the "higher" character of housing expenses linked to the qualitative dimensions of the good, which require income elasticity higher than the unit.

choice models is always less than the preceding section (reason for moving) and generally less effective than housing choice, which introduces high volatility of motivations and less behavioral predictability. We also see that two criteria tend to be exclusive, namely access to schools and attachment to the neighborhood. Although we have verified several combinations of explanatory variables, the neighborhood choice models are generally very simple (few factors) and the respondent's gender is significant for models E (schools) and G (cachet). Women are more inclined to mention proximity to schools, and men are more liable to stress the attraction of cachet.

Variable (reference category)	Category (measurement unit)	Model D Access to public transport-tation	Model E Access to schools	Model F Access to workplace	Model G Neighbor-hood cachet	Model H Attachment to neighbor-hood
Age * Income <30 years old * income >80K)	50 and older * income <50K					9.90 ***
Income	(per $10,000)				1.15 ***	
Household type (couple with children)	Single person			3.04 ***		
Children (no children)	With children	* 40–49 years old 3.52 ***	4.42 ***			
Prior status (owner)	Renter		0.647 **			
Respondent (male)	Female		1.47 **		0.670 **	
Number of criteria (>4)	1–2 criteria	0.035 ***		0.389 ***	0.478 ***	
	3–4 criteria	0.350 ***		0.550 ***	0.439 ***	
Model adjustment	Chi-square	47.3	43.0	36.0	29.0	19.2
	Sig	0.000	0.000	0.000	0.000	0.000
	-2LL	288.2	706.6	691.4	626.0	836.7
	Nagelkerke R^2	0.171	0.087	0.077	0.066	0.038

Table 7.5. *Odds ratios and adjustment of conjoint models of neighborhood choices (significance levels: ** = 5%, *** = 1%)*

When mentioned, access to public transportation (Model D) is complementary and appears especially important for households with children and parents in their 40s (3.52). Adolescent children's parents appreciate not having the burden of driving them, thanks to a public transport service. As expected, proximity to schools (Model E) is appreciated by parents, especially mothers. The 1996 Canadian social survey indicates that 21.4% of men and 29.8% of women in Québec City – 36% more women – stated they dedicated five hours or more per week to take care of their children [STA 96]. Although proximity to schools may be a common concern for both parents at the time of purchase, it is forgotten sooner by men. However, seemingly restricted in their budget, purchasers of their first residence (0.647) place less importance on this factor than households that are already property owners. In addition, with all else being equal, it is single people who choose to live in the neighborhood where they work, thus greatly reducing their commute times. Moreover, it would appear that this motive often heads the list of priorities among certain purchasers (0.389 and 0.550).

Appreciation of neighborhood cachet is intimately connected to income, often mentioned by men who place it at the top of their priority list. Despite the frequency of mentions (27%), attachment to the neighborhood was very difficult to model (R^2 of 0.038), although this criterion was very often mentioned alone, especially by purchasers older than 50 with limited revenue (<$50,000). Moreover, their propensity to mention this criterion is exceptional (9.9 to 1) and translates into either comfort linked to familiarity with their immediate environment, or an *a posteriori* rationalization of their financial or physical limitations in changing neighborhoods. Placed in perspective with aging of the population and the desire of retirees to stay in their home as long as possible, this question merits review through in-depth interviews with those households concerned. A study held in the Québec City region among seniors with limited mobility showed that they gradually restrict their range of activities (contraction of expanded space) as their motor skills diminish, yet with coping strategies differing from one person to the next [LOR 09].

Table 7.6 presents three conjoint models of residential choice:

– Model I, which details the circumstances that led to the mention of sales price as a choice factor;

– Model J, which concerns the size of the home and lot; and

– Model K, which identifies responses about architectural style.

The three models are significant, although Model J is better adjusted than the other two. However, criteria for models I (price) and K (style) clearly seem more exclusive than that of size, which is rarely mentioned alone. Moreover, women (0.67) are less affected by the price criterion than men.

The probability of mentioning the *price* criterion is inversely proportional to income, with an odds ratio of 0.81 (19% reduction) per $10,000 wage bracket. New buyers (renters) bring up the financial aspect more often than previous owners (1.59). Inversely, the size criterion (size of house, number of rooms and lot dimensions) is positively associated with revenue (1.12) and negatively with first-time buyers (0.576), who are less affected by this objective. More specifically, the size of the home is mentioned three times less frequently by low-income households (<$50,000) than by higher income households (>$80,000). The mention of architectural style is positively associated with revenue (14% increase per $10,000 wage bracket). It is the prior mention of attachment to the neighborhood (1.88), however, that is best correlated with an appreciation of the future home's architectural style. This illustrates a symbiotic effect between the search for a building that has character and selecting the appropriate location.

Variable (reference category)	Category (measurement unit)	Model I Sales price	Model J Size of residence	Model K Architectural style
Income	(per $10,000)	0.810 ***	1.12 ***	1.14 ***
Prior status (owner)	Renter	1.59 ***	0.576 ***	
Attachment (no mention)	Attached to the neighborhood			1.88 ***
Respondent (male)	Female	0.67 **		
Number of criteria (>4)	1–2 criteria 3–4 criteria	0.480 **	0.043 ***	0.315 ***
Model adjustment	Chi-square	69.7	176.2	56.3
	Sig	0.000	0.000	0.000
	-2LL	894.3	799.6	782.8
	Nagelkerke R^2	0.126	0.295	0.110

Table 7.6. *Odds ratios and adjustment of conjoint models of residence choices (significance levels ** = 5%, *** = 1%)*

7.7. Discussion and conclusion

This chapter explores both the motives that led to a move and the residential choice criteria among single-family properties in Québec City, by estimating the probability that a motive or choice criterion be mentioned according to the household's socioeconomic profile. Needs and aspirations in terms of housing strategies have evolved greatly following major societal transformations that have taken place in North American cities over the past four decades. These transformations include economic growth, an increased use of motorized vehicles (motorization), access to property, urban sprawl, women's integration in the labor market, changes to household structures, an aging population, etc. In this context, an understanding of the motives for moving and residential choice criteria is important both for understanding urban dynamics and to establish land management policies that are adapted to the wants and needs of the population.

Logistic regression has enabled us to define the relationships between household characteristics and the multidimensional aspect of residential behavior, the motives for moving and selection criteria. This chapter supports Rossi's [ROS 55] postulates on lifecycle relationships–residential mobility, as with other recent studies [DIE 02]. Various original aspects have been introduced herein, such as prior occupancy status (renter versus owner), which has a strong influence on the motives for moving. It is also interesting to observe the negligible role educational attainment plays in motives for moving. Education is, however, associated with the mention of a new job, once income is controlled for (model not presented). This result conforms to migratory macroadjustment and human resources theories, since individuals migrate to improve their situation or usefulness level, in relationship with a combination of personal attributes and qualitative perceptions of the destination. A longitudinal analysis of successive residential locations would shed more light on the subject, particularly with regard to the accumulation of wealth, education, adjustment to a change in the workplace and family context. A study of this type using the *event history analysis* method (Cox regression) focuses on the Québec City region, thanks to a longitudinal survey on housing, family and employment history [VAN 09].

Neighborhood selection criteria are primarily linked to age, revenue and family structure. In Québec City, the proximity of public transportation networks is of particular importance to parents in their 40s, concerned with their children's access to urban life. This finding falls within the context of a progressive decline in the use of public transportation, with 10% of trips being made in 2001 compared to 17% in 1991, in spite of the inauguration of two high-frequency bus services in 1992, aimed at meeting the demand and stimulating the use of public transportation. The particular context of Québec City must be mentioned, with its highway length/resident ratio being among the highest in North America (21 km per 100,000 residents). This is a result of the massive investment in road infrastructures that took place during the 70s, at the time partly justified by increasing demographic forecasts, which have since revealed themselves to be excessively optimistic. These projects promoted urban sprawl, hindering the introduction of a fully-efficient public transportation system.

The neighborhood selection grid and reasons for moving can be situated within the framework of hierarchical concepts of place–perspective, local and distant. Criteria referring to place–perspective are linked to the very structure of buildings and the property's immediate environment, such as the style or size of the house, the presence of trees, or landscaping characteristics. However, it is more difficult to attribute a criterion exclusively to one or other of the local and distant spaces. For example, family proximity is a very relative criterion. For some, this idea can represent a 40-minute trip by car, consequently referring to the perceptual concept of distant space. For others, family proximity can refer to a distance that can be reached on foot, which infers a spatial scale closer to the perceptual concept of local space.

The response frequency for some criteria varies according to the actual location chosen. This observation, *a priori* logical, opens the door to some apparent contradictions. In Québec City, the proximity criterion is cited more often by residents of new neighborhoods than by those who live in or near the downtown core. This anomaly may be linked to the perception of easy accessibility by those living in the suburbs who have easy access to a well-developed road network, or to the fact that accessibility varies greatly in the peripheral sectors. It may also be the result of a statement bias, with some implicit selection criteria not being verbalized. Buyers who choose the downtown area can make the implicit choice to move closer to services while at the same time shifting their preferences to other aspects, such as esthetics or safety. Is the recent dispersion of commercial areas to highway interchange districts [BIB 08], thus the network restructuring of local trip space, in the process of modifying mental maps?

The concept of identification with the area is also visible in the analysis of residential behavior: 27% of the respondents mentioned attachment among neighborhood selection criteria. Attachment is mentioned more often by residents of affluent neighborhoods than by those in more deprived areas, although the frequency is high for both groups[5]. The first group are often prior owners, have a higher education and salary level, and mention more choice criteria (1.4 times more for the neighborhood and 1.26 more for the property), with more responses expressed for proximity, esthetics, property size, architectural style and surrounding milieu (trees and landscaping). As concerns attachment, details that were not available on the characteristics of the former residence would have allowed us to measure the continuity–space congruence and observe which elements were carried over. This is an aspect we were not able to explicitly cover, given the absence of relevant data.

It is important to remember that the results of this study deal with a particular segment of Québec City's population (purchasers of single-family properties), over a given period (1993–2001), when real estate market conditions were characterized by a high availability of homes versus few potential buyers. This resulted in a wide choice for buyers in a rather atypical city, given the Canadian and North American context, combining a highly dispersed peripheral urban fabric with a densely-populated historical central sector. In addition, the socioeconomic make-up of Québec City is unique to Canada, with:

– twice as many families comprised of common-law couples than the national average (29% compared to 14%, compared with 8.2% in the US);

– approximately twice as many divorced persons (8.6% compared to 4.8%); and

– six times fewer immigrants than the national average (2.8% compared to 18.4%)

These are just a few aspects. Thus, the results of the study cannot be directly transposed to other Canadian cities without indicating the prevailing local conditions.

---

5 One notable exception, however, concerns older respondents (>50 years) with a low income who state feeling very attached to their neighborhood (odds ratio of 9.9, see Table 7.5).

In conclusion, several methodological biases characterize the type of study we conducted. Respondents' recollection of selection criteria taken into consideration at the time of purchase may vary from person to person. An analysis of the relationships between selection criteria mentioned and the true characteristics of the property and neighborhood would enable us to clarify this point. On one hand, longitudinal studies that consider individual trajectories of employment, family composition and health are required in order to better understand the causal factors and interactions that influence residential behaviors. On the other hand, other methods such as hedonic modeling allow us to estimate the variation of the marginal value of property and neighborhood attributes according to the buyers' household characteristics. Two papers that adopt the hedonic approach with data from this study have been published by Kestens *et al.* [KES 06] and Des Rosiers *et al.* [DES 07]. Although fragmentary, their results basically confirm those presented above.

7.8. Acknowledgments

Yan Kestens is a research scientist at the Montréal Public Health Directorate and at Montréal University's Department of Social and Prevantative Meidcine. Marius Thériault is a full professor at Laval University's *École Supérieure d'Aménagement du territoire et de Développement régional* (ESAD), in Québec City, and a regular member of the Center for Research in Regional Planning and Development (CRAD). François Des Rosiers is a full professor with Laval University's Faculty of Administrative Sciences, and is also a regular member of CRAD. This project was funded by the Social Sciences and Humanities Research Council and studies were authorized by Laval University's Research Ethics Committee. Real estate transaction data were obtained from the Communauté Urbaine de Québec.

7.9. Bibliography

[ADA 01] ADAMOWICZ V., BOXALL P., "Future directions of stated choice methods for environment valuation", *Choice Experiments: A New Approach to Environmental Evaluation Conference*, South Kensington, London, April 2001.

[ALT 92] ALTMAN I., LOW S. M., (eds.) *Place Attachment*, New York, Plenum Press, 1992.

[ART 78] ARTLE R., VARAIYA P., "Life cycle consumption and homeownership", *Journal of Economic Theory*, vol. 18, pp. 38-58, 1978.

[BEN 98] BENSON E.D., HANSEN J.L., SCHWARTZ A.L., SMERSH G.T., "Pricing residential amenities: The value of a view", *Journal of Real Estate Finance and Economics*, vol. 16, pp. 55-73, 1998.

[BIB 08] Biba G., Thériault M., Villeneuve P., Des Rosiers F., "Aires de marché et choix des destinations de consommation pour les achats réalisés au cours de la semaine. Le cas de la région de Québec", *Le Géographe Canadien*, vol. 52, no. 1, pp. 39-64, 2008.

[BON 99] Bonaiuto M., Aiello A., Perugini M., Bonnes M., Ercolani A.P., "Multidimensional perception of residential environment quality and neighborhood attachment in the urban environment", *Journal of Environmental Psychology*, vol. 19, pp. 331-352, 1999.

[BOX 02] Boxall P. C., Adamowicz W. L., "Understanding heterogeneous preferences in random utility models: a latent class approach", *Environmental and Resource Economics*, vol. 23, pp. 421-446, 2002.

[BRE 86] Breakwell G. M., *Coping with Threatened Identity*, London, Methuen, 1986.

[BRE 92] Breakwell G. M., "Processes of self-evaluation: Efficacy and estrangement", in: G.M. Breakwell (ed.), *Social Psychology of Identity and the Self-concept*, Surrey, Surrey University Press, 1992.

[CLA 00] Clark W.A.V., Mulder C.H., "Leaving home and entering the housing market", *Environment and Planning A*, vol. 32, pp. 1657-1671, 2000.

[CLA 04] Clark M., Deurloo C., Dieleman F.M., "Choosing neighborhoods: Residential mobility and neighborhood careers", *ENHR Conference Proceedings*, Cambridge, UK, 2004.

[CUM 86] Cummings R.G., Brookshire D.S., Schulze W.D., *Valuing Environmental Goods: A State of the Arts Assessment of the Contingent Method*, Totowa, NJ, Roman and Allanheld, 1986.

[DEN 03] Deng Y., Ross S.L., Wachter S.M., "Racial differences in homeownership: The effect of residential location", *Regional Science and Urban Economics*, vol. 33, no. 5, pp. 517-556, 2003.

[DES 00] Des Rosiers F., Thériault M., Villeneuve P.Y., "Sorting out access and neighborhood factors in hedonic price modelling", *Journal of Property Valuation and Investment*, vol. 18, no. 3, pp. 291-315, 2000.

[DES 07] Des Rosiers F., Thériault M., Kestens Y., Villeneuve P., "Landscaping attributes and property buyers' profiles: Their joint effect on house prices", *Housing Studies*, vol. 22, no. 6, pp. 945-964, 2007.

[DIE 01] Dieleman F.M., "Modelling residential mobility: A review of recent trends in research", *Journal of Housing and the Built Environment*, vol. 16, pp. 249-265, 2001.

[DIE 02] Dieleman F.M., Mulder C.H., "The geography of residential choice", in: J. Aragonés, G. Francescato, T. Gärling (eds.) *Residential Environments: Choice, Satisfaction, and Behaviour*, Westport, Connecticut, Bergin and Garvey, 2002.

[DUN 93] Duncan J., Ley D., *Culture/place/representation*, New York, Routledge, 1963.

[EAR 98] EARNHART D., *Combining Revealed and Stated Data to Examine Decisions of Housing Location: Discrete-choice Hedonic and Conjoint Analysis*, Lawrence, KS, University of Kansas, 1998.

[FEL 90] FELDMAN R.M., "Settlement-identity: Psychological bonds with home places in a mobile society", *Environment and Behavior*, vol. 22, pp. 183-229, 1990.

[FEL 96] FELDMAN R.M., "Constancy and change in attachments to types of settlements", *Environment and Behavior*, vol. 28, pp. 419-445, 1996.

[FIL 99] FILION P., BUNTING T., WARRINER K. "The entrenchment of urban dispersion: Residential preferences and location patterns in the dispersed city", *Urban Studies*, vol. 36, no. 8, pp. 1317-1347, 1999.

[FRI 81] FRIEDMAN J., "A conditional logit model of the role of local public services in residential choice", *Urban Studies*, vol. 18, no. 4, pp. 347-358, 1981.

[FRI 82] FRIED M., "Residential attachment: Sources of residential and community satisfaction", *Journal of Social Issues*, vol. 38, pp. 107-119, 1982.

[GAB 03] GABRIEL S., PAINTER G., "Pathways to homeownership: An analysis of the residential location and homeownership choices of black households in Los Angeles", *Journal of Real Estate Finance and Economics*, vol. 27, pp. 87-109, 2003.

[GAR 93] GÄRLING T., GOLLEDGE R.G., "Understanding behavior and environment: A joint challenge to psychology and geography", in: T. GÄRLING, R.G. GOLLEDGE (eds.) *Behavior and Environment: Psychological and Geographical Approaches*, Amsterdam, Elsevier Science Publishers, pp. 2-13, 1993.

[GIB 50] GIBSON J.J., *The Perception of the Visual World*, Boston, Houghton Mifflin, 1950.

[GIU 91] GIULIANI M.V., "Towards an analysis of mental representations of attachment to the home", *Journal of Architectural and Planning Research*, vol. 8, pp. 133-146, 1991.

[GOO 89] GOODMAN A.C., "Identifying willingness-to-pay for heterogeneous goods with factorial survey methods", *Journal of Environmental Economics and Management*, vol. 16, pp. 58-79, 1989.

[HEN 83] HENDERSON J.V., IOANNIDES Y.M., "A model of housing tenure choice", *American Economic Review*, vol. 73, pp. 98-113, 1983.

[HOE 97] HOESLI M, THION B., WATKINS C., "A hedonic investigation of the rental value of apartments in central Bordeaux", *Journal of Property Research*, vol. 14, no. 1, pp. 15-26, 1997.

[KES 04] KESTENS Y., THÉRIAULT M., DES ROSIERS F. "The impact of land use and surrounding vegetation on single-family property prices", *Environment and Planning B: Planning and Design*, vol. 31, no. 4, pp. 539-567, 2004.

[KES 06] KESTENS Y., THÉRIAULT M., DES ROSIERS F., "Heterogeneity in hedonic modelling of house prices: Looking at buyers' household profiles", *Journal of Geographical Systems*, vol. 8, no. 1, pp. 61-96, 2006.

[KIM 05] KIM J.H., PAGLIARA F., PRESTON J. "The intention to move and residential location choice behaviour", *Urban Studies*, vol. 42, no. 9, pp. 1621-1636, 2005.

[LOR 09] LORD S., JOERIN F., THÉRIAULT M., "La mobilité quotidienne de banlieusards vieillissants et âgés: Déplacements, aspirations et significations de la mobilité", *Le Géographe Canadien*, vol. 53, no. 3, pp. 357-375, 2009.

[LON 84] LONGLEY P.A., "Comparing discrete choice models: Some housing market examples", in: D.E. PITFIELD (ed.) *Discrete Choice Models in Regional Science*, London, Pion Limited, pp. 163-180, 1984.

[LOU 90] LOUVIÈRE J.J., TIMMERMANS H., "Hierarchical integration theory applied to residential choice behaviour", *Geographical Analysis*, vol. 22, pp. 127-144, 1990.

[MCF 78] MCFADDEN D., "Modeling the choice of residential location", in: A. KARLQUIST, L. LUNDQUIST, F. SNICKARS, J. WEIBUL (eds.) *Spatial Interaction Theory and Planning Models*, New York, Elsevier North-Holland, pp. 75-96, 1978.

[MEG 99] MEGBOLUGBE I., SA-AADU J., SHILLING J.D., "Elderly female-headed households and the decision to trade down", *Journal of Housing Economics*, vol. 8, pp. 285-300, 1999.

[MOL 01] MOLIN E.J.E., OPPEWAL H., TIMMERMANS H.J.P., "Analyzing heterogeneity in conjoint estimates of residential preferences", *Journal of Housing and the Built Environment*, vol. 16, pp. 267-284, 2001.

[MOL 03] MOLIN E.J.E., TIMMERMANS H.J.P., "Testing hierarchical information integration theory: The causal structure of household residential satisfaction", *Environment and Planning A*, vol. 35, pp. 43-58, 2003.

[MUL 99] MULDER C.H., HOOIMEIJER P., "Residential relocations in the life course", in: L.J.G. VAN WISSEN, P.A. DYKSTRA (eds.) *Population Issues, an Interdisciplinary Focus*, New York, Kluwer Academic/Plenum Publishers, 1999.

[NEC 84] NECHYBA T., STRAUSS R., "Community choice and local public services: A discrete choice approach", *Regional Science and Urban Economics*, vol. 28, pp. 51-73, 1998.

[ONA 83] ONAKA J.L., "A multiple-attribute housing desequilibrium model of residential mobility", *Environment and Planning A*, vol. 15, pp. 751-765, 1983.

[ORF 99] ORFORD S., *Valuing the Built Environment: GIS and House Price Analysis*, Ashgateh, Aldershot, 1999.

[PEL 02] PELLEGRINI P.A., FOTHERINGHAM A.S., "Modelling spatial choice: A review and synthesis in a migration context", *Progress in Human Geography*, vol. 26, pp. 487-510, 2002.

[PRO 78] PROSHANSKY H., "The city and self-identity", *Environment and Behavior*, vol. 10, pp. 147-170, 1978.

[QUI 76] QUIGLEY J., "Housing demand in the short run: An analysis of polytomous choice", *Explorations in Economic Research*, vol. 3, pp. 76-102, 1976.

[QUI 85] QUIGLEY J., "Consumer choice of dwelling, neighborhood, and public services", *Regional Science and Urban Economics*, vol. 15, pp. 41-63, 1985.

[REG 01] REGINSTER I., EDWARDS G., "The concept and implementation of perceptual regions as hierarchical spatial units for evaluating environmental sensitivity", *URISA Journal*, vol. 13, pp. 5-16, 2001.

[ROS 55] ROSSI P.H., *Why Families Move: A Study in the Social Psychology of Urban Residential Mobility*, Glencoe, Illinois, Free Press, 1955.

[ROS 74] ROSEN S., "Hedonic prices and implicit markets: Product differentiation in pure competition", *Journal of Political Economy*, vol. 82, pp. 34-55, 1974.

[ROU 89] ROURKE D.O., LAKNER E., "Gender bias: Analysis of factors causing male underrepresentation in surveys", *International Journal of Public Opinion Research*, vol. 1, pp. 164-176, 1989.

[STA 01] STATISTICS CANADA, Crime rates. Web. Statistics Canada, 2001. http://www12.statcan.ca/english/census01/, accessed 22 September 2010

[STA 96] STATISTICS CANADA, Canada's Workforce: Unpaid Work, 1996. http://www12.statcan.ca/english/census96/data/profiles/Index.cfm.

[SUN 96] SUNDSTROM E., BELL P.A., BUSBY P.L., ASMUS C. "Environmental psychology 1989-1994", *Annual Review of Psychology*, vol. 47, pp. 485-512, 1996.

[THÉ 03] THÉRIAULT M., DES ROSIERS F., KESTENS Y., VILLENEUVE P., "Modelling interactions of location with specific value of housing attributes", *Property Management*, vol. 21, no. 1, pp. 25-62, 2003.

[THÉ 05] THÉRIAULT M., DES ROSIERS F., JOERIN F., "Modelling accessibility to urban services using fuzzy logic: A comparative analysis of two methods", *Journal of Property Investment and Finance*, vol. 23, no. 1, pp. 22-54, 2005.

[THU 27] THURSTONE L.L., "A law of comparative judgment", *Psychological Review*, vol. 34, pp. 273-286, 1927.

[TIM 92] TIMMERMANS H., BORGERS A., VAN DIJK J., OPPEWAL H., "Residential choice behaviour of dual earner households: A decompositional joint choice model", *Environment and Planning A*, vol. 24, pp. 517-533, 1992.

[TIM 95] TIMMERMANS H., VAN NOORTWIJK L., "Context dependencies in housing choice behaviour", *Environment and Planning A*, vol. 27, pp. 181-192, 1995.

[TIM 96] TIMMERMANS H., VAN NOORTWIJK L., OPPEWAL H., VAN DER WAERDEN P. "Modeling constrained choice behaviour in regulated housing markets by means of discrete choice experiments and universal logit models: An application to the residential choice behaviour of divorcees", *Environment and Planning A*, vol. 28, pp. 1095-1112, 1996.

[TWI 96] TWIGGER-ROSS C.L., UZZELL D.L. "Place and identity processes", *Journal of Environmental Psychology*, vol. 16, pp. 205-220, 1996.

[VAN 99] VAN OMMEREN J., RIETVELD P., NIJKAMP P., "Job moving, residential moving and commuting: A search perspective", *Journal of Urban Economics*, vol. 46, pp. 230-253, 1999.

[VAN 09] VANDERSMISSEN M.H., SÉGUIN A.M., THÉRIAULT M., CLARAMUNT C., "Modelling propensity to move after job change using event-history analysis and TGIS", *Journal of Geographical Systems*, vol. 11, no. 1, pp. 37-65, 2009.

[WOL 00] WOLVERTON M.L., SENTEZA J., "Hedonic estimates of regional constant quality house prices", *Journal of Real Estate Research*, vol. 19, pp. 235-253, 2000.

Chapter 8

Distances, Accessibility and Spatial Diffusion

8.1. Introduction

The classical models of spatial analysis do have merit: they illustrate, in slightly differentiated spaces (or those judged as such through simplification), the tendancial spatial effect of a mechanism. In this sense, they play an irreplaceable pedagogical role. Despite this, they also bear numerous inconveniences: they are deductive models, reasoning on homogeneous behaviors, in invariant time, in an isotropic space that is frequently reduced in the equations to the sole Euclidian distance between locations.

But, space is differentiated – and scientists have become aware of the complexity of the world. The distance from one place to another or several others is a result of several factors that are objective (networks, simple or combined means of transportation, moments of mobility) and subjective (various perceptions and knowledge). Territorial accessibility is indeed a function of that contextual distance with differentiated space, differentiated time and differentiated perceptions. It may be approached in various ways. The theme of spatial diffusion is a good example of the structuring role of accessibility.

Chapter written by Pierre DUMOLARD.

8.2. Distance, distances?

The first way to estimate the proximity of a territory to a center of activity consists of determining the *physical distance* of each of its parts to that center of activity. In so doing, we implicitly use a *geometric* notion, which has become natural as it was assimilated, but is much less unequivocal than it appears at first sight.

8.2.1. *Definition of a mathematical distance between points*

We call a distance between two points, A and B, any measurement of spacing ($d(A,B)$) that conforms to the following three mathematical properties:

– separation $d(A,B) \geq 0$ two separate points are distant from each other;

– symmetry $d(A,B) = d(B,A)$ the distance from A to B equals that from B to A;

– uniqueness $d(A,A) = 0$ all locations are unique.

The property of symmetry implies that space, as a supporting element in the measurement of distances, is *isotropic* (having the same characteristics in any direction). If that property is not verified, as is frequently the case for real geographical spaces, we are no longer speaking of distances but of *separations*.

A space is said to be *metric* if the lengths that we can measure in it conform to the fourth property, that of triangular inequality: $d(A,B) \leq d(A,C)+d(C,B)$ all "paths" that are not in a straight line are longer than that straight line (unless C is found on that line).

8.2.2. *Various geographic distances between points*

The most common measurement of distance in a metric isotropic space is the *Euclidian* distance (distance as the crow flies). In Cartesian space, it is the minimum distance between two points: it can serve as a theoretical reference to which a distance traveled can be compared in order to estimate the "rugosity" of real space.

A number of other distances also exist:

– *Rectilinear* distance provides an example of the adaptation of a measurement of length to the particular context of a network of streets in which the segments intersect at right angles (therefore we also call it a Manhattan distance or a rectangular distance). Moreover, it can be generalized to include any space in which the segments intersect at any angle.

– *Geodesic* distance is the shortest curved distance between two points, used, for example, to measure a sphere, such as the distance between two airports.

– *Topological* distance between two points on a graph is the lowest number of cross-sections to travel through in order to move from one point to the other.

– *Ultrametric* distances conform to a "reinforced" version of the triangular inequality where $d(A,B) \leq \max[d(A,C),d(C,B)]$.

– Other distances also exist.

What is the distance between two ports in Brittany, when following the coastline? We know that Mandelbrot [MAN 73] illustrated the response in his theory on fractals, showing that this distance is dependent on the scale of the representation used to measure it, and thus highly variable.

8.2.3. *Distances between cartographic objects*

The points (with a 0 dimension in classical geometry) are not the only objects of cartographic representation: curves with 1D, surfaces have a 2D, and volumes have a 3D. Does the notion of distance still have meaning between these various types of objects? From an intuitive point of view yes, but what exactly is that meaning?

– the distance from a point to a right angle may be estimated by the distance to its orthogonal projection on that right angle;

– the distance from a point to a surface may be estimated by the distance to its average or median point;

– the distance between two surfaces may be estimated by the Hausdorff distance [BRI 98] between all the representative points of the two surfaces, which is the same as measuring the distance between point patterns so that they overlap.

In all these cases, we are brought back to a distance between points. What about the distance between curves, networks or surfaces? This will be discussed below.

8.2.4. *A generalized notion of distance*

Actual geographic spaces are not generally metric spaces in which we can automatically determine simple distances, such as Euclidian distances. This state of affairs exists for numerous reasons:

– The geographic space is very rarely isotropic (isotropic space is a geometric fiction, and a theoretical abstraction). If it were, spatial forms would be simple figures, for example the isochrones around any city would be perfect circles.

Geographic space is fundamentally *differentiated/differentiating*. It is heterogeneous in *texture* due to the relief, hydrography, build up and road networks. It is also heterogeneous in *structure* due to the modes of functioning, past and present, the spatial concentration of the activities (based on the principle of geographic rarity), specialization and territorial competition.

– The distances actually estimated or anticipated by users are often distances-times and distances-costs, rather than physical distances. This means that the referential is no longer only spatial but *spatiotemporal*, taking into account moments and durations.

– With the notion of distance, there is also a part that is perception and "lived" experience linked to cultural, psychological and social aspects. Such-and-such far away place will be perceived as closer than it is physically. Such-and-such a barrier to movement will be a comparative border for certain social and/or ethnic categories, etc. As seen in Chapter 6, space is also differentiated in the minds of individuals!

– The notion of geometric distance has also been generalized (in statistics for example) as measured by differences. It is no longer spatial, but relative to the characteristics of the subjects under study (which may be places).

We could just as well say that a relatively simple and apparently unambiguous notion, such as the notion of distance, becomes more complex when it is placed into a context. Accessibility is thus linked to a *contextual* distance.

8.3. Spatial accessibility

8.3.1. *Definition of spatial accessibility*

Accessibility from one place to another can be defined as the "optimal" path (in light of one criterion at any given moment) from between the two. For example, accessibility from one neighborhood to a central place is linked to the group of "optimal" paths to that center. This definition implies that we conceive space as a continuum that might have barriers to circumvent and brakes that stop movement (in the case of calculations, it is discrete and fragmented).

Accessibility is therefore a *concrete* situationalization of geographic kinetics. It is both a *generalization* and a *particularization* of the notion of distance.

It is a *generalization* as it is applied to space of types and dimensions: from point-to-point, from point-to-curve, from curve-to-point, from curve-to-surface, from surface-to-curve, from surface-to-surface. It is also a generalization as it can be applied to geometrical spaces as well as topological spaces[1]. Lastly, accessibility is a general notion as it concerns groups of places and mobile agents.

It is also a *particularization* as, rather than always being the same like a metric distance, accessibility results from the conjunction of elements that are modifiable in time and space. It supposes:

– a group of subjects in movement;

– a means of transportation (from the very simple to the very sophisticated); characterized by a speed (average or instantaneous) and an expenditure of energy;

– a transport infrastructure (from the most elementary to the most complex multimodal networks);

– an awareness or a perception of the path to take and the place to be reached;

– a moment of travel and anticipation of its duration.

All modes of travel have their own characteristics in a particular spatiotemporal context (which may be repetitive in time – yielding cyclical patterns, such as rush hours). Thus it is that central cities and city centers, described in the literature as places of maximum accessibility, are at certain moments places of minimum accessibility in terms of access time. Thus, accessibility varies according to numerous parameters.

8.3.2. *Measuring spatial accessibility*

There are cases in which the use of a simple metric distance may be appropriate: for example, when we consider large spaces that are weakly differentiated from the point of view of their texture, and for which we want structural inequalities. In most cases, however, seeking recourse in better adapted methods is necessary so that the results may have some degree of plausibility. In the other cases, various methods for calculating accessibility may be mobilized. Here, we briefly present methods arising from graph theory (which are compatible with vector-based geographic information systems or GIS), methods deployed in raster-based GIS, *ad hoc* modeling and methods linked to multiagent systems.

8.3.2.1. *Methods linked to graph theory*

All places visited are accessible through one or several networks. From the point of view of the mobility of persons and goods, geographical space is a network space,

---

1 Topology deals with the qualitative properties and relative positions of objects, independent of their shapes and dimensions.

which is to say a space in which the networks constitute the innervating system. As recognized for any plot of land, the right to access (or even right of passage on other plots), all useful portions of territory are accessible by the road network, which justifies the calculations of accessibility on all levels by graph theory. Modeling a network on a graph must take into account the mode of transportation and nature of the infrastructure. We must also distinguish the modes of transportation "with a tunnel effect" (such as trains, highways, access from the air) in which the use has a limited number of points of entry/exit and those "with quasi-continual spatial service" like walking, cycling or driving a car, in which the points of entry/exit are innumerable. The second type of graph, in practice, is difficult to create (lack of complete data, multimodal systems of transportation, etc.).

An oriented and valued graph formally represents a transportation network: accessibility from a peripheral place to a center city will be defined in it as an "optimal" path from the former place to the latter, in view of one or several impedance criteria applied to the interstices on the graph (distances in kilometers, duration of trip, cost of trip, a combination of both duration and cost). The classical algorithms used to search for the "optimal" path are of two main types:

– Dijkstra's algorithm, from a node of origin to a destination node with a positive or null valuation of the segments of road; and

– Bellman-Ford algorithm, for any node pair with connection links that are possibly valued negatively.

Diverse developments in graph theory were proposed for solving concrete problems. We cite diverse adaptations here, for example, taking into account:

– timed graphs or graphs with windows of time [BAP 10, DES 08], in particular for traffic management and planning trips that are multimodal or have local time constraints;

– instantaneous problems of flows in a network;

– adaptation to urban morphology [HIL 99, HIL 07, JIA 02];

– graphic representation [LHO 10] in the form of a graph with reliefs or chronomaps of long-term urban modifications [MAT 10] induced through the evolution of an intra- or inter-urban traffic network.

Still, many others could be noted, for example within the context of transportation at the request of the city of Besançon (see Chapter 5): the optimized research of itineraries, planned in a second not yet operational step of the application, for recalculation in near-real time [BOL 06]. A more detailed presentation of the use of graphs to measure distances and travel times is presented in Chapter 9.

Among the work using vector graphs, an important place must be accorded to the works of Hillier [HIL 99, HIL 07]. Taking the counter-argument of models inferring the mobility of the interaction of gravity between urban zones, he begins with detailed urban morphology to induce a differential of potential accessibility. The founding idea is the following: the city is a collection of buildings interconnected by a network of roads and, through progressive self-organization, this spatial configuration shapes intra-urban mobility.

Beginning with the principle that individual movements take the shortest path from each building to all others, a map of the shortest minimum paths interconnecting all the blocks may be built (called a "*least line map*"). Two types of path trajectories are noted: those having a precise destination ("*to movement*") and those in transit ("*through movement*"), defining two separate maps of potential mobility. We can observe one constant form, metaphorically named the "deformed wheel" of which the axle (downtown) and the spokes (penetrating rectilinear angles, in a small number) have maximum potential accessibility (thus attracting the functions that necessitate high traffic). The interstices, with more local accessibility (dedicated more to the habitat) are composed of a large number of small segments situated at right angles.

How do individuals move around in a complex built system? To answer this question, a large number of studies were conducted, which distinguished "*to movements*" and "*through movements.*" For each of them, three measurements were defined: minimum distance trajectories, rectilinear trajectories, and trajectories with few curves. The confrontation with empirical observations still gives the best correlations with rectilinear trajectories (with the "*least angle change*"). "Natural" movements are thus defined as those that correspond to urban morphology and not to some specific attractivity. Multiplication of the two indices corresponding to the two types of movements (to and through) allows us to obtain a single map of potential accessibility, which is very revealing of the functional structure of a large number of cities. For Hillier, therefore, accessibility is *spatial* before it is *social*.

The basic shortest trip algorithms (e.g. Dijkstra) are often implemented like additional modules in the vector GISs (the best known example is the "Network Analyst" module in the ArcView software or the "Network" module in the ArcInfo software). In the latter case, integration is that much easier as the modeling of cartographic objects is achieved in the form of an oriented planar graph. With regard to more recent developments, implemented in the form of specific software packages, the results are easy to integrate as additional layers are almost all vectorial GIS.

Recourse to graph theory allows us to determine that, due to accessibility between the nodes in the network, on a fine scale, it is almost impossible to have an exhaustive representation of the secondary road sections and, especially, of the interstitial space. These spaces are, however, accessible in reality. Thus, we must move on to another exhaustive representation of the territory in the form of matrix images using raster GIS.

8.3.2.2. *Methods linked to surfaces of friction*

This second approach, still classical, is known as the calculation of spatial impedance or spatial friction. It implies that the matrix representation of the geographical information be mobilized (one or several layers) allows us to determine the accessibility of an entire territory (and not only a network) in relation to one or several central points. This method is implemented in numerous raster GISs, such as Idrisi or GRASS.

The idea is to quantify, for each pixel, the resistance that it gives to one or several types of movement; thus we can take into account road sections traveled at various speeds, with various grades and/or degrees of winding road, more or less permeable barriers, inaccessible zones, etc. We can determine an "average" level of accessibility for a mode of transportation or for several modes of transportation in sequence, accessibilities at various moments, competition for accessibility, etc. In all cases, the resulting image or images are valued according to differential accessibility. In order to translate accessibility into kilometers, time or cost of the trip, one additional step of calibration and empirical validation is necessary.

The method of spatial friction has been proven and is simple to implement. It is best for rural territories, loose peri-urban spaces or large spaces that are finely served; whereas the approach through graph theory is much better suited for dense urban areas considered on a large scale or for large spaces in which the network is essential. These two approaches call on infrastructures: we can also envision approaches founded on the interaction between users and infrastructure.

8.3.2.3. *Methods founded on the interaction between users and infrastructure*

We can characterize, in a city, the various types of users' unequal access to u various types of public services or businesses. The notion of accessibility thus assumes a broader meaning, calling on two large classes of parameters, related to:

– their "social value" (their differential attractivity) for various segments of the population in view of the goals aimed for (work, leisure activities, etc.);

– their physical accessibility through various transportation modes.

The work conducted by researchers from Laval University [THE 03, THE 05] on the physical and behavioral components of accessibility to the services of Québec

City clearly belong to this type of research and an example of an application is presented in Chapter 10. Another example of *ad hoc* modeling, this one being less well-known, is found in the work conducted to report on the population's accessibility to green spaces in the city of Padua [ZUL 06]. The attractivity of green spaces was defined, for various age classes, in view of their surface and equipment. Physical accessibility to green spaces, through various modes of travel, was approximated through an inverse logistic function from the distances to the places of residence (it could also have been approximated by reference to the traffic networks). One development project to transform the riverbanks into a park was also simulated and similarly the zones with insufficient access to the green spaces of Padua were highlighted.

Here the interest lies in the definition of broadened accessibility to a sociospatial interaction of the types of equipment (uses and demands) that do not only take into account infrastructures, but also behaviors. Determination of this type of interaction can be also achieved through a "distributed artificial intelligence" (DAI) type approach.

8.3.2.4. *An approach to accessibility by DAI*

Approaches founded on DAI are supported by the concept of emergence based on interactions, in a specified spatial environment, of individuals who can be pixels in an image, persons and/or institutions, itineraries, etc. *Auto-organization* is the key word here. The DAI methods that are useable for calculations of accessibility are drawn from three major types: cellular automata, multiagent systems and genetic algorithms.

We can imagine solutions founded on cellular automata paired with a model of mobility behaviors, where the automata represent spatial constraints and rules of propagation over a territory. Then, we should unify formalism by moving to multiagent systems. Such systems include agents (acting "objects") with rules of behavior, "demographic" (appearance, disappearance, and possibly aging), interaction with other agents (implying modes of communication and rules of behavior) in an environment (presenting constraints and opportunities).

In the search for accessibility from one or several centers, the environment of action is composed of spatial objects that comparatively authorize mobility, with various speeds and conditions of access (networks of various types, space outside of the network, more or less permeable barriers, etc.). The agents are "mobile" in this space and are capable of interacting and minimizing their costs [TOU 95]. Temporal data can also be introduced so as to modify constraints linked to the environment of action of the agents (cumbersome on some days, at certain peak hours).

A group of researchers [BAN 05, BAN 06] (see Chapter 4) have proposed simulating, in time and space, diverse forms of intra-urban mobility and accessibility with the help of a meta-multiagent model (a model that allows us to implement various models of mobility). It is composed of:

– mobile agents equipped with perception (of other agents, of the environment) that is variable over time. These agents have a program of activities to accomplish and are organized into groups;

– an environment composed of the buildings, road network and instantaneous traffic;

– a module of initialization and control by the model's user.

The Miro project defined a prototype. A displacement resulted from environmental awareness by agents, tasks to be accomplished (a sequence of programmed activities) and interaction between agents in the environment. The mobile components are localized agents at moment, t, moving according to behavioral rules (reactive or cognitive), their instant knowledge of the environment (a cognitive map that varies over time) and their programs of activities. The environment creates a network with queues, and in the system plays an active role as it generates and receives traffic. The current implementation is applied to the mobility of pedestrians in downtown Besançon.

In his research on improving transportation at the request of City of Besançon, Bolot [BOL 06] designed a prototype that includes the modification of pickup itineraries in near-real time. The algorithm is initialized by a group of randomly simulated itineraries drawn from statistical data and calculated through a genetic algorithm. The algorithm includes "improvised" (when possible from the point of view of the delays it implies) new steps in the pickup itineraries underway.

DAI thus allows us to kinetically and dynamically simulate a territory by spreading the access costs out. Thus, accessibility is one of the components of the processes of spatial diffusion and the distribution of their mechanism.

8.4. Accessibility and spatial diffusion

As a process, any diffusion (of innovations, epidemics, forest fires, pioneer frontiers, etc.) necessarily has a *dynamic* dimension but also a *kinetic* dimension that is intrinsically linked to it. The contact between the elements in play implies mobility, and therefore differential accessibility.

8.4.1. *Chronology of diffusions*

For a long time, the processes of diffusion was mainly considered (particularly in epidemiology) from the angle of their chronology, where we can classically distinguish various phases:

– *beginning* of the process: in the case of a pandemic, it is the appearance of the disease. In the case of the diffusion of an innovation, it is adoption of the innovation by pioneers;

– *expansion:* the number of individuals reached increases rather quickly;

– *condensation:* the phase of generalization, where the rates of penetration become homogenized; and

– *saturation:* the rate of penetration reaches its maximum point.

These phases, which are continuous in time, are classically represented (on a Cartesian graph with a cumulative number of elements attained placed in ordinate and time in abscissa) by an S curve. This curve is generally modeled by a logistic function, but can also be modeled by other types of functions.

8.4.2. *The spatiotemporal diffusion process*

The chronological approach implicitly implies reducing the importance of the spatial conditions and nature of the diffusion and that which they can modify in the process. From a geographical point of view, we can differentiate diverse forms of diffusion: through *expansion* (i.e. the progression of a forest fire), *migration* (i.e. the relocalization of production), or *contagion* of proximity (spatial surroundings or hierarchical diffusion).

Hägerstrand [HÄG 67] was the first to build a formal model of the spatiotemporal diffusion of an innovation. It was a question of the progressive adoption of a public grant by farmers to transform a forest into pasture, observed and simulated in the Asby district in Sweden between 1928 and 1932. The zone was cut evenly into square grids of 5 km on each side, and each cell contained n_i farmers. At one given moment, each cell contained k_i adopters and $n_i - k_i$ potential adopters. A "*spatial field of contact*" is defined thus: from each cell i, the probability of drawing a cell j at random decreases with its distance to i. The sending of a message between cells i and j takes place as many times as there are adopters in the cell i (k_i times). The procedure is applied to all the cells in the zone and iterated. At the end of the iterations, we are left with a map of the numbers of adopters and chronology of the adoption process.

In a study of the diffusion of AIDS in the USA, Gould [GOU 92, GOU 93] pointed out two spatial modalities: *hierarchical* diffusion (from the large centers along the urban hierarchy toward smaller centers); and diffusion by *contagion* (as a watermark in the surrounding areas). Moreover, we can ask ourselves whether the diffusion in surrounding areas is not also a hierarchical diffusion but at a finer scale, in studies of large territories. Besides this, taking long-distance rapid transportation (airplanes, interstate trains) undoubtedly reinforces the weight and strength of diffusions between large centers. The methodological originality of Gould's study resides in the use of filtering by local operators, adapting the functions and their parameters to local conditions.

In her study, Saint-Julien [SAI 85] distinguishes the mechanisms of diffusion for innovations according to their geographical modalities: effects of barriers stopping movement, an adaptation to the system of populating receptor spaces, the role of new techniques decreasing the friction of time and space, territorial differentials in the speeds of propagation and their irreversible spatial effects. We can extend this work by asking ourselves several questions: what are the little or non-innovative territories? what are the (cumulative) dynamics of the innovation in other territories?

8.4.3. *Simulating diffusion through DAI*

Among the many current research projects that use this type of approach, we will briefly describe a few that are linked to cellular automata or the multiagent system.

8.4.3.1. *Scenarios of peri-urban diffusion*

Langlois and Phipps [LAN 97] developed the SIMURB model in order to simulate, through cellular automata, the spatiotemporal diffusion of the use of urban and peri-urban soil under various hypotheses of the growth of five types of population, four types of employment and two types of habitat. The SIMURB model, applied to a raster grid, contains the following elements:

– a matrix of probabilities of the transition of states (MTS) of use of the soil, with intangible values over time;

– four functions of spatial distribution which, combined with the values of the MTS, allow us to calculate, for the cells of the grid, local probabilities in view of their centrality, accessibility in relation to the road network, the state of their surrounding area and the availability of the soil. It also allows us to calculate cells changing state between two iterations modifying local conditions;

– a dynamic module (DM) "*à la Forrester*" [FOR 71] with interactions and retroactions between population, employment and habitat causes the probabilities of changing states to vary in the course of the iterations in view of the MTS and functions of spatial distribution;

– a "regulatory zoning" module is added to the equation in order to prohibit certain changes of state.

At each step in the simulation, a spatial configuration is produced in relation to the hypotheses (formulated in the MTS) of growth (or not) in population, employment and habitat, so that they interact in the DM. Figure 8.1 summarizes the overall architecture of SIMURB.

Let us note in passing that, close to the road network, the function of "accessibility" represents a disadvantage (negative externalities linked to proximity – noise, air pollution, etc.), and becomes an advantage beyond some threshold distance from it (progressively decreasing), leading to the function presenting the approximate aspect of an asymmetrical Gauss curve.

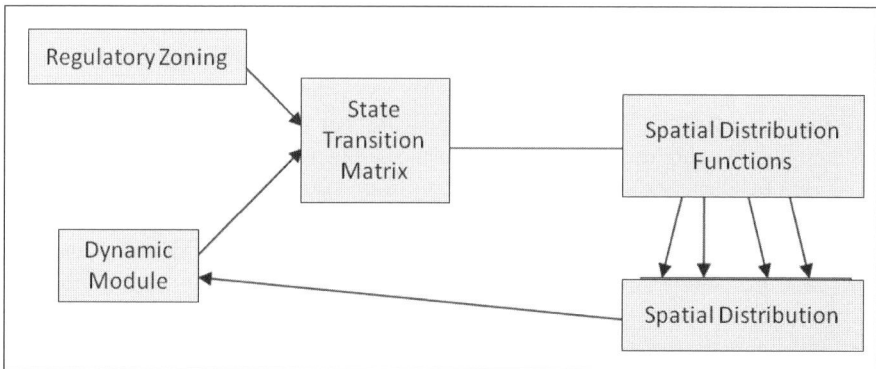

Figure 8.1. *General architecture of SIMURB [LAN 97]*

In order to adapt it to the parallel calculation, a SIMURB modification has allowed us to test four scenarios of spatial diffusion of land use for the agglomeration of Ottawa-Gatineau:

– scenario 1 (pursuit of current tendencies) produces a very moderate peri-urban diffusion;

– scenario 2 (strong demographic growth), yields a dense diffusion along the axes of communication and more dispersed among them;

– scenario 3 (economic consolidation) displays no demographic push and very little dispersion; finally

– scenario 4 (opening of the green belt) projects the appearance in it of a few aggregates.

8.4.3.2. *Examples of simulation by multiagent systems*

In a process of diffusion founded on at least three types of interaction (agent-agent, agent-environment, environment-environment), the cellular automata, lacking differentiation between agents and environment, reproduce internally only those of the environment-environment type. A large amount of research has thus used multiagent systems with the goal of simulating spatial diffusion by distinguishing between agents and the environment. We can cite, as an example, the rough outline of a multiagent model of spatial diffusion [DUM 99], the SMArtURB module of Banos [BAN 07], the works of Barros and Alves on "*peripheralization*" of the "*slum cities*" of the metropoles of Latin America [BAR 03], and a number of others.

Supplementary step: Daudé and Langlois [DAU 08] propose a general conceptual model of spatiotemporal diffusion and provide some examples of its possible application. First, they describe three types of models with various behaviors, then they attempt to define a generic model that distinguishes two types of elements: *hosts*, both emitters and receptors, connected by an oriented graph (channel of diffusion) either having acquired or not (at moment t_i) newness; and *vectors*, concretizing the interaction between the hosts. Two processes act on those two types of elements: (i) a process of *acquisition* complete with a rule that allows us to change the state of the hosts; (ii) a process of *propagation* determining the vectors emitted by each host. Several examples of application are then described (diffusion of a toxic cloud, an epidemic, or an innovation). The generic model is then formalized in order to cover all cases between the logistic model (within global surroundings) and the head-on functioning (within local surroundings). Finally, the authors of that study introduce a rule of propagation of the vector into it, which decreases over time.

8.5. Conclusion

In this chapter, we discussed the relationships betwen concepts of distance, proximity, access, mobility, accessibility and diffusion. We have seen various ways of modeling and simulating their spatial and temporal structure using statistical and computer science tools. Subsequent chapters will give several applications of these concepts integrated in the geographical analysis of urban landscapes. However, there is still much progress to be made in urban research in order to fully reconcile perception, behavior and physical features of the urban networks, with the aim of better understanding of how they jointly influence the long-term evolution of cities.

8.6. Bibliography

[BAN 05] BANOS A., CHARDONNEL S., LANG C., MARILLEAU N., THÉVENIN T., "Simulating the swarming city: a MAS approach", *Proceedings of the 9th International Conference on Computers in Urban Planning and Urban Management*, London, 2005.

[BAN 06] BANOS A., "Geosimulating the swarming city: a bouquet of alternatives", *Geoinformatics*, vol. 8, pp. 58-60, 2006.

[BAN 07] BANOS A., *SMArtURB*, University of Pau, http://arnaudbanos.perso.neuf. fr/geosimul.smarturb.smarturb.html, 2007.

[BAP 10] BAPTISTE H., "Determination of optimal paths in a time-delay graph", in: MATHIS P. (ed.), *Graphs and Networks*, ISTE-Wiley, London-New York, 2010.

[BAR 03] BARROS J., ALVES J.R., "Simulating rapid urbanisation in Latin American cities", in LONGLEY P., BATTY M. (eds.), *Advanced Spatial Analysis: the CASA Book of GIS*, ESRI Press, 2003.

[BOL 06] BOLOT J., Le transport à la demande, une piste pour le développement urbain durable, Thesis, Franche Comté University, 2006.

[BRI 98] BRISSAUD M., "Proches, voisins, adhérents", in: BELLET M., KIRAT T., LARGERON C. (eds) *Approches Multiformes de la Proximité*, Hermès, pp. 125-147, 1998.

[DAU 08] DAUDÉ E., "Multi-agent systems for simulation in geography", in GUERMOND Y. (ed.), *The Modeling Process in Geography, from Determinism to Complexity*, ISTE-Wiley, London-New York, 2008.

[DES 02] DESROSIERS J., "Gestion des opérations dans les réseaux de transport" in: FINKE G. (ed.), *Recherche Opérationnelle et Réseaux*, Hermès, Traité IGAT, pp. 141-166, 2002.

[DES 08] DESROSIERS J., "Operations management in transportation networks" in: FINKE G. (ed.), *Operations Research and Networks*, ISTE-Wiley, London-New York, 2008.

[DUM 99] DUMOLARD P., "Accessibilité et diffusion spatiale", *L'Espace Géographique*, vol. 3, pp. 205-214, 1999.

[FOR 71] FORRESTER J.W., "Counterintuitive behavior of social systems", *Technology Review*, MIT, vol. 73, no. 3, pp. 52-68, January 1971.

[GOU 92] GOULD P., "Epidémiologie et maladie", in BAILLY A., FERRAS R., PUMAIN D. (eds), *Encyclopédie de Géographie*, Economica, pp. 949-969, 1992.

[GOU 93] GOULD P., *The Slow Plague, a Geography of the AIDS Pandemic*, Blackwell, 1993.

[HÄG 67] HÄGERSTRAND R.T., *Innovation Diffusion as a Spatial Process*, Chicago University Press (original book in Swedish: 1953), 1967.

[HIL 93] HILLIER B., PENN A., HANSON J., GRAJEWSKI T., XU J., "Natural movement", *Environment and Planning B*, vol. 20, pp. 29-66, 1993.

[HIL 99] HILLER B., "The hidden geometry of deformed grids", *Environment and Planning B*, vol 23, pp. 169-191, 1999.

[HIL 07] HILLIER B., TURNER A., YANG T., PARK H.T., "Metric and topo-geometric properties of urban street networks: some convergences, divergences and new results", *Proceedings of the Space Syntax Symposium*, no. 1, pp. 1-21, Istanbul, 2007.

[JIA 99] JIANG B., CLARAMUNT C., BATTY M., "Geometric Accessibility and Geographic Information: extending desktop GIS to space syntax", *Computers, Environment and Urban Systems*, vol. 23, pp. 127-146, 1999.

[JIA 02] JIANG B., CLARAMUNT C., "Integration of space syntax into GIS: new perspectives for urban morphology", *Transactions in GIS*, vol. 6, no. 3, pp. 295-309, 2002.

[LAN 97] LANGLOIS A., PHIPPS M., *Automates Cellulaires, Application à la Simulation Urbaine*, Hermès, 1997.

[LAN 08] LANGLOIS P., "Cellular automata for modeling spatial systems", in GUERMOND Y. (ed.), *The Modeling Process in Geography, from Determinism to Complexity*, ISTE-Wiley, London-New York, 2008.

[LHO 10] L'HOSTIS A., "Graph theory and representation of distances, chronomaps and other representations", in: MATHIS P. (ed.), *Graphs and Networks*, 2nd edition, ISTE-Wiley, London-New York, 2010.

[MAN 73] MANDLEBROT B., *Les Objets Fractals: Forme, Hasard, et Dimension*, Paris, Flamarion, 1973.

[MAT 10] MATHIS P., "Dynamic simulation of the urban reorganization of the city of Tours", in MATHIS P. (ed.), *Graphs and Networks*, 2nd edition, ISTE-Wiley, London-New York, 2010.

[SAI 85] SAINT JULIEN T., *La Diffusion Spatiale des Innovations*, GIP Reclus, Montpellier, 1985.

[THÉ 03] THÉRIAULT M., Définition et Mesure de l'Accessibilité aux Services Urbains: Aspects Physiques, Comportements de Mobilité et Perceptions, Colloque Géopoint, 2003.

[THÉ 05] THÉRIAULT M., DES ROSIERS F., JOERIN F., "Modelling accessibility to urban services using fuzzy logic: A comparative analysis of two methods", *Journal of Property Investment and Finance*, vol. 23, no. 1, pp. 22-54, 2005.

[TOU 95] TOURET A., "Agripa: un modèle de calcul de courbes isochrones fondé sur un système multi-agents", *Revue Internationale de Géomatique*, vol. 3-4, pp. 299-314, 1995.

[ZUL 06] ZULIAN G., SECCO G., "Parks and users, where and how inhabitants can profit by urban parks", in: CARREIRO M.M., WU J. (eds.), *Ecology and Management of Urban Forests: an International Perspective*, Elsevier, 2006.

Chapter 9

Accessibility to Proximity Services in Poor Areas of the Island of Montreal

9.1. Introduction

The question of accessibility to services is not new, as witnessed in particular in the work done on spatial segregation in the Paris region in the 1970s and 1980s [PIN 86, PRE 75]. More recently, a number of authors revisited the concept of spatial accessibility to public and private services and facilities in cities, and indeed to urban resources, from the angle of spatial equity [TAL 98a, TAL 98b, TRU 93]. Indeed, accessibility to collective services and facilities constitutes a key issue in spaces where poverty is concentrated: low accessibility contributes to exacerbate the deficit of resources possessed by poor people. Good accessibility, however, compensates, at least partially, for the low level of individual resources [APP 06b, SEG 04].

The development of geographic information systems (GIS) and, in particular, the network analysis modules in them, allows us to calculate measures of accessibility for a given population with relative ease. The objective of this chapter, in a methodological vein, is to describe how it is possible, thanks to GIS, to operationalize the concept of spatial accessibility to urban resources in terms of social equity. In concrete terms, it is a question of verifying whether the areas of poverty on the Island of Montreal benefit from accessibility that is comparable to the more affluent zones on the Island.

Chapter written by Philippe APPARICIO and Anne-Marie SÉGUIN.

9.2. Data

In order to assess accessibility to proximity services in poor areas on the Island of Montreal, we have used two sets of spatial data: the census data of 2001 from *Statistics Canada* [STA 01] conducted on the level of the census tracts, and data on a series of services and facilities collected in 2005.

The first set of spatial data allows us to locate the deprived spaces on the Island. In order to do this, we have used the variable *frequency of low income units in the total population*, which is to say "the percentage of persons in private households who must allocate 20% more than the general average to food, housing and clothing"[1] [STA 02]. This variable thus refers to the notion of the relative poverty of individuals, which is to say that a person is defined as poor when his or her living conditions are well below those attained by the majority of the population of a given society [LEL 05, SEG 98, STR 96].

The second set of data lists 35 types of proximity services and facilities grouped into seven major categories (see Table 9.1). In total, almost 4,700 sites were integrated in GIS by geocoding using street addresses.

9.3. Methodology for measuring accessibility to services

With regard to methodology, the assessment of accessibility to the proximity services and facilities of a given population requires that we apply parameters to the following four elements:

– the spatial unit of reference to which the population is attached;

– the method of aggregation;

– the accessibility measures;

---

1 According to the approach based on the thresholds of low income adopted by Statistics Canada, a person is considered as having a low income if his/her actual household income is below a threshold established on the structure of Canadian household expenses. The thresholds of low income are established based on the data from the *Survey of Household Spending*. They are expressed in terms of the percentage of revenue before taxes. The thresholds of low income for the year are established based on households who must allocate 20 percentage points more than all of Canadian households for these three items. The data from the study are then analyzed in order to determine the levels of revenue of households who must allocate 20 percentage points more than the Canadian average for those three basic needs. These levels of revenue are calculated for various sizes of regions of residence and for various sizes of households. They correspond to the thresholds of low income for the basic year. These thresholds are then updated each year, based on the consumer price index [STA 97].

– the type of distance used to calculate the measures of accessibility.

Although certain notions have already been discussed in Chapter 8, in this section we will outline in detail the various ways of calibrating parameters on those elements, while specifying and justifying our methodological choices for this study.

Services and facilities	N	Services and facilities	N
Cultural facilities		**Sports and recreation facilities (continued)**	
Municipal library	55		
Movie theater	28	Park (one hectare or more)	481
Cultural center	37	Park (five hectares or more)	126
Educational facilities		Outdoor rink	252
Elementary school (Francophone)	242	Indoor rink	48
Elementary school (Anglophone)	73	Outdoor pool	100
High school (Francophone)	44	Indoor pool	54
High school (Anglophone)	25	**Other facilities**	
Adult Ed Center (Francophone)	59	**and services**	
Adult Ed Center (Anglophone)	19	Shopping center	122
Health facilities and services		Local employment office	17
Community health center (CLSC)	29	Postal counter	131
Dental clinic	653	Daycare	542
Medical clinic	209	Police station	49
Large pharmacy	196	Metro station	64
Hospital	28	Supermarket	167
Sports and recreation facilities		**Bank branches**	
Youth center	47	Caisse Populaire Desjardins	133
Municipal or community sports center	54	Laurentian Bank	59
Community parks/gardens	100	National Bank of Canada	75
Wading pool	166	Other banks	210

Table 9.1. *Selected proximity services and facilities*

9.3.1. *The spatial unit of reference*

In most of the Canadian and US studies on accessibility to urban facilities, the spatial unit of reference selected is the census tract to which variables are attached, in particular sociodemographic and socioeconomic variables [APP 07a, HEW 02, LIN 01] [OTT 94, TAL 98b, TRU 93]. Within the framework of this study, we have also chosen the census tract[2] as a spatial unit of reference, and have done so in order to examine the relationships that exist between the percentage of persons with low income in the total population and the indicators of spatial accessibility to the various public and private services and facilities selected.

9.3.2. *The aggregation method*

In order to assess accessibility to a given service for a population residing in a census tract, three methods of aggregation can be envisioned [HEW 02]. The first method consists in calculating the distance between the centroid of the census tract and the service (see Figure 9.1a). By far the simplest and most commonly used – see, in particular, [OTT 94, TAL 98b, TRU 93] – this method has the inconvenience of not taking into account the spatial distribution of the population within the census tract. In order to remedy that shortfall, we must seek recourse in a spatial geographic level nested in that of the census tracts, such as that of the dissemination areas or dissemination blocks[3].

The second method consists of calculating the weighted average point by the total population of the lower level spatial entities included in a census tract, then calculating the distance between that weighted centroid and the service. The coordinates x and y of the average weighted point i, are calculated as follows:

$$\left(\overline{x_i}, \overline{y_i} \right) = \left(\frac{\sum\limits_{k \in i} w_k x_k}{\sum\limits_{k \in i} w_k}, \frac{\sum\limits_{k \in i} w_k y_k}{\sum\limits_{k \in i} w_k} \right) \qquad [9.1]$$

where w_k represents the total population of the spatial unit k entirely included in the census tract i; and x_k and y_k are, respectively, coordinates x and y of the spatial entity k.

Finally, the third and last method of aggregation consists of calculating the distance between the service and each centroid of the spatial units of the lower level, which are included in the census tract, then calculating the weighted average of the total population of those distances (see Figures 9.1b and 9.1c). Compared to the first two methods, this last method is obviously more accurate as it takes into account,

---

2 According to Statistics Canada [STA 02], a census tract usually includes between 2,500 and 8,000 persons. On average, in the territory occupied by the metropolitan region of Montreal, a census tract includes close to 4,000 inhabitants.

3 A dissemination area includes between 400 and 700 inhabitants [STA 02]. On average, in the territory encompassed within the metropolitan region of Montreal, close to 600 inhabitants reside in a dissemination area, and a little more than 100 inhabitants per dissemination block.

more precisely, the distribution of the population inside the census tract, as is clearly shown in Figure 9.1. Thus, we opted for this method of aggregation for our work.

In Figure 9.1, the gaps between the values of distance for a census tract and the closest service according to cases (a) and (c) clearly illustrate the problem of errors in aggregation. Let us take the example showing the highest variations – that of the census tract **iv** in Figure 9.1. The distance separating the centroid of this census tract and the closest service is 160 m (a). This is compared to 449 m if we calculate the weighted average of the total population in relation to the distances between the dissemination areas included in census tract **iv** and the closest service (b). Lastly, this compares to 414 m if we calculate the weighted average of the total population in relation to the distances between the dissemination blocks included in census tract **iv** and the closest service (c).

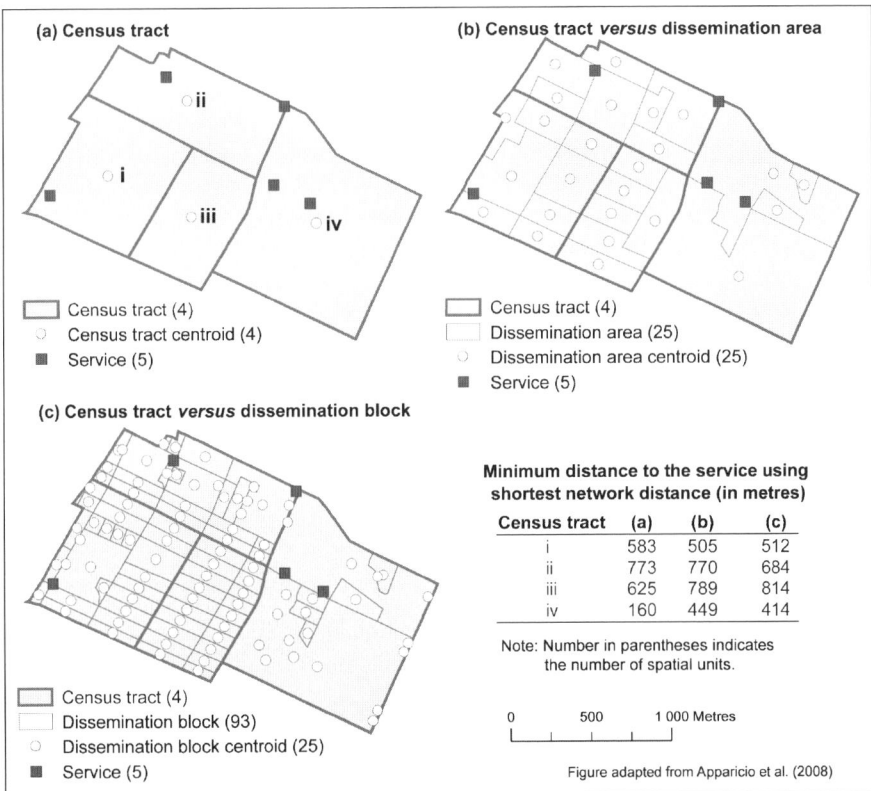

(a) Census tract

ii
i
iii
iv

Census tract (4)
Census tract centroid (4)
Service (5)

(b) Census tract *versus* dissemination area

Census tract (4)
Dissemination area (25)
Dissemination area centroid (25)
Service (5)

(c) Census tract *versus* dissemination block

Census tract (4)
Dissemination block (93)
Dissemination block centroid (25)
Service (5)

Minimum distance to the service using shortest network distance (in metres)

Census tract	(a)	(b)	(c)
i	583	505	512
ii	773	770	684
iii	625	789	814
iv	160	449	414

Note: Number in parentheses indicates the number of spatial units.

0   500   1 000 Metres

Figure adapted from Apparicio et al. (2008)

Figure 9.1. *Methods of aggregation for assessing the distances separating the populations of census tracts from the proximity services*

9.3.3. *The measures of accessibility*

The most commonly used measures of accessibility in the studies on accessibility to urban resources are:

– the distance separating the resident from the closest service;

– the number of services within n meters or minutes;

– the average distance to all services or to n closest services; and finally

– the gravity model [CER 99, HAN 97, TAL 98a, TAL 98b].

Several authors have shown that the choice of these measures is decisive as the accessibility maps vary from one indicator to another [APP 07a, SMO 04, TAL 98a, TAL 98b]. In addition, each of these refers to a different conceptualization of accessibility: the minimum distance allows us to assess *immediate proximity*; the number of services in a radius of n meters or minutes of travel allows us to evaluate *the supply of services in the immediate environment*; the average distance to the group of services refers to the *average cost from one origin to reach all destinations*; the average distance to the closest services n allows us to assess *the average cost to reach a diversified supply of urban resources*; whereas the gravity model measures *the potential attractiveness of the services in view of their size* from a given point of origin (a census tract, for example).

If we apply the principle of the method of aggregation described previously, which takes into account the distribution of the population in the census tract calculated on the basis of dissemination blocks, the equations used to calculate the measures of accessibility may be expressed as follows:

$$Z_i^a = \frac{\sum_{b \in i} w_b (min|d_{bs}|)}{\sum_{b \in i} w_b}, \qquad [9.2]$$

where Z_i^a is the distance between the census tract i and the closest service; d_{bs} is the distance between the centroid of the dissemination block b completely within census tract i and the service s; and w_b is the total population of the dissemination block b.

$$Z_i^b = \frac{\sum_{b \in i} w_b \sum_{j \in S} S_s}{\sum_{b \in i} w_b}, \text{ with } S_s = 1 \text{ when } d_{bs} \le n \text{ and } S_s = 0 \text{ when } d_{bs} > n, \qquad [9.3]$$

where Z_i^b is the number of services within n meters or minutes in the census tract i (1,000 meters, for example); S is the total number of services in the study area; S_s is the service s; and w_b is the total population of the dissemination block b.

$$Z_i^c = \frac{\sum\limits_{b \in i} w_b d_{bs}}{\sum\limits_{b \in i} w_b},$$

[9.4]

where Z_i^c is the average distance between census tract i and all services in the study area; d_{bs} is the distance between the centroid of the dissemination block completely within census tract i and the service s; and w_b is the total population of the dissemination block b.

$$Z_i^d = \frac{\sum\limits_{b \in i} w_b \sum\limits_{S} \dfrac{d_{bs}}{n}}{\sum\limits_{b \in i} w_b},$$

[9.5]

where Z_i^d is the average distance to the n closest services; d_{bs} is the distance between the centroid of the dissemination block b completely within census tract i and the service s (the values of d_{bs} are sorted in ascending order); and w_b is the total population of the dissemination block b.

$$Z_i^e = \frac{\sum\limits_{b \in i} w_b \sum\limits_{S} S_{ws} d_{bs}^{-\alpha}}{\sum\limits_{b \in i} w_b},$$

[9.6]

where Z_i^e is the value of the gravity potential; d_{bs} is the distance between the centroid of the dissemination block b within census tract i and the service s; S is the total number of services in the study area; S_{ws} is the weight attributed to the service s, such as its size or the number of commodities available there (for example, for a cinema, it could be the number of seats or movie theatres that it includes); and α is the parameter of friction that usually assumes the value of 2 or 1.5.

Regarding the choice of accessibility measures, we are very specifically interested in the immediate proximity to urban resources here. Thus, we have opted for the minimum distance for each of the 35 types of services selected.

9.3.4. *Types of distances*

In the field of urban studies, the four most commonly used types of distances to calculate the accessibility measures are:

– Euclidian distance (distance measured as the crow flies);

– Manhattan distance which forms a right angle and which is relatively simple to use to calculate distance in the cities that have a rectilinear street network [FOT 00, PUM 97];

– the shortest path between two points through a network of streets; and finally

– the shortest travel time (or distance time), which is the quickest route between two points through a network of streets [APP 03] (see Figure 9.2).

From the coordinates x and y of points i and j, the Euclidian [9.7] and Manhattan [9.8] distances can easily be calculated by GIS software, such as ArcGIS or MapInfo, or even by Excel or statistics software (SPSS, SAS, Stata, etc.). To do so, the coordinates must be projected in a plane system projection (transverse Mercator or Lambert projection, for example) and not in a spherical system (latitude/longitude):

$$d_{ij} = \sqrt{(x_i - x_j)^2 + (y_i - y_j)^2} \qquad\qquad [9.7]$$

$$d_{ij} = |x_i - x_j| + |(y_i - y_j)| \qquad\qquad [9.8]$$

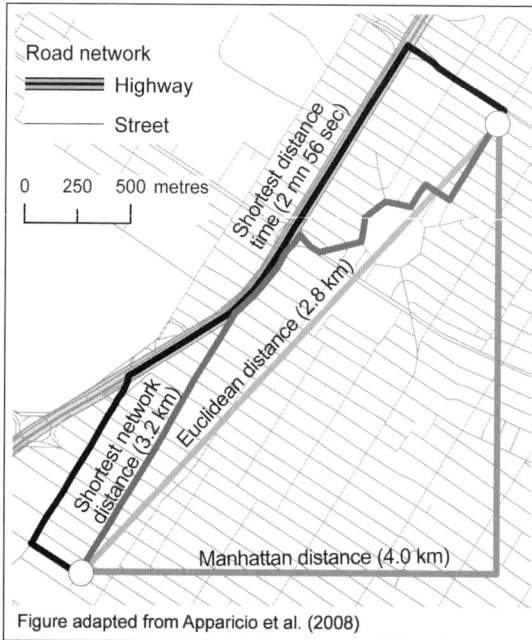

Figure 9.2. *The types of distances*

If the coordinates are projected in a spherical system, it is possible to calculate the distance as the crow flies between two points based on the following spherical trigonometry formula:

$$d_{ij} = 2R\,arcsin\left(\sqrt{sin^2\left(\delta_i - \delta_j\right) + cos(\delta_i)cos(\delta_j)sin^2\left(\frac{\varphi_i - \varphi_j}{2}\right)}\right)$$ [9.9]

in which R is the path of the Earth, and $\delta_i, \delta_j, \varphi_i, \varphi_j$ are, respectively, the coordinates of latitude and longitude of points i and j.

The network distances are, however, more complex to calculate as they suppose that there is a structured network of streets available with directions for the lanes, impedances for crossing the road segments and intersections, which must be integrated into a GIS. They also suppose a network analysis module or software program (for example, the *Network Analyst* extension for ArcGIS or ArcView, or the *TransCAD* software program).

The shortest network distance is very useful for assessing the distance of a route as if we were to walk that route; thus, it is often used in studies on proximity to services [APP 07a, APP 06a, APP 06b, OTT 94, TAL 98b, WIT 03]. Therefore, we have selected the shortest network distance within the framework of this study. In order to calculate it, we must take out the road segments of the integrated network that have restrictions for pedestrians beforehand, in the GISs or, more simply, prohibit the use of those segments, implementing restriction rules.

The shortest distance time (minimum travel time) – the quickest route between two points in a network of streets – requires a more precise modeling of the road network than is allowed, in particular, by the GISs [APP 03, LAU 95, THE 99, ZEI 99]. This type of distance requires us to determine beforehand the mode of transportation used before making the calculations with specific impedances and rules (e.g. one-way restrictions for cars). If several modes may be used according to individual choices, then the choice of mode slightly complicates the modeling of multimodal trips (see Chapter 5).

Let us take the example of modeling a road network in order to calculate the travel times as if they were traveled by car (see Figure 9.3). Most importantly, we must remember that a network is formed of lines (segments of streets) connected by nodes (the junctions between the segments: street intersections). Therefore, it is possible to associate information with both the segments and the nodes. Regarding the segment, the direction of traffic flow in the lanes may be shown in a field with, for example, the values of 0 for traffic flowing in both directions, 1 for the same direction the line was digitized, and -1 for a reverse situation (see Figure 9.3). It is also possible to calculate in a field the time it takes to travel over a segment in view of the maximum authorized speed in the segment based on equation [9.10]. Here, T_{mn} represents the time in minutes to travel over the segment, L_m is the length of the segment in meters, and $V_{km/h}$ is the speed limit in the segment represented as kilometers per hour [MIT 99]:

$$T_{mn} = \frac{L_m * 60}{V_{km/h} * 1,000}$$
[9.10]

Several parameters can also be attached to the nodes: to integrate the stop signs, the traffic lights and the "yield" signs, and to specify authorized turns at the segment junctions. In order to do this, we will create a table called "Turntable", which will usually have four fields: the field ID of nodes; the field ID of the origin road segment (also called origin section); the field ID of the destination road segment (destination section); and the time necessary to cross the intersection expressed in minutes or in seconds, which is the penalty (or in the worst case prohibition) for making a turn.

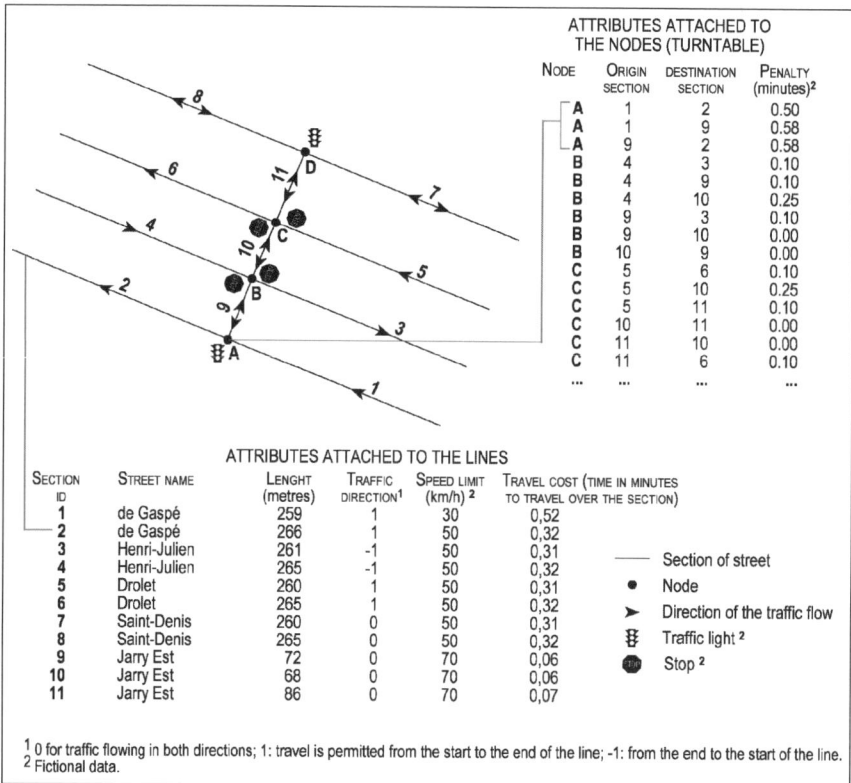

ATTRIBUTES ATTACHED TO THE NODES (TURNTABLE)

NODE	ORIGIN SECTION	DESTINATION SECTION	PENALTY (minutes)[2]
A	1	2	0.50
A	1	9	0.58
A	9	2	0.58
B	4	3	0.10
B	4	9	0.10
B	4	10	0.25
B	9	3	0.10
B	9	10	0.00
B	10	9	0.00
C	5	6	0.10
C	5	10	0.25
C	5	11	0.10
C	10	11	0.00
C	11	10	0.00
C	11	6	0.10
...	...	...	...

ATTRIBUTES ATTACHED TO THE LINES

SECTION ID	STREET NAME	LENGHT (metres)	TRAFFIC DIRECTION[1]	SPEED LIMIT (km/h) [2]	TRAVEL COST (TIME IN MINUTES TO TRAVEL OVER THE SECTION)
1	de Gaspé	259	1	30	0,52
2	de Gaspé	266	1	50	0,32
3	Henri-Julien	261	-1	50	0,31
4	Henri-Julien	265	-1	50	0,32
5	Drolet	260	1	50	0,31
6	Drolet	265	1	50	0,32
7	Saint-Denis	260	0	50	0,31
8	Saint-Denis	265	0	50	0,32
9	Jarry Est	72	0	70	0,06
10	Jarry Est	68	0	70	0,06
11	Jarry Est	86	0	70	0,07

—— Section of street
● Node
➤ Direction of the traffic flow
Traffic light [2]
Stop [2]

[1] 0 for traffic flowing in both directions; 1: travel is permitted from the start to the end of the line; -1: from the end to the start of the line.
[2] Fictional data.

Figure 9.3. *Example of modeling a road network*

For the last field, the penalties are attributed in view of the presence or absence of a traffic light, stop sign or yield sign, but also in view of the type of corner (straight ahead, left turn, right turn, U-turn). As an example, in Figure 9.3, due to the presence of a traffic light at intersection A, the penalty – time required to crossing the intersection – between road segments 1 and 2 is estimated at 0.50 minutes, or 30 seconds. Still at the same intersection, this penalty is raised to 35 seconds (0.58 minutes) between road segment 1 and 9, as it is a question of a right turn here where the possible presence of pedestrians crossing Jarry Street East should slow drivers down. Another example, at intersection B, is that the absence of a mandatory stop sign on Jarry Street East justifies the fact that no penalty will be attributed to the pairs of road sections 9-10 and 10-9. However, the presence of a stop sign on Henri-Julien Avenue in road segment 4 explains the attribution of a penalty of six seconds (0.10 minutes) to go straight ahead or turn left (toward segments 3 and 9), and of 15 seconds (0.25 minutes) to turn left on Jarry Street East (toward segment 10).

9.4. Methodological approach: designing an accessibility indicator

Once the accessibility measures have been computed in the GISs, we will have a table of data at our disposal including the census tracts of the Island of Montreal in rows and the 35 services and facilities selected in columns. The value of each cell in the table corresponds to the shortest route between the census tract i, and the closest service j. The analysis of this table rests on two classical, multivariate analyses that pair a principal component analysis (PCA) with a K-means classification [ESC 98] [EVE 01, LEB 04]. First, the PCA allows us to understand the structure of the table of accessibility measures and to reduce it to a few explanatory factors (components). Second, the K-means operated on the initial factors n of the PCA allows us to obtain a typology of the census tracts in view of their proximity to the selected services and facilities. Finally, based on a classical bivariate analysis (analysis of variance, or anova), it will be possible to show the existing relationships between this indicator and the variable of poverty.

9.5. The findings

9.5.1. *The identification of spaces of poverty*

Several Canadian and US studies have kept the threshold of 40% in order to locate high-concentration areas of urban poverty [JAR 91, KAS 90]. Yet for the census tracts in Montreal we must note that this threshold also allows us to locate 25% of the poorest census tracts of the Island of Montreal, as the values of the 75th and 76th percentiles are raised, respectively, to 39.8% and 40.14%. Consequently, we have kept this threshold in order to identify the spaces where poverty is concentrated on the Island of Montreal. As several authors have already shown [APP 07b, APP 06c, DRO 96, LAN 01, REN 97, SEG 98], the cartography of the percentage of persons with low income in the total population by census tract clearly indicates that the spaces in which poverty is concentrated are especially localized in the center of the Island of Montreal (see Figure 9.4).

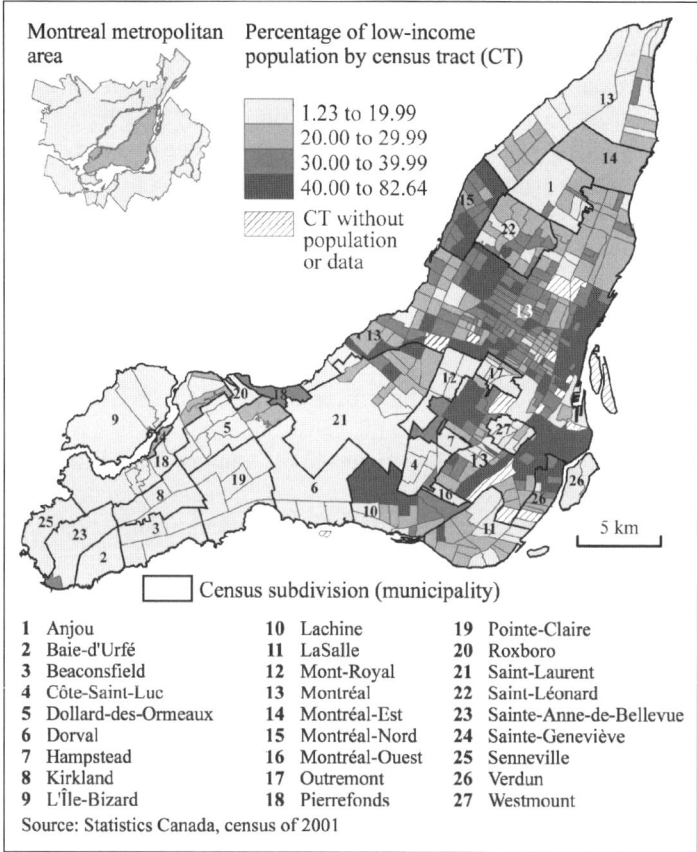

Montreal metropolitan area

Percentage of low-income population by census tract (CT)

1.23 to 19.99
20.00 to 29.99
30.00 to 39.99
40.00 to 82.64

CT without population or data

5 km

Census subdivision (municipality)

1	Anjou	10	Lachine	19	Pointe-Claire
2	Baie-d'Urfé	11	LaSalle	20	Roxboro
3	Beaconsfield	12	Mont-Royal	21	Saint-Laurent
4	Côte-Saint-Luc	13	Montréal	22	Saint-Léonard
5	Dollard-des-Ormeaux	14	Montréal-Est	23	Sainte-Anne-de-Bellevue
6	Dorval	15	Montréal-Nord	24	Sainte-Geneviève
7	Hampstead	16	Montréal-Ouest	25	Senneville
8	Kirkland	17	Outremont	26	Verdun
9	L'Île-Bizard	18	Pierrefonds	27	Westmount

Source: Statistics Canada, census of 2001

Figure 9.4. *Poverty on the Island of Montreal in 2000 [STA 01]*

9.5.2. *The indicator of proximity to urban resources*

Arising from a concern for simplification, we will present here only the final results of the multivariate analyses, which are those of the K-means calculated according to standardized coordinates from the census tracts on the first seven factors of the PCA, which represent 69% of the total variance. The final results of the K-means indicate nine types of proximity to urban resources[4]. Overall, the first four classes – A to D – group together census tracts with a very low proximity to collective services and facilities (see the values of the centers of gravity in Table 9.2). Those census tracts are mainly located in affluent suburban municipalities on the west and the east side of the Island of Montreal (see Figure 9.5). The next two classes – E and F – group together census tracts with a relatively average proximity, similar to that observed for the entire group of census tracts. The census tracts of classes G and I, located especially in the central neighborhoods of the Island, benefit from good proximity to services and facilities; the best situation can be observed for the 135 census tracts in class I (see Table 9.2).

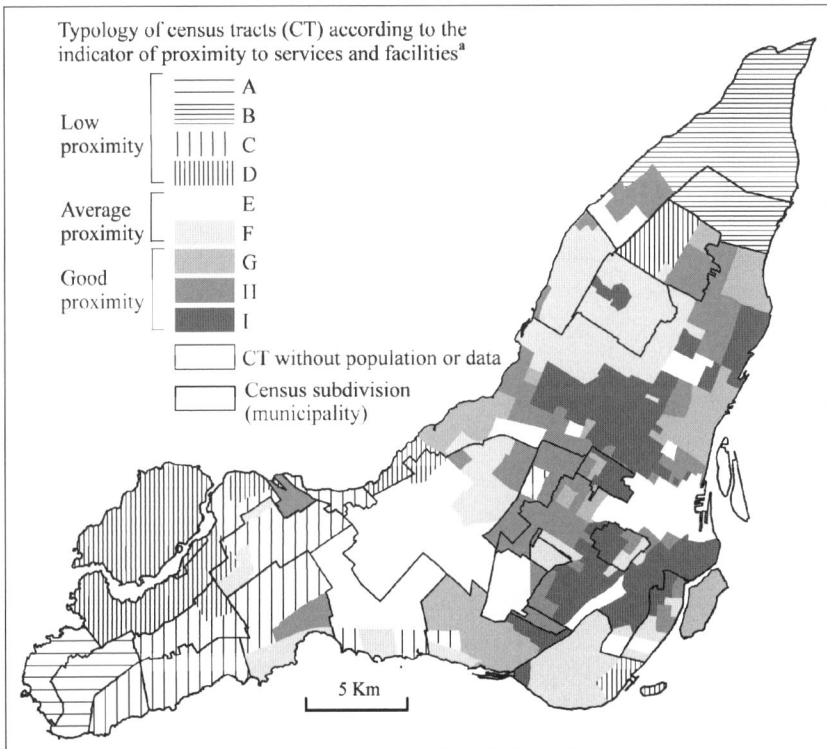

[a] See the centers of gravity of the classes in Table 9.2.

Figure 9.5. *The indicator of proximities to collective services and facilities, Island of Montreal, 2005*

---

4 We must note that we have calculated several classifications for K-means in SAS (Statisitical Analysis Software): from four to 15 classes. We finally opted for the classification with nine classes as it obtained the strongest values of *Pseudo-F* and *cubic clustering criterion* (74.01 and 12.69, respectively).

Class	A	B	C	D	E	F	G	H	I	ALL
Census tracts (N)	3	16	26	23	29	109	71	94	135	506
Municipal library	2239	1865	1896	2195	2091	1692	**1110**	**1378**	**1071**	1451
Movie theater	6321	8568	3506	3165	**1504**	2784	2911	**2181**	**2182**	2714
Cultural center	8175	2306	2405	3177	**1704**	2442	**1588**	**1665**	**1288**	1889
Elementary school (F.)	4197	876	1625	1557	1536	**706**	**501**	**671**	**556**	790
Elementary school (A.)	4373	2672	**1254**	2406	1906	**1080**	2874	**1329**	**1396**	1649
High school (F.)	2154	1892	2588	4826	2661	**1686**	**1297**	**1443**	**1368**	1756
High school (A.)	**2589**	11113	**2811**	3528	**2429**	**1890**	4194	3163	**1963**	2918
Adult ed center (F.)	9615	3339	3232	3041	2383	**1618**	**1304**	**1588**	**1204**	1751
Adult ed center (A.)	9354	12294	4758	**3864**	4854	**2320**	5077	4262	**3032**	3956
Community health center	12563	2459	4677	4075	**1790**	**1906**	**1455**	**1542**	**1277**	1922
Dental clinic	1304	2025	1121	856	**464**	**567**	**498**	**476**	**441**	593
Medical clinic	2383	2192	2252	1575	943	867	657	787	608	912
Large pharmacy	2424	2229	1510	1654	963	**744**	**637**	**741**	**640**	851
Hospital	11417	11684	5483	5315	**1878**	**2448**	**2363**	**2404**	**1914**	2884
Youth center	2579	**1730**	5029	3680	**1976**	2681	**1037**	**1435**	**1432**	1981
Sports center	3537	1860	2462	2901	**1415**	2613	**1158**	**1292**	**1196**	1704
Community parks/gardens	18002	1988	8151	7838	**1319**	**1614**	**885**	**1007**	**966**	1937
Wading pool	1222	1246	1284	1050	**750**	1059	**952**	974	**815**	963
Park (1 ha or more)	491	429	624	460	**337**	**350**	**340**	397	417	397
Park (5 ha or more)	1087	967	2047	988	**845**	**764**	**714**	996	**728**	879
Outdoor rink	1223	774	834	1053	1408	**672**	**640**	777	**635**	752
Indoor rink	3001	2081	2403	3248	2015	1637	**1330**	**1316**	**1283**	1596
Outdoor pool	1887	1486	1415	**1192**	2515	**956**	**1108**	1652	1467	1389
Indoor pool	3001	1888	2941	4401	1860	**1563**	**1416**	**1565**	**969**	1620
Shopping center	3482	**1048**	**994**	**1053**	**1155**	**904**	**905**	**903**	2227	1302
Local employment office	3539	2767	5799	4562	**2631**	**2670**	**2323**	**2370**	**2271**	2712
Postal counter	2374	1667	1829	1627	**864**	1045	**793**	**818**	**793**	984
Daycare	1992	821	1214	1029	**514**	**566**	**530**	**436**	**501**	587
Police station	9660	3272	2813	2583	**1464**	**1408**	**1467**	**1179**	**1082**	1523
Metro station	25255	8741	14281	12407	**2347**	3588	**2099**	**1766**	**1058**	3536
Supermarket	5445	1522	1720	1465	1040	**885**	**672**	**831**	**649**	907
Caisse pop. Desjardins	11223	**1132**	3028	2146	1853	**986**	**700**	**1176**	**705**	1179
Laurentian Bank	10017	2183	2657	1823	1642	**1055**	**992**	**1149**	**1038**	1299
National Bank of Canada	14060	3626	6683	3397	**1351**	**1169**	**1269**	**1231**	**976**	1692
Other banks	1928	2288	1534	2853	**765**	**1009**	1219	**773**	**929**	1116

Numbers in bold indicate the average values of groups that are lower than those observed for all census tracts (ALL column).

Table 9.2. *Average values of the classes of the indicator of proximity to urban resources*

9.5.3. *Poverty and proximity to urban resources: is there any relationship?*

The output from the analysis of variance and the boxplots reported in Figure 9.6 clearly indicate that there is a significant relationship between the level of poverty and the indicator of proximity to services by census tract. However, all things considered, this relationship remains very limited as the value of the R square is only 0.206. Despite that, it is possible to isolate two major elements.

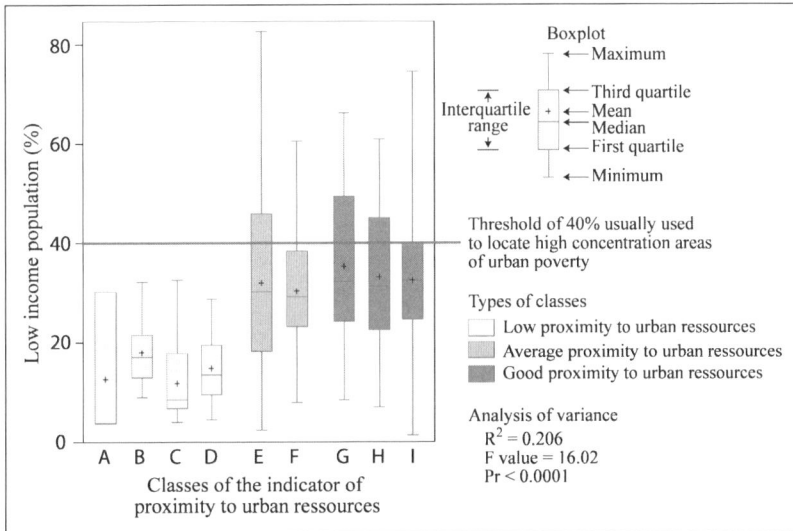

Figure 9.6. *The level of poverty in the census tracts belonging to the various classes of the indicator of proximity to services and facilities*

First, all the census tracts characterized by a very low proximity to services (A to D) had very low poverty rates. These are census tracts of suburban municipalities located in the east and west regions of the Island of Montreal, where very motorized family households reside. Consequently, in those spaces, the low proximity to services does not constitute a real problem for the residing populations.

Second, poverty is more significant in the census tracts that benefit from either average proximity (classes E and F) or good proximity to services (classes G to I). In addition, among the 124 census tracts of poverty concentration (a rate higher than or equal to 40%), 32 (25.8%) benefit from average proximity (E and F) and 92 (74.2%) from good proximity (G to I). There are two possible reasons for this situation. On one hand, the Ville de Montréal has taken care to "properly equip" these less well-to-do neighborhoods with public services and facilities, in particular by building libraries, *maisons de la culture* (cultural centers), numerous parks, municipal and community sports centers, etc. On the other hand, a significant proportion of census tracts with a high concentration of poverty nonetheless keep up a certain social mix [SEG 98]. They are also spaces with high population densities, which contribute to attracting services offered by the private sector.

Lastly, only the 32 census tracts with high concentrations of poverty that benefit from only average proximity to services can be considered as problematic: they are located in the boroughs of Montréal-Nord, Saint-Michel, Center-Sud and Verdun (see Figure 9.5).

Classes of the indicator of proximity to services	Total population		Population above the low-income thresholds		Low-income population	
	N	%	N	%	N	%
Total	**1,808,871**	**100.00**	**1,257,005**	**100.00**	**513,765**	**100.00**
Very low proximity	**300,093**	**16.59**	**252,265**	**20.07**	**42,805**	**8.33**
A	6,032	0.33	4,595	0.37	845	0.16
B	73,316	4.05	58,050	4.62	12,985	2.53
C	127,990	7.08	110,995	8.83	15,910	3.10
D	92,755	5.13	78,625	6.25	13,065	2.54
Average proximity	**541,116**	**29.91**	**370,375**	**29.46**	**161,305**	**31.40**
E	90,182	4.99	59,905	4.77	26,165	5.09
F	450,934	24.93	310,470	24.70	135,140	26.30
High proximity	**967,662**	**53.50**	**634,365**	**50.47**	**309,655**	**60.27**
G	217,247	12.01	139,685	11.11	70,405	13.70
H	356,656	19.72	234,500	18.66	113,395	22.07
I	393,759	21.77	260,180	20.70	125,855	24.50

Table 9.3. *Distribution of various populations according to the classes of indicator of proximity to services and facilities*

9.6. Conclusion

According to a number of authors, the question of accessibility to public and private services and facilities is crucial for social equity and, more particularly, for the poorest populations [SEG 04, TAL 98a, TAL 98b, TRU 93]. In Montreal, our results indicate that the situation is far from dramatic. Quite on the contrary, the sectors with the best proximity to services and facilities have the highest rates of poverty and the inverse is also true. In fact, only 25.8% of the census tracts with a high concentration of poor population – which is to say with percentages of low-income people higher or equal to 40% – benefit from a proximity to services that is judged to be average. All the other poor sectors enjoy good proximity to public and private urban resources. In addition to this, only 8.3% of persons with low income reside in spaces characterized by a very low proximity to services as compared to 16.6% for the entire population (see Table 9.3).

Research work conducted on the *urban underclass* and on neighborhood effects in US metropolises indicates that less affluent central neighborhoods are most often deficient with regard to the level of collective public and private services and facilities [KAS 90, MAS 94, SMA 01]. Montreal thus distinguishes itself on this level from its US counterparts. In order to explain this state of affairs, we can refer to the significant differences that we observe in the terms and conditions of financing of social policies and of collective urban facilities and services [SEG 02].

9.7. Bibliography

[APP 03] APPARICIO P., SHEARMUR R., BROCHU M., DUSSAULT G., "The measure of distance in a social science policy context: Advantages and costs of using network distances in eight Canadian metropolitan areas", *Journal of Geographic Information and Decision Analysis*, vol. 7, no. 2, pp. 105-131, 2003.

[APP 06a] APPARICIO P., SÉGUIN A.M., "Measuring the accessibility of services and facilities for residents of public housing in Montreal", *Urban Studies*, vol. 43, no. 1, pp. 187-211, 2006.

[APP 06b] APPARICIO P., SÉGUIN A.M., "L'accessibilité aux services et aux équipements: un enjeu d'équité pour les personnes âgées résidant en HLM à Montréal", *Cahiers de Géographie du Québec*, vol. 50, no. 139, pp. 23-98, 2006.

[APP 06c] APPARICIO P., SÉGUIN A.M., "L'insertion des HLM montréalaises dans le milieu social environnant", *L'Espace géographique*, vol. 1, pp. 63-85, 2006.

[APP 06d] APPARICIO P., "L'identification et la qualification des espaces de pauvreté à Montréal: quelques pistes de recherche", *Cahiers de Géographie du Québec*, vol. 50, no. 141, pp. 471-487, 2006.

[APP 07a] APPARICIO P., CLOUTIER M.S., SHEARMUR R., "The case of Montreal's missing food deserts: Evaluation of accessibility to food supermarkets", *International Journal of Health Geographics*, vol. 6, no. 4, 13 p., 2007.

[APP 07b] APPARICIO P., SÉGUIN A.M., LELOUP X., "Modélisation spatiale de la pauvreté urbaine à Montréal: Apport méthodologique de la régression géographiquement pondérée", *The Canadian Geographer/Le Géographe canadien*, vol. 51, no. 4, pp. 412-427, 2007.

[CER 99] CERVERO R., "Tracking accessibility: Employment and housing opportunities in the San Francisco Bay area", *Environment and Planning A*, vol. 31, no. 7, pp. 1259-1278, 1999.

[DRO 96] DROUILLY P., *L'Espace Social de Montréal 1951-1991*, Montréal, Les éditions du Septentrion, 1996.

[ESC 98] ESCOFIER B., PAGÈS J., *Analyse Factorielles Simples et Multiples: Objectifs, Méthodes et Interprétation*, Paris, Dunod, 1998.

[EVE 01] EVERITT B., LANDAU S., LEESE M., *Cluster Analysis*, Arnold Publishers, 2001.

[FOT 00] FOTHERINGHAM A.S., BRUNSDON C., CHARLTON M., *Quantitative Geography. Perspectives on Spatial Data Analysis*, London, Sage Publications, 2000.

[HAN 97] HANDY S.L, NIEMEIER D.A., "Measuring accessibility: An exploration of issues and alternatives", *Environment and Planning A*, vol. 29, no. 7, pp. 1175-1194, 1997.

[HEW 02] HEWKO J., SMOYER-TOMIC K.E., HODGSON M.J., "Measuring neighbourhood spatial accessibility to urban amenities: Does aggregation error matter?", *Environment and Planning A*, vol. 34, no. 7, pp. 1185-1206, 2002.

[JAR 91] JARGOWSKY P.A., BANE M.J. "Ghetto poverty in the United States, 1970-1980", in: JENKS C., PETRESON P.E., *The Urban Underclass*, Washington, The Brookings Institution, 1991.

[KAS 90] KASARDA J.D., "Structural factors affecting the location and timing of urban underclass growth", *Urban Geography*, vol. 11, no. 3, pp. 234-264, 1990.

[LAN 01] LANGLOIS A., KITCHEN P., "Identifying and measuring dimensions of urban deprivation in Montreal: An analysis of the 1996 census data", *Urban Studies*, vol. 48, no. 1, pp. 119-139, 2001.

[LAU 95] LAURINI R., THOMPSON D., *Fundamentals of Spatial Information Systems*, London, Academic Press, 1995.

[LEB 04] LEBART L., MORINEAU A., PIRON M., *Statistique Exploratoire Multidimensionnelle*, Paris, Dunod, 2004.

[LEL 05] LELOUP X., APPARICIO P., SÉGUIN A.M., *Le Concept de Relative Déprivation: Survol des Définitions et des Tentatives de Mesure Appliquées à l'Urbain*, Document de recherche, Les Inédits, INRS-Urbanisation, Culture et Société, 2005.

[LIN 01] LINDSEY G., MARAJ M., KUAN S., "Access, equity, and urban greenways: An exploratory investigation", *The Professional Geographer*, vol. 53, no. 3, pp. 332-346, 2001.

[MAS 94] MASSEY D.S., "America's apartheid and the urban underclass", *Social Service Review*, pp. 471-487, 1994.

[MIT 99] MITCHELL A., *The ESRI Guide to GIS Analysis. Volume 1: Geographic Patterns & Relationships*, Redlands, ESRI Press, 1999.

[OTT 94] OTTENSMANN J.R., "Evaluating equity in service delivery in library branches", *Journal of Urban Affairs*, vol. 16, pp. 109-123, 1994.

[PIN 86] PINÇON-CHARLOT M., PRETECEILLE E., RENDU P., *Ségrégation Urbaine, Classes Sociales et Équipements Collectifs en Région Parisienne*, Paris, Éditions Anthropos, 1986.

[PRE 75] PRETECEILLE E., PINÇON M., RENDU P., *Équipements Collectifs, Structure Urbaine et Consommation Sociale, Introduction Théorique et Méthodologique*, Paris, Center of Urban Sociology, 1975.

[PUM 97] PUMAIN D., SAINT-JULIEN T., *L'Analyse Spatiale. Localisation dans l'Espace*, Paris, Armand Colin – Cursus Géographie, 1997.

[REN 97] RENAUD J., CARPENTIER A., LEBEAU R., *Les Grands Voisinages Ethniques dans la Région de Montréal en 1991: une Nouvelle Approche en Écologie Factorielle*. Québec, Collection Études et recherches, Ministère des Relations avec les citoyens et de l'Immigration, 1997.

[SEG 98] SÉGUIN A.M., "Les espaces de pauvreté", in MANZAGOL C., BRYANT C.R. (eds.), *Montréal 2001. Visages et Défis d'une Métropole*, Montréal, University of Québec Press, 1998.

[SMA 01] SMALL M.L., NEWMAN K., "Urban poverty after the truly disadvantaged: The rediscovery of the family, the neighborhood, and the culture", *Annual Review of Sociology*, vol. 27, pp. 23-45, 2001.

[SEG 02] SÉGUIN A.M., DIVAY G., Pauvreté Urbaine: la Promotion de Quartiers Socialement Viables, Rapport de Recherche Réalisé pour le Réseau Canadien d'Analyse des Politiques Publiques/Canadian Policy Research Network, Discussion paper F/27, Family Network, Réseau Canadien d'Analyse des Politiques Publiques/Canadian Policy Research Network 2002.

[SEG 04] SÉGUIN A.M., DIVAY G., "Lutte territorialisée à la pauvreté: examen critique du modèle de revitalisation urbaine intégrée", *Lien Social et Politiques*, vol. 52, pp. 67-79, 2004.

[SMO 04] SMOYER-TOMIC K.E., HEWKO J.N., HODGSON M.J., "Spatial accessibility and equity of playgrounds in Edmonton, Canada", *The Canadian Geographer*, vol. 48, no. 3, pp. 287-302, 2004.

[STA 97] STATISTICS CANADA, Seuils de Faible Revenu, Ottawa, Statistique Canada, 1997.

[STA 01] STATISTICS CANADA, Census of Canada 2001, Ottawa, Statistics Canada, http://www.12.statcan.ca/english/census01/home/Index.cfm, 2001.

[STA 02] STATISTICS CANADA, Dictionnaire du Recensement de 2001, Ottawa, Division des Opérations du Recensement, 2002.

[STR 96] STROBEL P., "De la pauvreté à l'exclusion: société salariale ou société des droits de l'homme ?", *Revue Internationale des Sciences Sociales*, vol. 148, pp. 201-218, 1996.

[TAL 98a] TALEN E., "Visualizing fairness: Equity maps for planners", *Journal of American Planning Association*, vol. 64, no. 1, pp. 22-38, 1998.

[TAL 98b] TALEN E., ANSELIN L., "Assessing spatial equity: An evaluation of measures of accessibility to public playgrounds", *Environment and Planning A*, vol. 30, no. 4, pp. 595-613, 1998.

[THÉ 99] THÉRIAULT M., VANDERSMISSEN M.-H., LEE-GOSSELIN M., LEROUX D., "Modelling commuter trip length and duration within GIS: Application to an O-D survey", *Journal of Geographic Information and Decision Analysis*, vol. 3, no. 1, pp. 40-56, 1999.

[TRU 93] TRUELOVE M., "Measurement of spatial equity", *Environment and Planning C*, vol. 11, no. 1, pp. 19-34, 1993.

[WIT 03] WITTEN K., EXETER D., FIELD A. "The quality of urban environment: Mapping variation in access to community resources", *Urban Studies*, vol. 40, no. 1, pp. 161-177, 2003.

[ZEI 99] ZEILER M., *Modeling our World: The ESRI Guide to Geodatabase Design*, Redlands, ESRI Press, 1999.

Chapter 10

Accessibility of Urban Services: Modeling Socio-spatial Differences and their Impacts on Residential Values

10.1. Introduction

Since its origins, the concept of urbanization has been based on the desire to share and provide common access to public services and facilities. While cities and towns had, for a long time, fulfilled a defensive role, from the beginning of the industrial revolution, agglomeration economies have become the recurring theme of urban growth [POL 05]. With globalization, and despite delocalization facilitated by the internet (e.g. telework), it is still service diversity, offered within agglomerations, that determines their place in the urban hierarchy. For Québec City in Canada, an average size city (seventh largest Canadian agglomeration by population), its historic position as a provincial capital and tertiary service city (government, insurance, education, etc.) has secured an undeniable quality of life. This quality varies greatly, however, over its urban landscape as a function of the structure and age of its various neighborhoods.

This chapter presents the concepts and implementation of original approaches recently employed to measure intra-urban service accessibility variations in the Québec City region. The accessibility indices that were created have been validated in local residential price econometric models [DES 00, THE 03, THE 05, THE 07a,

Chapter written by Marius THÉRIAULT, Marion VOISIN and François DES ROSIERS.

THE 07b] or as forecasting components in the travel behavioral models (see Chapter 3) [BIB 06, BIB 07, BIB 08]. Theoretical approaches range from the gravitational potential model used to measure urban centrality and tailored for monocentric cities, to the theory of "time-geography" proposed by Hägerstrand [HÄG 70], based on the concept of a spatiotemporal prism used in a circumstantial evaluation of accessibility.

On the functional level, the distance or duration of trips are measured as a function of various metrics, which range from a simple calculation of the Euclidian distance to a full simulation of travel routes in a network graph with relatively sophisticated features in terms of optimization of itineraries and impedance constraints (see Chapters 8 and 9). These physical measurements, which are associated with network structures (see Chapter 5), are interpreted in the light of observed mobility behaviors (see Chapters 2, 3 and 4) in order to construct accessibility indices based on the concept of "preferences revealed by action" (Chapter 6) rather than those that are explicitly declared (Chapter 7). The pertinence of revealed preferences is then validated by means of hedonic models (see Chapters 11 and 12) that allow us to assess the price internalization assumption and measure their marginal contribution to the formation of real estate values.

As a result of its history, Québec City presents a complex accessibility pattern. In four centuries it has grown from a colonial capital to an economic metropolis, then a national capital with an industrial framework, to that of a provincial capital specializing mainly in services. The agglomeration of Québec City, after the disappearance of a large number of activities which have marked its history, presents a spatial structure (distribution of activities) that is not very compatible with the classical model of the monocentric city. Instead, it corresponds roughly to the polycentric model often opposed to it (Chapter 1). The activity distribution has an axial structure that was initially based on topography but became consolidated during the last few decades of the 20th Century: e.g. the Saint-Charles River and Charlesbourg axes in the lower town, which provide local services, and the La Cité-Sainte-Foy axis in the upper town which concentrates most of the regional services. Currently, we are assisting in the rapid development of a new market-based city – called Lebourneuf – which profits from an especially dense highway network built during the 1970s in anticipation of demographic growth that never materialized. For this reason, accessibility (or proximity) to the historic center constitutes little more than a minor influence on the formation of real estate values, especially when compared with more sophisticated measurements of accessibility to schools, jobs, shopping, health services, etc. [DES 00].

Over the last 10 years, we have developed an original approach to express these relative accessibilities in order to compare, without the need for simplifying assumptions with regard to the form of the expected effect (whether gravitational or not), the marginal utility of each type of accessibility in a real estate market. This also enables us to weigh the average importance assigned by households to each accessibility component when they are purchasing new residences. Moreover, recent studies concerning the effect of ease of access to public transportation have facilitated the measurement of its impact on residential values in Québec City [DES 06a, DES 06b]. In this North American city the modal share of public transportation was 10% when considering all trips (origin-destination survey) in 2001, as opposed to 16% 10 years earlier.

The remainder of this chapter is divided into four sections. Section 10.2 provides a brief review of the literature to justify, in addition to the concepts already presented in the other chapters, the importance of integrating the social and behavioral dimensions into the construction of accessibility indices. Section 10.3 briefly presents the methodological process used to estimate the local variations in centrality and accessibility to urban services. It also deals with problems that complicate the calculation of additional indices for various categories of services (schools, shopping, hospitals, etc.), a necessary step to properly measure the marginal value of each one. Section 10.4 briefly presents some empirical attempts to model the effects of centrality and accessibility in the setting of residential values for a Québec City market made of detached single-family residences. Auto-regressive (SAR – *spatial auto-regression*) [ANS 88] and space-weighted regression methods [FOT 02] are used to counter the problems of spatial autocorrelation. Finally, the last section reviews the difficulties inherent to this type of process and proposes some additional research topics.

10.2. The perceptual and social components of accessibility

Due to lack of space, we cannot present a review of the literature on GIS and/or spatial analysis in the transportation and mobility domains here. We prefer to direct the reader to some recent works that will help him or her to evaluate the relevance of this domain of expertise [HEN 00, HEN 04, THI 00] which this research belongs to. The accessibility theme already constitutes in itself a vast research subject that we do not pretend to fully cover, redirecting the reader to a particular issue of the *Geographical Systems* review published in 2003 [KWA 03a, KWA 03b]. This review discusses the state of the art of the operational modeling of this phenomenon that has always interested urban studies specialists.

Since the evaluation modalities for distance and network travel times are abundantly discussed elsewhere in this book (Chapters 3, 4, 5, 8 and 9), we will focus here on the social and perceptual dimensions of accessibility that are too often disregarded in urban studies in spite of their crucial importance. In fact, although the distance between locations and travel time constitute phenomena whose physical reality is indubitable (though moderated by transportation technologies), the concept of accessibility also depends on behavior patterns, preferences and social values. It can be shown that a financially-impoverished individual does not have the same perception of accessibility to urban facilities as someone who is more fortunate; a pedestrian does not have the same accessibility as a motorist. The perception of accessibility is a function of the time periods and continents, the rhythm of life and the aspirations of people, as well as transportation technologies that are available.

The perception of accessibility depends on many personal and sociological factors that include individual aspirations, time constraints, the perception of space and time (in public transportation, a 10-minute wait at a station is often perceived as being longer than 10 minutes of travel). Accessibility is the result of an afterthought, a subjective evaluation of a person's ability to control space and time to accomplish things. To model it based only on physical variables, such as time, is to create an abstraction of the relative utility of each project that determines the total time individuals are prepared to spend in order to reach an activity site. It may be acceptable for some to spend six hours crossing the Atlantic for a vacation but nobody would accept that kind of travel as a necessary part of their daily activities. As a result, accessibility is a contemplated compromise (evaluated on a project-by-project basis) between the time required to travel to an activity site and the quality (or diversity that increases satisfaction from a range of choices) of the opportunities that this investment provides (an aspect of agglomeration economies). Unlike transportation simulations, in real life, and, especially when making a critical decision such as a place of residence, travel time is not always minimized. Instead, preference is often given to obtaining a maximum range of choices (satisfaction of needs and preferences), while at the same time considering the travel constraints inherent to this approach. The final decision is, therefore, based on a range of criteria and multidimensional; operational accessibility modeling must include all these aspects.

In addition to the trip geosimulation principles presented elsewhere in this book, our accessibility modeling is based, at the design level, on two fundamental concepts:

1) a behavioral approach which is encapsulated in "time geography" developed by Hägerstrand [HÄG 53]; and

2) the theory of utility [WIE 92].

Kim and Kwan [KIM 03] presented the principles of a spatial analysis algorithm capable of determining the set of accessible destinations to perform a given activity, under the constraint of the travel impedance (graph theory), individual schedule constraints and schedule of operation for targeted destinations (available time intervals, activity duration, windows of operation). It is based on the concept of the spatiotemporal prism developed by Hägerstrand to determine the feasibility of travel, but also discriminates between establishments as a function of the "marginal utility" principle. Longer travel time reduces the available activity completion time (as well as time for other activities) and, as a result, diminishes the utility of the latter. By identifying and prioritizing the overall *opportunity set* of satisfying choices, a much more reliable result can be obtained when compared to the normal approach that merely identifies the closest establishments.

The accessibility indices developed for this study are founded on the same principles. Nevertheless, adaptations were necessary since we wanted to model the set of possible activities even though the spatiotemporal windows are not defined. The allocation method varies from one person to another, but usually follows certain general rules that are associated with personal and family-related constraints. Adults without children have, on average, more time than fathers and mothers, but this does not necessarily mean that they will accept longer travel times when they want to go out to dinner. To verify this, the mobility behavior of these groups can be examined. If it is true, there are elasticity differences that characterize their propensity to travel [JOE 01]. All significant elasticity differences between activity types and people categories provide us with general (sociological) indications of their revealed preferences. These can be analyzed at an individual level by studying their action space (Chapter 6), but they can also be studied in a synthetic manner by defining groups with significantly different behavior, for example men and women (Chapter 2) or consumer groups (Chapter 3).

By determining the relative travel time acceptability thresholds (from total to none), preference functions can be created [JOE 01] to simulate all possible trips in a GIS (on a network with impedances) for a particular location (for example, a residence) and assign an acceptability rating to each service point by using a fuzzy logic function [THE 05]. The sum of the ratings provides information concerning the number of suitable opportunities available from that location. This constitutes a sufficiently substantial basis for creating accessibility indices that define the link between demand (propensity to travel to reach the service as a function of its rarity, mean activity duration and preferences) and supply (distribution of facilities and transportation network structure).

10.3. Centrality, relative and differential accessibilities

Several studies, performed to examine the impact of accessibility on real estate values have proposed a monocentric urban structure with a William Alonso bid-price curve and use the general cost involved in reaching that center (distance, time, travel cost) as an accessibility indicator. In Québec City, the structure is not definitively monocentric, and it is more appropriate to use a gravitational model for the actual facilities to measure the centrality. In a critique of the deficiencies of the monocentric model, Tiefelsdorf [TIE 03] proposes an alternative that evaluates the potential interaction between various locations in order to measure the position of each in terms of the balance between services (supply of opportunities) and demand (population) for services while considering their relative location (distance). We are going to use this index in section 10.3.1 in order to perform a classical measurement of urban centrality. Section 10.3.2 presents some distinguishing components of mobility behaviors in Québec City, whereas section 10.3.3 briefly explains that they lead to travel elasticity thresholds. The following sections (10.3.4 to 10.3.6) briefly discuss the procedures for calculating the urban service accessibility indices. More detailed explanations are provided elsewhere [THE 04, THE 05, THE 07a].

10.3.1. *Centrality as an indicator of interaction potential*

For each of n residential locations, the potential flow (M_i) generated from location i is calculated by considering the set of m available opportunities:

$$M_i = \sum_{j=1}^{m} \mu_{ij} \quad , \quad i \in \{1,...,n\}, \quad j \in \{1,...,m\} \qquad \mu_{ij} = \frac{e^{\lambda} P_i^{\lambda_o^p} O_j^{\lambda_d^o}}{D_{ij}^{\lambda^d}} \qquad [10.1]$$

where:

μ_{ij} is the expected traffic (car) between locations i and j;

P_i represents the total population residing in location i;

O_j is the sum of opportunities (activities or services) provided in location j;

D_{ij} represents the network travel time between i and j (minutes); and

$\lambda, \lambda_o^p, \lambda_d^o, \lambda^d$ are the gravitational model coefficients adjusted using regression on travel data recorded during the OD survey, in accordance with a procedure proposed by Tiefelsdorf [TIE 03] and equation [10.1].

Next, the ratings are redistributed by comparing each residential location to that which obtained the best potential, in order to define a centrality index (M_i^*) whose values have a controlled variation between 100 (best centrality) and 0 (extremely peripheral location – inhabited or disconnected from transportation networks):

$$M_i^* = 100 \left(M_i \Big/ max(M_1,...,M_n) \right) \quad , \quad i \in \{1,...,n\} \qquad [10.2]$$

Figure 10.1 presents the results calculated for Québec City with a complete road network (more than 20,000 directional links with impedance values), the residential locations and the service points being grouped in a matrix composed of hexagonal cells (500 meters per side). This is an essential compromise to reduce the number of simulated trips to reasonable limits. It can clearly be seen that the axial structure is aligned, in part, on the parliament hill between Cap-Rouge and Old-Québec (La Cité) and, in part, on an axis transverse to the Saint-Lawrence River which links Charlesbourg to La Cité. Finally, additional axes are drawn from Val-Bélair to Vanier and South of Beauport.

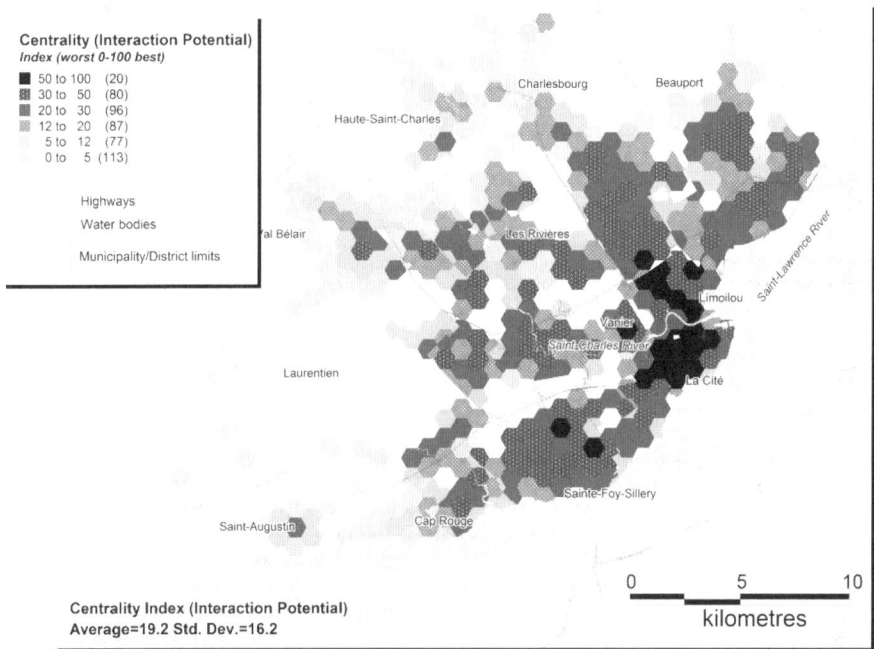

Figure 10.1. *Centrality index of Québec City, 2001*

10.3.2. *Differentiating among mobility behaviors*

In Chapter 2 we observed significant differences between the commuting times for men and women. But what are the differences between the various transportation modes (corresponding to different services used) and between different types of people? Table 10.1 presents some results for Québec City based on calculations made for car trips declared by Québec City residents during the OD survey of 2001. We consider individual-level trips (origin and destination) geo-referenced at the building- and urban block-levels (approximate 100-meter precision), which permits accurate travel distance and duration estimations.

Trip purpose/ *types persons/households*	No. of trips	Average (minutes)	Variance (minutes)	t Student	p
Work					
Women	6,371	10.30	39.09		
Men	6,576	10.76	42.06	4.048	0.000
Mothers	2,796	10.28	38.43		
Fathers	2,716	10.83	41.37	3.259	0.001
School					
Women	1,234	7.93	42.75		
Men	1,152	7.15	38.86	2.956	0.003
Adults (16 and older)	407	9.89	42.52		
Children (15 and younger)	1,979	7.07	39.36	8.196	0.000
Shopping (large chains)					
Women with no children	1,440	6.78	20.91		
Mothers	418	7.37	23.91	2.295	0.022
Men with no children	1,081	6.84	23.16		
Fathers	183	7.92	26.78	2.778	0.006
Households with no children	2,521	6.81	21.87		
Families	601	7.54	24.80	3.414	0.001
Grocery					
Households with no children	2,229	4.52	14.01		
Families	669	5.15	21.08	3.612	0.000
Healthcare					
Adults (16 and older)	859	8.26	35.15		
Children (15 and younger)	84	5.76	15.61	3.781	0.000

Table 10.1. *Differences in travel time by car per trip purpose and type of person/household; OD 2001 Survey, Québec City [THE 04]*

Previous research [THE 04, THE 05] has shown very significant differences between several activities, the travel elasticity being greater for work and greatly diminished for less important activities such as shopping. In addition, significant differences appear between various categories of people, thus translating the variable constraint thresholds as a function of age, household composition and gender. Other significant differences have been brought to light [THE 04]; Table 10.1 shows the most interesting. It is, therefore, not sufficient to model accessibility; it is also necessary to determine which social group is being evaluated and to specify the transportation mode as elasticity, even temporal, can vary greatly from one mode to another. Thus, individuals who do not have cars but can use public transportation are forced to accept higher tolerance thresholds. Their perception of accessibility is different and their action space is greatly reduced, but not necessarily in proportion to the increase in travel time. Everyone has fixed time budgets and, as a result, this leads to inequality. However, for the purpose of this chapter, we restrict our view to car travel.

10.3.3. *Determining elasticity thresholds and opportunities*

Differences in average travel times also translate into differences in distributions of duration (from the lowest to the highest). By ordering travel times, frequency distributions can be created and percentile ranking can be calculated. As mentioned in Chapter 5, frequency distributions follow a gamma-type shape with strong clustering towards the right (skewness > 0) and strong concentration near the origin (kurtosis > 0). For this study, we have arbitrarily chosen the median thresholds of (C_{50}) and 90th percentile (C_{90}) to distinguish:

– fully-satisfactory destinations in terms of travel time ($t \leq C_{50}$);

– facilities too distant to be satisfactory ($t \geq C_{90}$); and

– those that are in an intermediate position ($C_{50} < t < C_{90}$) for which a degree of satisfaction is linearly interpolated, leading to a fuzzy logic assessment.

The principle is simple: opportunities that are more accessible than most of the trips are deemed "satisfactory". An activity located at a distance that has been "accepted" in less than 10% of the actual trips is deemed to be "non-satisfactory".

Purpose	Persons/ households	C_{50} Median (minutes)	C_{90} (minutes)	Skewness	Kurtosis
Work	Women	9.5	18.6	0.868	0.873
Work	Men	9.8	19.3	0.949	1.247
School	Children	4.8	16.1	1.302	1.547
Shopping	Families	6.5	14.1	1.307	1.575
Grocery	Families	3.5	11.2	1.958	5.077
Healthcare	Adults	6.8	16.6	1.197	1.750

Table 10.2. *Satisfaction thresholds for car trips based on purpose and type of persons/household [THE 04]*

Table 10.2 presents the thresholds used for five accessibility indices (for cars) used in this study:

– accessibility to employment (average rating obtained for jobs filled by women and men);

– accessibility to schools using thresholds for children;

– accessibility to shopping for families;

– accessibility to grocery stores for families; and

– accessibility to healthcare for adults.

It is necessary to mention that these social categories were targeted as a function of the real estate market used in section 10.4 to validate the indices, i.e. detached single-family homes that are regularly purchased by families with children, with most benefiting from two incomes.

10.3.4. *Modeling accessibility of urban services*

As a function of the aforementioned considerations, accessibility of a particular location corresponds to the ability that persons living in that area can, within an acceptable travel time, reach a satisfactory diversity of activity locations that are important to them. This definition adds a human dimension (preferences and constraints) into the very definition of the concept of accessibility. It clearly entails that people can exercise choices between several destinations competing for the same activity and that tolerance to travel can also vary as a function of the context. The best accessibility rating corresponds to the site offering access to a maximum of satisfactory choices [THE 07a].

The detailed procedure for calculation of accessibility indices is based on the same geosimulations as that for centrality, to which is associated a summation of satisfactory opportunities for each category of service targeted and each category of person [JOE 01, THE 04, THE 05]. For each of the n residential locations, the sum of satisfactory opportunities is calculated (A_i) while considering the specific opportunities provided (O_j) in the set of m locations where the service exists while including their acceptability (S_{ij}) for a resident in location i. This acceptability is determined independently for each class of service as a function of the satisfaction thresholds presented in Table 10.2. In other words, the median (C_{50}) and the 90th percentile (C_{90}) of the distribution of observed trips, for the given category of persons, during the OD survey:

$$A_i = \sum_{j=1}^{m} S_{ij} O_j \quad , \quad i \in \{1,...,n\}, \quad j \in \{1,...,m\}$$ [10.3]

where $S_{ij} = 1 \ \forall \ D_{ij} \le C_{50}$; $S_{ij} = 1 - \left(\dfrac{D_{ij} - C_{50}}{C_{90} - C_{50}} \right) \ \forall \ C_{50} < D_{ij} < C_{90}$; $S_{ij} = 0 \ \ \forall \ D_{ij} \ge C_{90}$

where D_{ij} represents the travel time by car (in minutes). Next, the rating is reorganized by comparing each residential location to that with the best ranking. This defines an accessibility index (A_i^*) whose values have a theoretical variation between 100 (best accessibility) and 0 (location that has no access to services within an acceptable travel time or less than C_{90}):

$$A_i^* = 100 \left(A_i \Big/ max(A_1, ..., A_n) \right) \ , \ \ i \in \{1, ..., n\}$$
[10.4]

Figures 10.2 and 10.3 present maps of two accessibility indices for Québec City in 2001: access to the job market for adults and access to schools for children. The two maps look similar because they both use the same road network but they differ in detail because the opportunity distributions differ (number of jobs and number of students in school) and the acceptability thresholds (choice range) are also different.

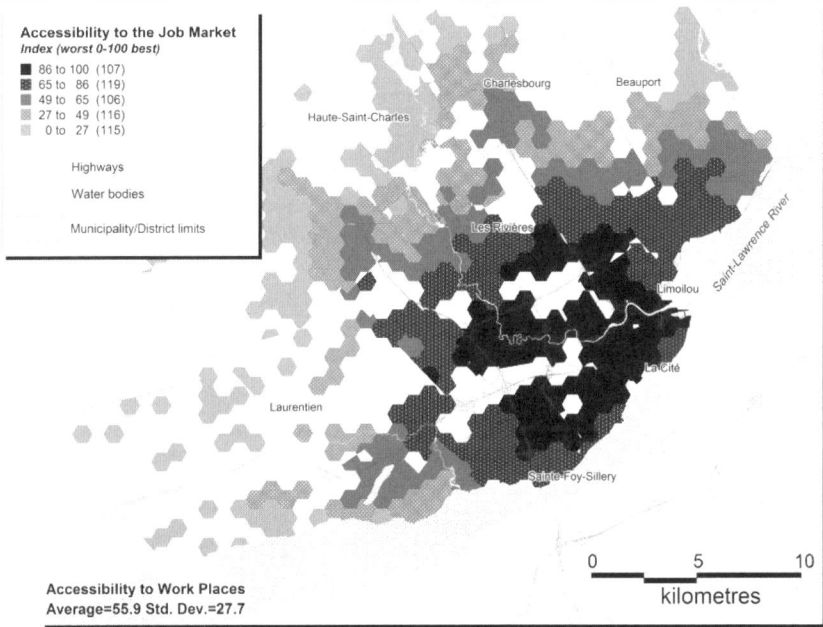

Figure 10.2. *Accessibility indices for the job market, Québec City, 2001*

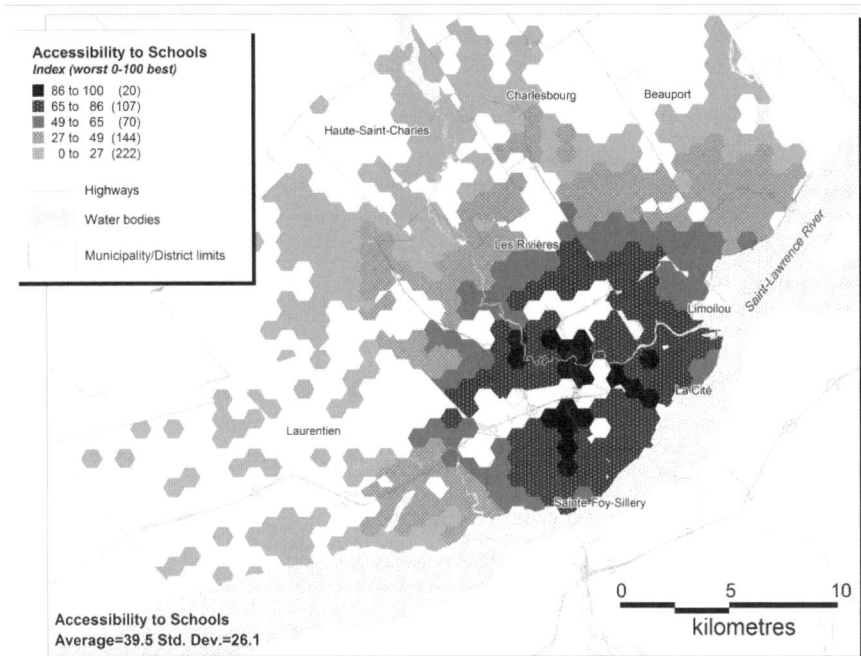

Accessibility to Schools
Index (worst 0-100 best)

■ 86 to 100 (20)
▓ 65 to 86 (107)
■ 49 to 65 (70)
▓ 27 to 49 (144)
□ 0 to 27 (222)

Highways

Water bodies

Municipality/District limits

Accessibility to Schools
Average=39.5 Std. Dev.=26.1

Figure 10.3. *Accessibility indices for schools (children), Québec City, 2001*

10.3.5. *Multicollinearity between indices and accessibility differentials*

The combination of several of these accessibility indices into a single regression model poses a serious multicollinearity problem because the accessibility maps are strongly correlated ($r > 0.8$). We have already shown [THE 05], however, that these indices are significantly different from one another (Wilcoxon test). They therefore, provide additional information to measure the overall accessibility when all types of facilities are combined.

In a subsequent paper [THE 07a], we propose a technique that helps to isolate the specific part of each accessibility index by comparing them, two-by-two, in order to define differential outranking relationships. This technique consists of centering each index in its regional average in order to identify locations with higher (positive) and lower (negative) ratings than the regional average. The differences between two accessibilities are then calculated to determine which outclasses the other by relative advantage.

For each residential location (i), the accessibility differential ($\Delta_i^{\alpha \to \omega}$) is calculated by determining the rating difference between a service class (α) which outclasses ($\to$) another one (ω):

$$\Delta_i^{\alpha \to \omega} = \left[A_i^{*\alpha} - \overline{A}^{*\alpha} \right] - \left[A_i^{*\omega} - \overline{A}^{*\omega} \right] = -\Delta_i^{\omega \to \alpha} \ , i \in \{1,...,n\} \tag{10.5}$$

The initial indices must be centerd in their respective averages ($\overline{A}^{*\alpha} = \frac{1}{n}\sum_{i=1}^{n} A_i^{*\alpha}$ and $\overline{A}^{*\omega} = \frac{1}{n}\sum_{i=1}^{n} A_i^{*\omega}$) to avoid multicollinearity problems.

This precaution removes any structural effects from the differential that are common to the two indices (same road network, common urban format) and only retains that which is unique to each distribution, by conducting a relative comparison (better or worse than the regional average) of the services at each location. It should be noted that this method of evaluating accessibility is relatively natural since households that are looking for homes will compare potential locations against one another (this also applies to services). Thus, accessibility is measured in a contextual manner rather than an absolute one.

As an example, accessibility to major employment locations within 45 minutes would be considered "excellent" in Toronto (a highly-congested city) but not satisfactory in Québec City, where congestion is a rare occurrence. Accessibility to shopping cannot be evaluated in the same manner as that for employment because the travel elasticity is different (see Tables 10.1 and 10.2). What is important in the choice of location is essentially an evaluation of which one is better than the others and, in the case of a conflict between services (shopping versus work versus school), to choose the most acceptable compromise. This leads to the justification of using the outranking concept.

Figure 10.4 presents an example of an accessibility differential in Québec City in 2001. The positive ratings indicate the sectors where accessibility to schools outranks accessibility to employment (Sainte-Foy, Sillery, etc.); while the negative sectors (Beauport center, Loretteville, etc.) have a relatively deficient accessibility to schools which is partially offset by easy access to the job market. These relative accessibility indices are clearly less correlated to one another than the preceding ones [THE 07a] ($|r| < 0.4$). They generate very little multicollinearity when juxtaposed in a regression model [THE 07b] and this allows them to be combined into a single hedonic model. With five accessibility indices, 10 (5 x 4/2) distinct differentials can be created but it can be shown (by analysis of the principal components) that a single accessibility index combined with four, well-chosen, differentials contain all information.

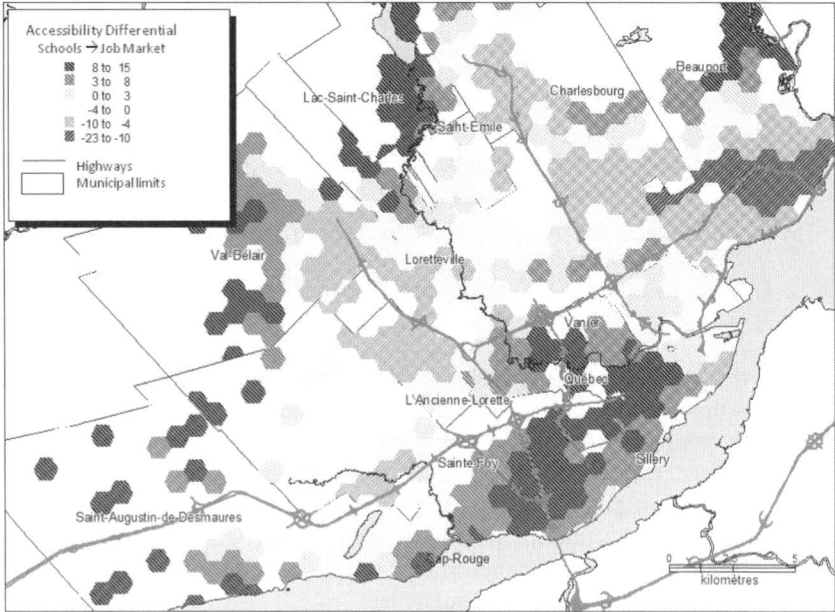

Figure 10.4. *Schools versus job market accessibility differentials, 2001*

10.4. Modeling the impact of accessibility on residential values

This section establishes the relevance of the indices presented in section 10.3 by verifying their internalization in real estate values and by measuring their impact on residential housing prices. The basic reasoning is the following: although centrality and accessibility indices are based on people's behavior, the procedure applied (enumeration of opportunities) to construct them is based on a hypothesis whose plausibility we will prove. This will be done by establishing a link between these indices and the spatial (and marginal) variations of single-family sales prices for houses sold in Québec City during the same period. If our assumption is correct, families who purchase a home should consider accessibility parameters for all urban facilities (employment, schools, shops, etc.) when selecting their residence. As the best locations are more desirable, competition should be higher to acquire them, which, by virtue of the law of supply (inelastic) and demand, would result in a local increase in sales prices.

A database comprising 3,453 detached single-family home transactions (2,495 bungalows – one floor – and 958 cottages – two or more floors) compiled in 1995 and 1996 in Québec City allows us to verify this hypothesis[1]. The database describes each transaction (selling price, date, etc.) and the characteristics of the property (size, construction quality, accessory buildings, etc.). It was drawn from the municipal assessment role that indicates location at the centroid of each lot. Population census data (Statistics Canada) for 1991 by enumeration area are then added in order to consider the socioeconomic characteristics of the surrounding neighborhood. The real estate market remained very stable in Québec City during this period, marked by virtually no appreciation in house prices.

As we shall see in Chapter 11, hedonic modeling uses multiple regression techniques to identify contributive elements of the aggregate price of a composite good. In this case, it is possible to use several regression methods, such as:

– ordinary least-squares (OLS) method, which is not very relevant, as the resulting residuals are highly autocorrelated (see Table 10.3, Moran's I);

– a spatial autoregressive variable applied on the dependent variable (SAR-LAG), which produces more robust estimators by correcting for spatial autocorrelation among sales prices;

– a spatial autoregressive variable applied on error terms (SAR-ERR), which adjusts autocorrelation on the model's residuals; and

– geographically-weighted regression (GWR), which generates estimators for each coefficient in all points of space, thus allowing spatial autocorrelation to express itself directly in the model rather than in error terms.

Table 10.3 presents a summary of the results of the four methods. Akaike's information criterion, based on the theory of information, allows us to compare the performances (it must be minimized) of these four models whose mathematical structure differs slightly (see the following sections). The GWR and SAR-ERR models outrank the two others in terms of overall performance (Akaike, adjusted R^2, standard error of the estimate): spatial autocorrelation (very weak and insignificant Moran's I) and significance (F of marginal contribution and maximum likelihood ratios are significant). Thus, we only use the latter two forms for the following sections. The OLS method is inappropriate because of the scope of spatial autocorrelation in its residuals, and the SAR-LAG model displays similar problems. Overall, the SWR and SAR-ERR account for more than 83% of price variations recorded. The coefficients are presented in Table 10.5.

---

1 Atypical transactions were removed, and only *bona fide* sales with prices between $40,000 and $300,000, lot size of 3,000 square meters or less, and living space of 40 square meters or more were considered.

Method	Adjusted R^2	Error type	Akaike	Moran I	Log maximum likelihood	F / ML
OLS	0.8115	0.140	-3,762	0.0636	1,909	F: 572.82 (27; 3426)
SAR-LAG	0.8247	0.135	-3,837	0.0230	1,946	ML: 73.9 (1)
SAR-ERR	0.8306	0.132	-3,921	-0.0022	1,987	ML: 156.0 (1)
GWR	0.8504	0.124	-4,312	-0.0019	2,183	F: 5.04 (221; 3205)

Table 10.3. *Comparison of OLS (ordinary least squares), SAR-LAG (dependent autoregressive), SAR-ERR (error term autoregressive) and GWR (geographically-weighted) regression models [THE 07b]*

10.4.1. *The spatial autocorrelation issue*

Spatial autocorrelation is a typical issue that needs to be resolved in order to calibrate this type of real estate price model. It is present and significant in most of the attributes and indices considered (including the sales price), which often results in the near impossibility of measuring and integrating all the geographical phenomena that have an influence on price variation. The omission of important factors can be felt in the strongly autocorrelated residuals, which negatively impact the evaluation of coefficients and especially the estimation of their degree of significance (by way of the standard error of the coefficient). Such models may be biased because they might attribute a variation to a given element, through its coefficient, which is related to another element (unmeasured) with an analog spatial structure.

Spatial index	Index Moran's I	GWR coefficients	Mean % impact on residential value
Centrality	0.4217	0.8451	0.6925
Accessibility to the job market	0.8493	0.7746	0.7914
School-job market differential	0.6967	0.9542	0.8215
Job market-health differential	0.7639	0.8555	0.6995
Health-businesses differential	0.7167	0.9530	0.5217
Businesses-grocery differential	0.7629	0.9684	0.4939
Accessibility to the five services	n/a	n/a	0.8245

Table 10.4. *Spatial autocorrelation (Moran's I at 2,500 meter radius) of accessibility indices, GWR coefficients and accessibility impacts; all coefficients are significant (p<0.001) [THE 07b]*

In this example, the spatial autocorrelation between the natural sales price logarithms (dependent variable) is quite significant (I = 0.34, p<0.001). The same can be said for all independent variables [THE 07b] with a maximum for municipal tax rates (I = 0.81, p<0.001), the apparent age of properties (0.28 – in relation to historical urban development) and residence size (0.16 – floor space). The other property-specific attributes display Moran coefficients between 0.01 and 0.12, all significant at the 5% threshold. Table 10.4 presents the measured spatial autocorrelation coefficients in centrality, accessibility and differential indices used for hedonic modeling. Ranging between 0.42 and 0.85, they are all very significant (p<0.001), which justifies the use of SAR and GWR techniques that seek to control the spatial autocorrelation.

10.4.2. *Autoregressive models of externalities derived from accessibility*

Based on econometrics principles [ANS 88], autoregressive models exert explicit control of spatial autocorrelation when evaluating the parameters. The relationships between neighboring observations are incorporated by means of a spatial interaction matrix ($\mathbf{W}$). This matrix has $n \times n$ location pairs that introduce a weighting effect to which a particular coefficient (ρ, λ) is assigned during regression to specify its effect. In other words, we can say that these methods essentially remove "*spatial noise*" from the model to subtract bias from the coefficients. Spatial noise can be modeled either using an independent variable (SAR-LAG; equation [10.6]), or by using an error term (SAR-ERR; equation [10.7]) which establishes control over all the model variables. Using such an approach, we obtain an approximation of the dependent variable as well as ($\mathbf{u}$), a vector with $n \times 1$ error terms distributed in an identical and independent manner.

The SAR-LAG model adjusts a vector with $n \times 1$ coefficients (β) that measure the relationship with an ($n \times k + 1$) matrix with k independent variables ($\mathbf{X}$) to produce an approximation of an ($n \times 1$) vector with values dependent on variable ($\mathbf{y}$). The additional variable has a value of 1 in the entire column and is used to calculate the origin of the regression. A fundamental difference with respect to ordinary regression – the dependent variable is present on both sides of the equation:

$$y = X\beta + \rho Wy + u \qquad [10.6]$$

The SAR-ERR model uses a slightly different expression but is based on the same components:

$$y = X\beta + \lambda W(y - X\beta) + u = X\beta + \lambda Wy - \lambda WX\beta + u \qquad [10.7]$$

The implementation of these models, using actual data, is greatly facilitated by GeoDa software, developed by Luc Anselin's team [ANS 06]. It supplies the resources necessary for the exploratory data analysis as well as implementing many hypothetical tests required to validate the pertinence of the OLS and SAR functional forms. In this example (see Table 10.5), we used a SAR-ERR model that evaluates the relationship between the natural logarithm of the sale price (multiplicative form) and a set of independent variables:

1) the specific attributes of the property (from the Property Size to a Stand-alone Garage);

2) a municipal tax variable (Tax rate);

3) indices of centrality and accessibility; and

4) certain socioeconomic and family composition indices obtained from the census (from Income to Poor households).

The multiplicative form allows us to normalize the price distribution and provides the advantage of evaluating the value contributions in percentage variation per unit of independent variable (some are binary [B] i.e. presence-absence). Here, the independent variable is itself expressed as a logarithm (Property Size, Apparent Age), and an elasticity coefficient is obtained instead (see Chapter 11 for further explanation of the method).

Only the significant variables have been used in these models; the potentially pertinent attributes were distilled using the OLS method. In any case, it is necessary to note that centrality and accessibility indices were used by default. By controlling the effects in an explicit manner, this approach allows us to measure the marginal contribution (*ceteris paribus*) of accessibility as well as the magnitude of its impact on price variations. Thus, in the SAR-ERR model, maximum centrality (100) has the effect of increasing the prices by 9%; whereas maximum accessibility to employment increases it by 14%, with the two effects being cumulative. The differentials also increase the raw indices, but their interpretation is more complex because they are bipolar (negative-positive) and their scales vary over the variable intervals, although they are always much less than |100|. In any case, they are all significant to the 5% significance threshold, except for the healthcare-shopping differential, which is significant to the 10% threshold.

Although this model is practically exempt from spatial autocorrelation when measured at the overall scale of the property (see Table 10.3), it is, however, possible that land values in some neighborhoods are systematically under-estimated (positive residuals) or over-estimated (negative residuals). This gives rise to the concept of "local spatial autocorrelation" which is concentrated in certain hot spots of a spatial autocorrelation that may be linked to some peculiarities (e.g. a nice view of the surroundings). The LISA (*local indicator of spatial autocorrelation*) exploratory analysis method, developed by Anselin [ANS 95], allows us to detect and map these areas of poorer model performance. It was applied to the residuals in the SAR-ERR model and the results are presented in Figure 10.5.

Variable	SAR-ERR model		SWR model		
	Coefficient	z (Sig.)	Median coefficient	Q1 – Q3 coefficients	Spatial drift
Constant	10.5820	122.7 ***	10.4027	10.138–10.607	***
Lot size [Ln(m^2)]	0.0719	8.5 ***	0.0621	0.0382– 0.0853	***
Living space [m^2]	0.0044	30.9 ***	0.0048	0.0040–0.0048	
Living space * Cottage	-0.0007	-10.1 ***	-0.0007	-0.0010–-0.0004	
Apparent age [Ln(years)]	-0.1362	-35.5 ***	-0.1349	-0.1500–-0.1175	***
Municipal services [B]	0.1112	5.3 ***	0.0988	0.0284–0.2033	*
Finished basement [B]	0.0571	10.6 ***	0.0493	0.0451–0.0550	
Number of bathrooms	0.0287	5.6 ***	0.0292	0.0229–0.0360	
Number of fireplaces	0.0363	6.5 ***	0.0357	0.0297–0.0425	
Hardwood stairs [B]	0.0374	5.4 ***	0.0321	0.0229–0.0432	
Quality indicator [B]	0.1064	11.5 ***	0.1020	0.0871–0.1355	**
Outside finishing [B]	0.0328	5.9 ***	0.0283	0.0154–0.0360	
Wood floors [B]	0.0274	5.4 ***	0.0276	0.0203–0.0330	
In-ground pool [B]	0.0665	6.4 ***	0.0646	0.0513–0.0827	
Attached garage [B]	0.0708	7.4 ***	0.0882	0.0551–0.1077	
Stand-alone garage [B]	0.0400	5.9 ***	0.0455	0.0288–0.0629	
Tax rate [$/100$]	-0.0833	-4.9 ***	-0.0402	-0.0894–0.0594	***
Centrality	0.0009	4.0 ***	0.0006	0.0003–0.0011	
Accessibility to job	0.0014	3.7 ***	0.0021	0.0014–0.0026	***
School-job differential	0.0042	2.7 **	0.0029	-0.0024–0.0139	***
Job-health differential	0.0042	2.0 *	0.0028	-0.0013–0.0078	***
Health-businesses differential	0.0027	1.9	-0.0001	-0.0037–0.0074	***
Businesses-grocery differential	0.0022	3.0 **	-0.0001	-0.0033–0.0040	***
Household income [$]	0.0022	8.9 ***	0.0028	0.0016–0.0033	***
Households with preschool child/children [%]	-0.0008	-2.2 *	-0.0001	-0.0007–0.0004	
Households with school-age child/children [%]	0.0010	2.0 *	0.0006	-0.0008–0.0017	*
Poor households [%]	-0.0021	-3.2 **	-0.0015	-0.0036–0.0003	***
λ Autoregressive coefficient	0.9237	44.0 ***			

Significance: * $0.05 \geq p > 0.01$; ** $0.01 \geq p > 0.001$; *** $p \leq 0.001$

Table 10.5. *Autoregressive (SAR-ERR) and GWR models of the impact of marginal accessibility on residential values; Dependent variable: natural logarithm of sales price ($) [THE 07b]*

In addition to not producing significant overall spatial autocorrelation, a good regression model must be distinguished by a minimum number of sites where the local spatial autocorrelation is significant. In Figure 10.5, the problematic sites are fewer in number (2,958 transactions are located in sites where the local spatial autocorrelation is not significant). Areas of under-evaluation can still be identified, however, especially in some neighborhoods of Sillery, Sainte-Foy and Cap-Rouge, while the model over-estimates the values of some sectors of Saint-Émile and Québec City (Duberger).

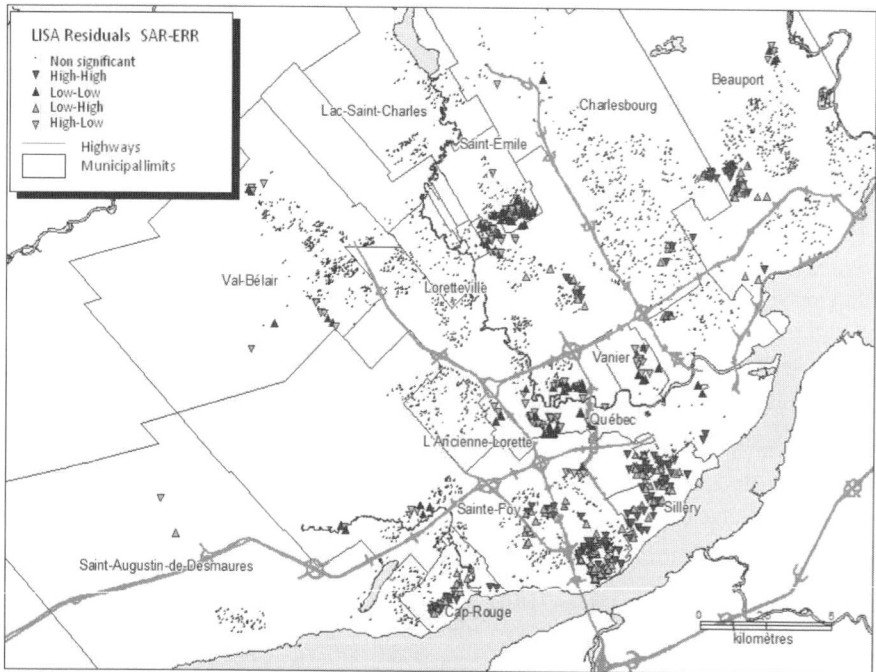

Figure 10.5. *Local spatial autocorrelation indicators (LISA) for residuals, autoregressive model in the error terms [THE 07b]*

10.4.3. *The issues of spatial drift on coefficients*

It might be acceptable for an economist to consider the space as noise in order to extract unbiased coefficients, but it is much less acceptable for a geographer to do so. In fact, the SAR methods postulate that there is a unique coefficient – applicable uniformly over the entire region – that is capable of modeling the entire phenomenon, meaning there is a single uniform process in action. If this reasoning is applied to the accessibility effects, it raises the question as to whether residents in neighborhoods with poor accessibility assign the same marginal value to this amenity as residents who have chosen to live in places with high centrality/accessibility.

More precisely is it not reasonable to consider that, in order to access property, less fortunate households must accept a compromise between the type of residence that they can afford (land size, conveniences) and better accessibility? If this is the case, there would be geographic factors in the accessibility evaluation coefficients that would vary as a function of buyers' sociologic characteristics (this effect is not seen in the current model). This means that the marginal price paid for accessibility and the choice of residential neighborhood are interdependent, such that buyers who choose less interesting locations value this attribute less (or are forced to do so). If such is the case, a significant spatial drift in the evaluation phenomena would be observed and the unique constraint coefficient (SAR) would be inadequate. In technical terms, it can be said that the phenomenon is non-stationary, whereas the autoregressive methods are based on the postulate that the effects are stationary.

GWR methods have recently been developed in order to model non-stationary phenomena. While there are many of them, the most well-known were proposed by Brunsdon, Fotheringham and Charlton [BRU 96, FOT 02]. A procedure that uses Monte Carlo simulations was then developed by Leung [LEU 00] in order to verify the existence of *spatial drift* and was incorporated into the GWR 3.0 software created by Fotheringham *et al.* [FOT 02].

The last column of Table 10.5 identifies the independent variables in our model that present a significant spatial drift (non-stationary effects). It is interesting to note that all the accessibility indices have non-stationary coefficients, which means that the intensity of the effect varies in space. A few other independent variables are also non-stationary. Furthermore, it is noteworthy that these are, in general, phenomena that are highly structured in space: property size – larger in the outskirts, smaller downtown; apparent age of the structures – including renovations; the general quality of the buildings – a socioeconomic indicator; tax rates – specific to each municipality; the household incomes and the proportion of poorer households. The spatial structure of all these phenomena is abundantly documented in the literature and it appears to be relatively unreasonable to model their effect using a single coefficient.

For example, in the case of land value, this means that the marginal value of one square meter of land should be similar to the downtown area (where the land is rare and highly-coveted) and in the remote suburbs (where space is relatively abundant and, consequently, cheaper); the economic theory has been teaching us the contrary for centuries (bid curve). This difference is evidently incorporated into the origin of the regression. With a SAR model the origin is constant over the complete region, while in a GWR model it varies over the region and, as expected, is non-stationary.

10.4.4. *The GWR model*

While the calculations required to establish them are relatively laborious, the principle of GWR models is relatively simple. After determining an optimal search radius to select observations, the local regression model is adjusted around each observation point by selecting the points that are located inside the latter and by weighting the effects using an inverse function of the Euclidian distance. The weighting depends on the nature of the regression being performed. In this example, we created a Gaussian model with a fixed search radius (*bandwidth*) of 2,385 meters. A distance, obtained by iteration, was used to optimise the Akaike criterion [FOT 02]. At the location of each house sold in the region, a regression is performed which, in the spatial drift case, yields maps of coefficients variation (see Figures 10.6 and 10.7).

The model specification takes the following form:

$$y_{(g)} = X\beta_{(g)} + \varepsilon \quad \text{and} \quad \hat{\beta}_{(g)} = (X^t W_{(g)} X)^{-1} X^t W_{(g)} y \quad g \in \{1..n\} \qquad [10.8]$$

where $\mathbf{y}_{(g)}$ is the local vector of observations in the search radius around point g; $\beta_{(g)}$ is the vector of the coefficients estimated with the observations; and $\mathbf{W}_{(g)}$ is the distance weighting matrix adjusted with the points. By making successive adjustments around each of the n points (sold properties), we obtain, for each independent variable, a map of coefficients for the entire region (see Figure 10.6).

It is possible that some variables have a significant effect on the overall model, even though they are not significant locally ($t < |1.96|$); this identifies areas of the city where the relationship is clearly less important. The black points in Figures 10.6 and 10.7 identify the places where accessibility to employment and the school→ employment differential become significant. It can clearly be seen that the coefficients vary in space and the effects are cumulative. Thus, sectors of Cap-Rouge, Sainte-Foy and Saint-Augustin, where access to employment is not locally determinative (accessibility is excellent but does not make a difference), are greatly favored by the ranking of schools over employment. This reflects a characteristic of the neighborhood but probably also the priority given to desirability of services by home buyers in these areas. In contrast, residences in Saint-Émile are greatly penalized by low accessibility to educational services.

Figure 10.6. *Geographically-weighted coefficients – accessibility index to employment*

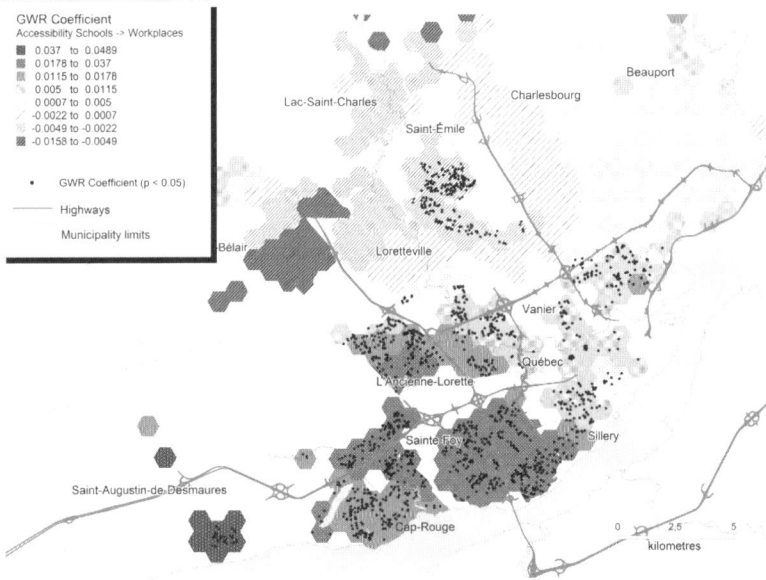

Figure 10.7. *Geographically-weighted coefficients – differential accessibility value assigned to schools versus employment*

Since these local coefficients are acquired for the complete set of accessibility indices, it is possible, by local application of the GWR model and by interaction with the values of each index, to calculate its marginal contribution to the real estate values. Then, by an accumulation of effects, it is possible to determine the local accessibility internalisation value for the set of five services in the residential prices (see Figure 10.8). Not surprisingly, the emerging pattern quite accurately mirrors the bid-rent curve that prevails in Québec City. It is also possible to measure the improvement of the local adjustment by using GWR instead of the SAR-ERR model by comparing its LISA map (see Figure 10.9) with the previous one (see Figure 10.5). In the GWR model, there are only 149 transactions in the significant local spatial autocorrelation areas as opposed to 495 in the SAR-ERR model.

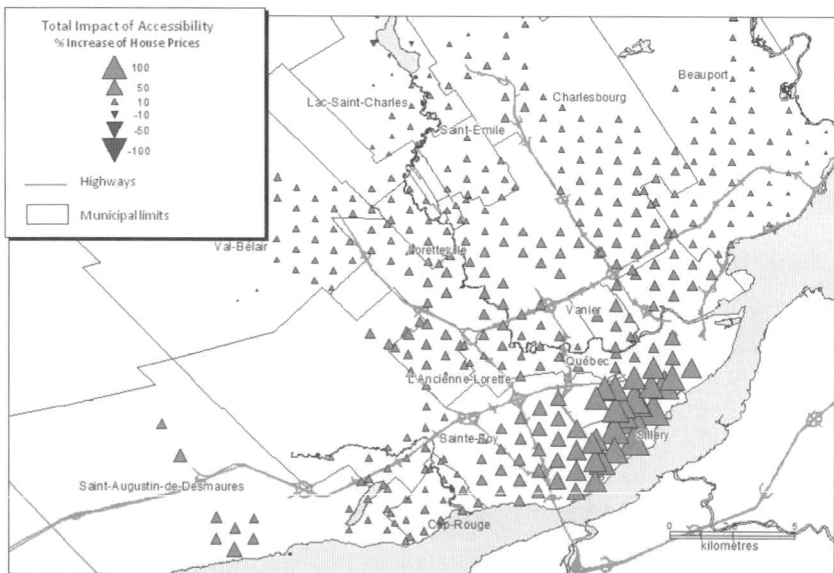

Figure 10.8. *Geographically-weighted cumulative contribution of accessibility to employment, schools, businesses, groceries and health care services in residential values*

According to the LISA, the only sector with a notable bias is located in the Duberger sector, where properties are overvalued by the model. Surrounded by four autoroutes, this neighborhood has excellent road access, but includes very low pedestrian access and, bordered for the most part by the St-Charles River, has been repeatedly flooded over the past three decades. These two factors are not taken into consideration in the current model. However, such a result illustrates the advantage of exploratory analysis in generating new hypotheses in order to progressively refine the model's specification.

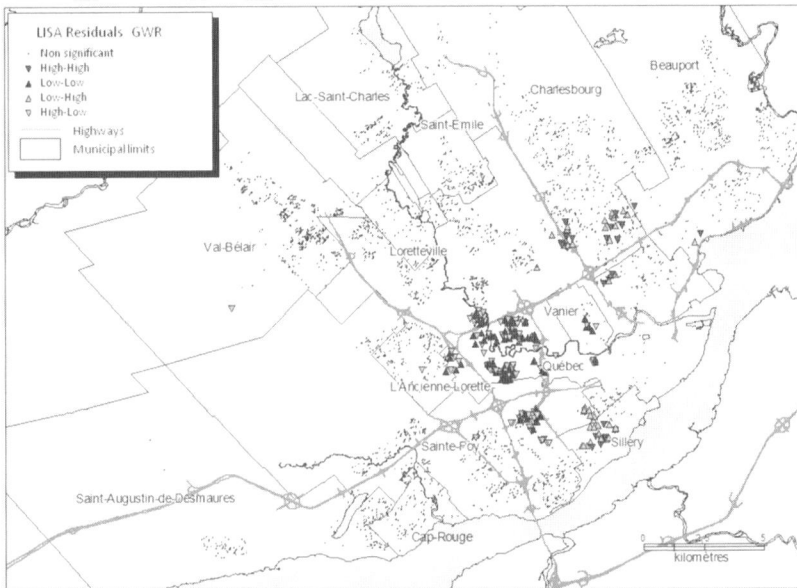

Figure 10.9. *Local indicators of spatial autocorrelation (LISA) of residuals, GWR model*

Table 10.6 presents a synthesis of the marginal contributions of centrality, job accessibility and overall accessibility to the five services (jobs, schools, businesses, healthcare and grocery). We can see major differences between SAR and OLS methods on one hand, and GWR versus SAR on the other hand. Being stationary, the effect of centrality is evaluated the same way by the four models: on average, from 2–3% increase in value. The most preferred neighborhoods score from 10–15%. Evaluations are much more contrasted for accessibility, with a maximum of 15% for those sectors having the best access to jobs, according to SAR-ERR, compared with a 100% increase in value according to GWR.

The two methods display an increase in value when accessibility points are cumulated, but in clearly lower proportions: a 15–20% increase (five services) according to SAR-ERR; from 100–190% according to GWR. Comparing residences of similar quality, the real estate value would be approximately three times more in neighborhoods providing the best accessibility in comparison with a hypothetical location that offered no accessibility. The difference in perspective between GWR and SAR is important. But are the GWR results robust? Why are the impact distributions practically Gaussian with the OLS and SAR methods, but skewed to the right and very leptokurtic with GWR? Which approach provides the most valid result? We will now attempt to answer these questions.

	Min (%)	Max (%)	Average (%)	Standard deviation (%)	Skewness	Kurtosis
Centrality						
OLS	0.0	15.0	3.6	1.9	0.6	0.0
SAR-LAG	0.0	13.8	3.3	1.8	0.6	0.0
SAR-ERR	0.0	9.4	2.3	1.2	0.6	0.0
GWR	-1.1	11.9	2.1	2.3	1.3	-1.1
Accessibility to jobs						
OLS	1.1	32.3	18.8	7.5	-0.1	1.1
SAR-LAG	0.6	15.0	9.0	3.4	-0.2	0.6
SAR-ERR	0.6	15.0	9.0	3.4	-0.2	0.6
GWR	-3.4	101.3	14.2	13.7	3.2	-3.4
Accessibility to the five services						
OLS	4.9	39.2	18.9	8.2	0.8	4.9
SAR-LAG	3.6	18.8	9.0	3.8	0.9	3.6
SAR-ERR	0.6	20.2	9.0	5.4	0.3	0.6
GWR	-18.7	190.1	19.4	23.7	3.9	-18.7

Table 10.6. *Marginal impact of centrality and accessibility on sales prices of single-family residences in Québec City (Value growth rate in %) [THE 07b]*

10.4.5. *The multicollinearity between regression parameters*

Recent literature already provides several answers to the questions posed above. Experiments with controlled data have shed light on problems of multicollinearity among local parameters generated by GWR [WHE 04, WHE 07]. In fact, the problem stems from the systematic reuse of the same measurement points during the estimation of $\beta_{(g)}$ at two neighboring localities, which produces highly correlated coefficients and generates a very high multicollinearity between estimators.

We conducted an experiment with the results of our model in order to test the sensitivity of the GWR method to sampling variations. We chose a random sampling of 50% of the transactions, ensuring that it was equally stratified in a cell grid to avoid creating a spatial sampling effect. We then built a GWR model on this base. The observations excluded from the first sample (complement) were then used to calibrate a second GWR model. By combining the coefficients from the two GWR models we obtain what we call the twin sample #1. We then repeat the same modeling procedure (two complementary GWR models) so as to obtain the twin sample #2. With this procedure, the coverage rate for the same neighboring observations is greatly reduced and the two twin samples are relatively different.

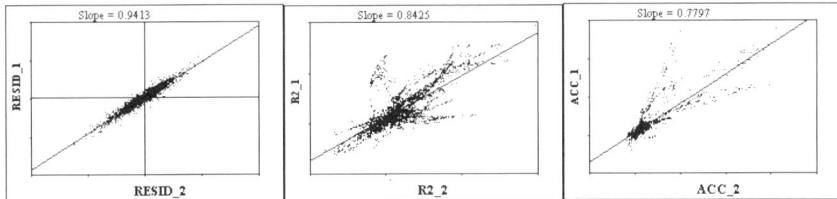

RESID_1: Residuals of twin sample #1; RESID_2: Residuals of twin sample #2
R2_1: Local R^2 of twin sample #1; R2_2: Local R^2 of twin sample #2
Acc_1: Impact of accessibility to the five services (%) with twin sample #1 (min 31, max 95)
Acc_2: Impact of accessibility to the five services (%) with twin sample #2 (min 36, max 102)

Figure 10.10. *Sampling sensitivity analysis in the*
space-weighted regression model [THE 07b]

Are the results of the GWR models comparable? Findings are reported in Figure 10.10. The residuals of the two sets of models conform perfectly among themselves and with the original model [THE 07b]. However, the situation is more problematic when we compare the local R^2 and the impacts of accessibility. Nonconformities are evident and we can even distinguish subsamples in the figure to the right. Maximum impact evaluations that were at 190% in the original model are now no more than 95% and 102% in the twin sample models. As there was very little multicollinearity in the initial accessibility indices, this artifact would appear to stem from the method itself. Obtained from different bases, this result confirms that of Wheeler and Tiefelsdorf [WHE 04].

10.5. Conclusion

Although quite sophisticated in terms of methodology, our application example ends on a slight impasse. We know that the internalization of accessibility in real estate values is usually greater than that of centrality, as defined by an interaction potential. We have also shown that accessibility defined based on the sum of satisfactory opportunities allows us to measure complementary (and significant) impacts to those of centrality. The accessibility differential method also allows us to cumulate effects without generating multicollinearity among indices (measured with variance inflation factor coefficients), thus allowing the comparison of specific impacts of various classes of urban services. Lastly, we were able to determine that, while robust, the SAR methods, because they impose global coefficients to measure an effect we know is non-stationary, are too rigid to allow us to measure the local impact of accessibility.

Despite the problem of the multicollinearity of estimators, GWR approaches could allow us to measure this type of impact, but they will have to be perfected in order to remove this undesirable effect and remain independent of sampling. Several approaches are possible. For example, a bootstrapping procedure during the evaluation of the local coefficients could improve the situation. What the price would be in terms of calculation time remains to be determined.

10.6. Acknowledgements

This research was funded by the Canadian Social Science and Humanities Research Council (SSHRC), the Canadian Natural Sciences and Engineering Research Council (NSERC) and the Canadian Centre of Networks of Excellence in Geomatics (GEOIDE). Data were obtained thanks to agreements with Québec province's ministry of Transportation, the *Réseau de Transport de la Capitale* and the Québec Urban Community. The authors are grateful to more than 30 collaborators and students who participated at various stages of improvement to the datasets used for this research.

10.7. Bibliography

[ANS 88] ANSELIN L., *Spatial Econometrics: Methods and Models*, Dordrecht, Kluwer, 1988.

[ANS 95] ANSELIN L., "Local indicator of spatial association – LISA", *Geographial Analysis*, vol. 27, pp. 95-115, 1995.

[ANS 06] ANSELIN L., SYABRI I., KHO Y., "GeoDa: An introduction to spatial data analysis", *Geographical Analysis*, vol. 38, pp. 5-22, 2006.

[BIB 06] BIBA G., DES ROSIERS F., THÉRIAULT M., VILLENEUVE P., "Big boxes versus traditional shopping centers: Looking at households' shopping trip patterns", *Journal of Real Estate Literature*, vol. 14, no. 2, pp. 177-204, 2006.

[BIB 07] BIBA G., THÉRIAULT M., DES ROSIERS F., "Analyse des aires de marché du commerce de détail à Québec: Une méthodologie combinant une enquête de mobilité et un système d'information géographique", *Cybergeo, Revue Européenne de Géographie*, article 382, http://www.cybergeo.eu/index2870.html, 2007.

[BIB 08] BIBA G., THÉRIAULT M., VILLENEUVE P., DES ROSIERS F. "Aires de marché et choix des destinations de consommation pour les achats réalisés au cours de la semaine. Le cas de la région de Québec", *Le Géographe Canadien*, vol. 52, no. 1, pp. 39-64, 2008.

[BRU 96] BRUNSDON C.F., FOTHERINGHAM A.S., CHARLTON M.E., "Geographically weighted regression: A method for exploring spatial non-stationarity", *Geographical Analysis*, vol. 28, pp. 281-298, 1996.

[DES 00] DES ROSIERS F., THÉRIAULT M., VILLENEUVE P., "Sorting out access and neighborhood factors in hedonic price modeling", *Journal of Property Investment & Finance*, vol. 18, no. 3, pp. 291-315, 2000.

[DES 06a] DES ROSIERS F., THÉRIAULT M., DUBÉ J., VOISIN M., *Does the overall quality in the supply of an urban bus service affect house prices? – A North-American case study*, 13th ERES Annual Conference, Weimar, Germany, June 2006.

[DES 06b] DES ROSIERS F., THÉRIAULT M., DIB P., DUBÉ J., *Public Transit Improvement and Property Values – A Canadian Case Study*, 22nd ARES Annual Meeting, Key West, Florida, April 2006.

[FOT 02] FOTHERINGHAM A.S., BRUNSDON C., CHARLTON M.E., *Geographically Weighted Regression: The Analysis of Spatially Varying Relationships*, Chichester, Wiley, 2002.

[HÄG 53] HÄGERSTRAND R.T., *Innovation Diffusion as a Spatial Process*, Chicago, University Press, 1953.

[HÄG 70] HÄGERSTRAND R.T., "What about people in regional science?", *Papers of the Regional Science Association*, vol. 4, no. 1, pp. 7-21, 1970.

[HEN 00] HENSHER D.A., BUTTON K.J. (Eds.), *Handbook of Transport Modelling*, Amsterdam, Pergamon, 2000.

[HEN 04] HENSHER D.A., BUTTON K.J., HAYNES K.E., STOPHER P.R. (Eds.), *Handbook of Transport Geography and Spatial Systems*, Amsterdam, Elsevier, 2004.

[JOE 01] JOERIN F., THÉRIAULT M., VILLENEUVE P., BÉGIN F., "Une procédure multicritère pour évaluer l'accessibilité aux lieux d'activité", *Revue Internationale de Géomatique*, vol. 11, no. 1, pp. 69-104, 2001.

[KIM 03] KIM H.M., KWAN M.P., "Space-time accessibility measures: A geocomputational algorithm with a focus on the feasible opportunity set and possible activity duration", *Geographical Systems*, vol. 5, no. 1, pp. 71-91, 2003.

[KWA 03a] KWAN M.P., JANELLE D.G., GOODCHILD M.F., "Accessibility in space and time: A theme in spatially integrated social science", *Geographical Systems*, vol. 5, no. 1, pp. 1-3, 2003.

[KWA 03b] KWAN M.P., MURRAY A.T., O'KELLY M.E., TIEFELSDORF M., "Recent advances in accessibility research: Representation, methodology and applications", *Geographical Systems*, vol. 5, no. 1, pp. 129-138, 2003.

[LEU 00] LEUNG Y., MEI C.L., ZHANG W.X., "Statistical tests for spatial nonstationarity based on the geographically weighted regression model", *Environment and Planning A*, vol. 32, pp. 9-32, 2000.

[POL 05] POLÈSE M., SHEARMUR R., *Économie Urbaine et Régionale: Introduction à la Géographie Économique*, 2nd edition, Paris, Economica, 2005.

[THÉ 03] THÉRIAULT M., DES ROSIERS F., VILLENEUVE P., KESTENS Y., "Modeling interactions of location with specific value of housing attributes", *Journal of Property Management*, vol. 21, no. 1, pp. 25-62, 2003.

[THÉ 04] THÉRIAULT M., DES ROSIERS F., "Modeling perceived accessibility to urban amenities using fuzzy logic, transportation GIS and origin-destination surveys", in: F. TOPPEN, P. PRASTACOS (eds.), *Proceedings of AGILE 2004 7th Conference on Geographic Information Science*, Heraklion Crete University Press, pp. 475-485, 2004.

[THÉ 05] THÉRIAULT M., DES ROSIERS F., JOERIN F., "Modelling accessibility to urban services using fuzzy logic: A comparative analysis of two methods", *Journal of Property Investment and Finance*, vol. 23, no. 1, pp. 22-54, 2005.

[THÉ 07a] THÉRIAULT M., DES ROSIERS F., DUBÉ J., "Testing the temporal stability of accessibility values in residential hedonic prices", *Scienze Regionali, Italian Journal of Regional Science*, vol. 6, no. 3, pp. 5-46, 2007.

[THÉ 07b] THÉRIAULT M., DES ROSIERS F., VOISIN M., "Assessing the marginal value of accessibility to urban amenities. Getting rid of spatial drift", *Proceedings of the European Regional Science Association Annual Conference*, Paris, 2007.

[THI 00] THILL J.C. (ed.), *Geographic Information Systems in Transportation Research*, Amsterdam, Pergamon, 2000.

[TIE 03] TIEFELSDORF M., "Misspecification in interaction model distance decay relations: A spatial structure effect", *Geographical Systems*, vol. 5, pp. 25-50, 2003.

[WHE 04] WHEELER D.C., TIEFELSDORF M., "Multicollinearity and correlation among local regression coefficients in geographically weighted regression", *Geographical Systems*, vol. 7, pp. 161-187, 2004.

[WHE 07] WHEELER D.C., CALDER C.A., "An assessment of coefficient accuracy in linear regression models with spatially varying coefficients", *Geographical Systems*, vol. 9, no. 2, pp. 145-166, 2007.

[WIE 92] VON WIESER F., "The theory of value", *Annals of the American Academy of Political and Social Science*, vol. 1891-1892, pp. 600-628, 1892.

Chapter 11

Hedonic Price Modeling: Measuring Urban Externalities in Québec

11.1. Introduction

One of the main difficulties that arise in estimating the market value of a real estate asset, in particular residential real estate, is in deciding how to establish the marginal value of each of its components. On one hand, some are "intrinsic" to the asset, and are comprised of structural attributes (such as type and architectural style, square meters, age, number of rooms, bathrooms, with or without certain extras, etc.), property characteristics (such as lot size, layout, quality of landscaping) and appurtenances on the property (such as swimming pools, garages, patios). On the other hand, there are "extrinsic" components that come from both the location of the property in an urban or regional space (proximity to, and accessibility of, activities and services) and neighborhood characteristics (such as the socioeconomic profiles of neighborhood residents, environmental quality, level of quietness and security).

But, why bother with this issue since market mechanisms do a good job in ascribing value to each individual real estate asset through the interaction of supply and demand, with value being defined as the "most likely price" that would be paid for the asset in a pure and perfect competition market context?[1] The reason is, firstly, that it is often necessary to reconstitute the market value of real property that

Chapter written by François DES ROSIERS, Jean DUBÉ and Marius THÉRIAULT.
1 Remember that *pure competition* implies that there are sufficient numbers of both buyers and sellers such that, individually, none may influence the price whereas *perfect competition* refers to the transparency of the market regarding the attributes of the transacted good, which are known to both parties.

has not been recently bought or sold on the basis of the quantity and nature of its attributes through the implicit prices obtained by statistical inference from a representative sample of local transactions. This is common practice in real estate appraisals used for property taxation purposes at the local level, in particular, in English-speaking countries where this type of expertise is widespread. Also, in a broader context, this occurs because a better understanding of urban and spatial dynamics and the behavior of households regarding mobility and choice of a home require having in-depth knowledge of the factors determining real estate values. This includes the willingness to pay the price economic agents assign to various attributes of the residential property. This knowledge is rather complex or heterogeneous.

Now, if the intrinsic characteristics of a home often go a long way to explaining some 60–70% of a sale price, the rate at which extrinsic characteristics contribute to that value tends to increase over time. This is particularly noticeable as the property becomes more centrally located, until location comes to represent more than half its market value, for example when located in central areas of a metropolitan region. This shows the heterogeneous nature of any real estate asset and, more specifically, of primary residences that are subject to a myriad of outside influences. These external influences, better known as "externalities", may have a positive effect (positive externality) or a negative one (negative externality) on prices depending on how these effects are perceived by users of the asset and utility – or degree of satisfaction – they generate. It is precisely this "marginal utility" of various attributes – both intrinsic and extrinsic – of a real estate asset that the hedonic approach makes it possible to capture, which is the subject of this chapter.

Since it was first applied to the automotive industry almost seven decades ago, [COU 39], using econometric modeling of hedonic prices – called the hedonic approach – has become the analytical framework of choice for estimating the implicit price of attributes for heterogeneous goods. Although this method is applicable to a large number of phenomena, for economists the residential market has fast become one of the prime fields for applying this method due to the very complex nature of a residential property.

It was, however, the computer processing revolution of the 1970s, with the advent of the microcomputer and user-friendly statistics software, which allowed the hedonic approach to take off. This first occurred in the US and then in the rest of the world. It is especially noticeable in places where property values were the basis of local financing systems. Its popularity derives from the robustness of the theoretical corpus underlying it, which combines, through *multiple linear regression analysis*, the properties of differential calculus with the potential of probability theory and statistical inference. So deriving hedonic or implicit prices through econometric modeling brings us directly back to what is essentially a statistical concept of market value as defined above. Furthermore, such an approach makes it possible to go beyond simple appraisal and explore the causal dimensions of value, the most significant determinants of which, as well as their marginal contribution to the phenomenon, may then be identified.

Despite the undeniable benefits of a hedonic approach based on the regression procedure called *ordinary least squares* (OLS). It requires that an overall rather restrictive series of conditions be adhered to:

– the relation between the dependent (endogenous) variable and the explanatory (exogenous) variables is linear in nature in the parameters (although various transformations may be applied to data so as to account for nonlinearities in the variables);

– the exogenous variables are independent of one another;

– a corollary to the previous condition is that the terms of the regression equation are considered to be additive, which implies that the marginal contribution of a given residence's attribute is unaffected by other variables in the model;

– the error terms (or residuals) of the model are also independent from each other and homoskedastic, that is, they show constant variance;

– the sales samples used to build the model are representative of the universe of residential units it is being used to estimate;

– the variables used (mainly the dependent variable, i.e. the sale price) as well as the residuals of the regression model follow a normal distribution.

Non-adherence to these conditions, due in particular to the presence of excessive multicollinearity among explanatory variables or of residuals that are heteroskedastic or correlated in space (spatial autocorrelation), would affect the standard error of the coefficients associated with exogenous variables. This would invalidate their statistical significance test (student's *t*-test). Moreover, the linearity of the hedonic relation *vis-à-vis* several residential attributes (age, living space, etc.) can be seriously questioned, as can the additivity of the terms of the regression equation. This is because the implicit prices of a property's quantitative attributes (e.g. living space) are often an offshoot of its qualitative elements (e.g. its condition or location).

These methodological limits to using the hedonic approach – which have successful workarounds, as we shall show below – have engendered the development of more flexible, parallel, non-econometric approaches since they require no *a priori* assumption regarding the form of the relationship being tested. These approaches are more heuristic and empirical than deductive, and are essentially geared to simple value estimation and in no way explain the underlying dynamics of value. This is because they have their own conceptual limitations, including the "black box syndrome", as well as a heavy dependence on both data and analytical procedures, while they often generate divergent results[2].

In the following sections, we present a few examples of hedonic modeling applied to measuring urban externalities in light of research conducted in Québec City and Montreal since 1995. Though they offer a very partial picture of all the subtleties and possibilities of adaptation of the hedonic price method, these examples should allow the reader to obtain a good idea of how it may potentially be used as a tool for aiding decision-making. This is particularly the case when used with a geographical information system (GIS) that allows the measuring of externalities to be considerably refined. Section 11.2 presents hedonic modeling within the framework of microeconomic theory, while section 11.3 deals with measuring urban externalities, notably, those features linked to the choice of a functional form. In section 11.4, there is a discussion of the various problems encountered with parameterization of the hedonic model and the setting of implicit prices. Then, in section 11.5, three examples of applying the method are presented to illustrate the complexity of the process of value formation, since the final impact of a phenomenon on the market value of neighboring properties is a function of the spatial overlapping of positive and negative effects. The chapter concludes with the findings on the pros and cons of the usefulness of the hedonic price method in light of possible extensions that would increase its efficiency.

11.2. Hedonic modeling and the microeconomic theory

Although hedonic theory gained acceptance with the work done by Rosen [ROS 74], its first empirical applications go back to Court [COU 39] who applied the concept to the automotive industry, followed by Stone [STO 56], Griliches [GRI 61] and Lancaster [LAN 66]. Since the work of these authors, the hedonic approach has been the subject of a great number of analyses in many research areas and disciplines. Among these areas and disciplines are the economics of the public sector and the environment, labor economics and, to a lesser extent, marketing and industrial organization [BEN 05, BER 95, GOE 01]. The subjects dealt with vary greatly and, to name but a few, range from measuring the value of air quality in urban environments [GRA 88] to an analysis of the power of the market [NEV 01] and quality differentials [GOR 80] in the food industry.

In real estate economics, the hedonic approach has become the most appropriate method for building price indexes [GRI 71, HOE 97a]. As the housing market is generally highly competitive and relatively transparent, it has fast become the field

---

2 Artificial neural networks are among the contemporary approaches critically analyzed in [WOR 95].

of choice for applying the hedonic approach [BAJ 05]. Though analyses focusing on single-family homes, whether stand alone or collectively owned, dominate the literature on the subject, rental housing [DES 96a, HOE 97b, JUD 91, SIR 89, SIR 91] and commercial real estate [BEN 90, DES 05, MEJ 02, SIR 93] have not been overlooked.

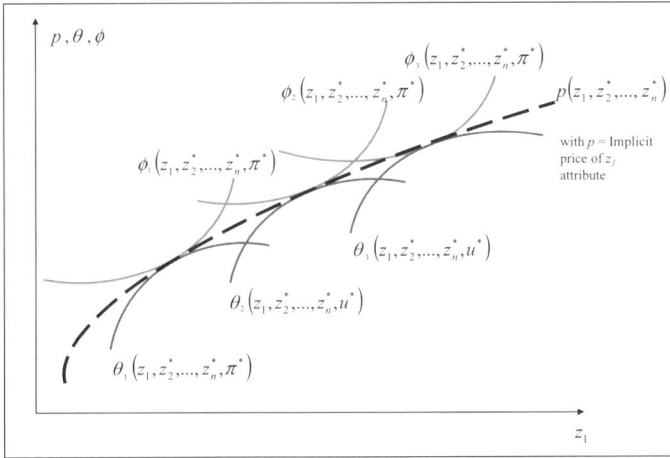

Figure 11.1. *Determination of the implicit price of the* z_1 *attribute. Adapted from [ROS 74]*

Conceptually, hedonic theory is applied to heterogeneous (or complex) goods and rests on the assumption that the market price for a complex good is a function of the utility the buyer derives from consuming a given quantity of each of the n attributes comprising it. On the other, it is a function of the profit the seller gets from it. Z's market price thus results from the equilibrium of supply and demand, on the basis of known characteristics. Each consumer or buyer sees him or herself attributed a higher bid price function θ that reflects a willingness to pay for the good. This depends on the number and quality of n attributes comprising Z (z_1, ..., z_n) as well as the level of utility v generated, given an income y and preference structure α. This higher bid price function may be expressed as:

$$\theta = \theta\left(z_1,...,z_n,\upsilon_{.y,\alpha}\right) \qquad [11.1]$$

Similarly, the individual supply function ϕ defines the lowest price at which a producer, or seller, is prepared to give up the Z good, given its attributes, level of anticipated profit π, and given a production level M and a cost function β:

$$\varphi = \varphi\left(z_1,...,z_n,\pi_{.M,\beta}\right) \qquad [11.2]$$

The market reaches equilibrium, for each attribute, at the point of tangency of the price bid and supply functions. This is illustrated in Figure 11.1, where all the dimensions other than the z_1 attributes are kept constant, and the dotted line of the envelope curve represents the hedonic or implicit price for attribute z_1. By generalization, we obtain a family of envelope curves in a hyperplane with n dimensions that constitute the hedonic prices at equilibrium of n attributes of the Z good.

Rosen's contribution [ROS 74] specifically made it possible to determine two problematical aspects of hedonic theory. In the first place, as the hedonic price function is the product of the combination of supply and demand factors, a problem of identification arises. The equilibrium price is observable, whereas the supply and demand functions are not. Furthermore, the linearity of the hedonic function, which implies that hedonic prices are constant whatever the number of attributes considered, may be seriously challenged in light of empirical evidence. This suggests that several phenomena – particularly in real estate – take on a nonlinear structure. For example, the marginal contribution of each square meter of living space (or plot) to the value of a residence tends to decrease according to the property's size. The same is true for the annual depreciation of a building, where the rate goes down the older it is. Lastly, a third dimension, brought elucidated by Tyrvainen [TYR 97], refers to the necessity of using the hedonic approach for homogeneous markets with regard to the structure of household preferences. When this is not the case, the implicit prices derived through modeling do not faithfully reflect the contribution of the attributes to the phenomenon being studied, and may even be a source of tax inequity if the goal of the analysis is to develop a land valuation model for purposes of local taxation [EPP 98].

Despite this, as we shall come to see, the problems in applying the hedonic approach, whether conceptual (identification of the functional form of the hedonic relationship) or methodological (multicollinearity, heteroskedasticity, and autocorrelation) are not without solution. This is because refinements to the method over the decades – most notably, due to technological innovations in the area of GIS and the development of new procedures of econometric and spatial analysis – have made it possible to attenuate the negative effects they have on the setting of implicit prices. Lastly, regarding the identification problem raised earlier, Straszheim [STR 87] emphasizes that the practice of hedonic modeling has made it possible to clearly delineate the determinants of supply and demand.

11.3. Measuring urban externalities: market segmentation and functional form issues

As a number of authors have rightly pointed out [BER 79, HOC 93, RIC 77], the level of real estate income and values at any location are the result of various overlapping influences. These influences include proximity effects operating on a property's immediate surroundings and accessibility effects that may be felt at greater distances – even over an entire urban region. This explains the emphasis on accessibility factors [COL 85, DES 96b, DES 00, GUN 83, THE 03, THE 05], on neighborhood attributes [COL 90, DES 01, DES 02a, DES 05, GRE 80] and on the quality of the local environment [DES 99b, DES 02b, KES 04, SIM 06]. Despite its limitations, the hedonic method remains both conceptually and methodologically the most appropriate and reliable method for untangling these often complex, overlapping influences, and isolating their contribution to real estate values. To achieve this, however, a number of conditions must first be satisfied, including an appropriate delimitation of the market being analyzed and the identification of a functional form that can adequately capture the hedonic function of the attribute on its own.

11.3.1. *Market segmentation*

Due to the importance market segmentation occupies in the real estate modeling process, this topic has been widely discussed in literature [ADA 96, BAJ 85, GOO 03]. Though there is no formal rule to this effect and the modalities of segmentation are often dictated by technical constraints – particularly the size of the housing stock and related volume of transactions – there are some principles that must be adhered to:

– The price modeling sample must be representative of the universe targeted by the model, which is the "*target stock*". This implies that sufficient information on all the properties in that stock are available, for which we may then set a statistical profile for purposes of comparison. In principle, the choice of a sufficiently large sample makes this objective achievable but prior verification of it is useful.

– The selected segment must be somewhat homogeneous. Thus, including "high end" and "low end" properties in the same regression equation should be eschewed, since the valuations by households of various residential attributes may vary significantly according to income and education. Consequently, care should be taken so that the distribution of selling prices is not excessively broad in scope. This can be made sure of in advance by performing a careful analysis of the sample's descriptive statistics (mean and median values, standard deviation[3], inter-quartile range, skewness and kurtosis coefficients). For the same reasons, a motley mixture of architectural styles and customers (for example young households with children and seniors) should also be avoided. The effect of adhering to these conditions is to reduce the risks of heteroskedasticity of the model's residuals.

---

3 Based on the experiment, a reasonable standard deviation is about 20%.

– Should the choice of segment selected contain several identifiable spatial sub-markets, binary-type (0/1) variables should be used to capture the economic behavior specific to them. Although there are several ways to delimit these spatial entities, in general we would use elements that structure:

- the urban fabric (major roadways, railways, rivers/streams, industrial parks and so on) or even socioeconomic criteria (average household income, educational level);

- architectural styles (that match the development phases of the specific housing stock) or historical features (old neighborhoods versus new ones).

That said, the analyst's market segmentation choice should account for the aim of the modeling process. The construction of an explanatory model of the urban and real estate dynamics of the overall area should be supported by a relatively heterogeneous sample – without deviating from the main determinants of value (square meters, age, condition, location and so on). This implies using a large number of explanatory variables so the regional dynamic of the formation of land and real estate prices may be revealed. If the objective is to measure the specific contribution of a given externality to the market value of a residence (for example, the impact of the proximity of a highway, hydroelectric transmission lines, a metro station and so on), care must be taken to place a boundary around the study. This boundary should be placed to avoid potentially "parasitical" dimensions in the analysis and enable us to focus on the modalities of measuring the phenomenon being studied and the choice of functional form that enables us to capture its dynamic in an optimal manner.

11.3.2. *Measure of the phenomenon and choice of the functional form*

Due to the complexity of the interactions of various determinants of value, measuring capitalization of urban externalities in real estate prices brings us back to the modalities of measuring the phenomenon and to the problems mentioned earlier of the nonlinearity of the hedonic function. By extension, they lead us to the non-additivity of the hedonic function's terms and to the choice of an adequate functional form [LIN 80].

With regard to the modalities of measuring the phenomenon, three types of variables are generally used in hedonic models:

– *Numerical variables (N)* are the quantitative variables used for numerically measuring a given attribute. They are measured on a ratio scale and can be *continuous* if they can assume all possible values within a given interval (e.g. selling price, square meters) or *discrete* if they can only assume a whole number (e.g. age of the property, number of rooms or bathrooms, number of parking spaces, number of months since the transaction).

– *Dichotomous or binary variables (B)* – also referred to as *categorical variables* – use a nominal scale and basically serve to designate the presence (1) or absence (0) of an attribute. Much used in econometric analyses, the benefit of these variables is that they only require partial information on residential characteristics (e.g. presence of a swimming pool or a finished basement). To measure the contribution of some attributes, a multi-category binary variable will be used where each category is viewed as a separate variable by the model. Hence, it is essential to determine the category of reference (generally, the dominant category) that is excluded from the model and which is parameterized by default in function of this category (e.g. Type: single-storey house; Area 1).

– *Rank variables (R)*, lastly, are on an ordinal scale and present the variables ranked subjectively along a scale (e.g. 1, 3, 5, 7, 9), where different values are assumed by a given attribute. Very useful in describing the qualitative level of a residential characteristic (e.g. the quality of the landscaping), this type of variable has the disadvantage – unlike the binary variable – of generating a marginal contribution that is directly proportional to the rank assigned to the variable. This is because the proportionality ratio is not necessarily in keeping with the one the market attributes to different quantities (or levels of quality) of the attribute.

It should be pointed out that the linear regression method (LRM) often ends up being substituted for nonlinear regression procedures, which are less user-friendly and more difficult to interpret in terms of results. Despite this, the *nonlinearity in the data* may, for its part, be suitably handled using appropriate mathematical transformations that may be applied, if required, both to the model's exogenous and endogenous variable. Several transformations can be used, depending on the nature of the phenomenon being analyzed. Some of those most frequently applied to hedonic modeling are described in the following:

– *Reciprocal transformation*, which consists of replacing the values attributed to a variable (X) by its inverse ($1/X$). Based on the principle of decreasing marginal contribution, this transformation is well suited for taking into account, for example, the effect of distance to the town-center on market value.

– Also based on the principle of decreasing marginal contribution, the *exponential transformation* consists of raising the values of a variable to some power in order to emphasize the gap between them (exponent >1) or compress them (exponent is between 0 and 1). The most frequent exponential transformations consist of raising the X variable to some power (X^y) or in taking its square root ($\sqrt{X}$ or $X^{0.5}$). The latter is often used on the square meters of lots when the sample contains lots of greatly varying sizes.

– *Quadratic transformation* is often used in economics. It involves inserting a second- or third-degree polynomial in the regression equation so as to capture the changes of direction in the hedonic function and identify its maximum and minimum limits.

– *Logarithmic transformation* of numerical-type variables, one of the frequently most used, makes it possible to linearize relations proper to highly skewed statistical distributions and may be described on the basis of decimal logarithms (*log*) or natural logarithms (*ln*). When applied to the dependent variable, i.e. the selling price, a logarithmic transformation standardizes its distribution and, by the same token, generates a model whose terms are multiplicative rather than additive. The resulting regression coefficients express adjustment factors that are higher or lower than the unit, rather than fixed unit prices[4]. If the transformation is applied at the same time to the endogenous variable and the appropriate independent variables, the regression coefficients obtained express price elasticities *vis-à-vis* the attributes in question.

– *Multiplicative transformation*, for its part, relies on the principle of interaction among attributes and consists of creating an interactive variable from the product of two separate variables, one of which is often a binary or rank variable (for example, the living space * quality category). Transformations like these may be used to provide more detail towards finding the marginal contribution of a major attribute, such as living space, when the model has various types of properties, such as:

– single-storey houses and two-storey houses; or

– several building qualities (such as co-ownership apartments in low-, medium- or high-density properties).

From this, an implicit price specific to each tested category is obtained rather than a constant average contribution. Interactive variables may also result from the product of two continuous variables so as to measure the marginal contribution of their joint variation [THE 05].

– *Ratio transformation* is a variant of the aforementioned transformation and consists of dividing one variable by another. A unit price ($/m²), often used as the dependent variable to counter the heteroskedasticity effects is a good example of this.

These transformations may be used on their own or in combination with one another. Using them on variables is thus an elegant way of taking into account nonlinear relationships between prices and property characteristics. This improves

---

4 Multiplicative models obtained by logarithmic transformation of the dependant variable (i.e. the selling price) – also called the "semilogarithmic model" – is the most commonly encountered functional form in hedonic analysis.

the performance and predictive granularity of the model without recourse to cumbersome, nonlinear regressive methods. In some cases, measurement of the effects of an externality on values requires applying transformations that can capture not only the nonlinearity of the phenomenon, but also the non-monotonous nature of the hedonic relationship. This is particularly due to the overlapping of negative proximity effects stemming from noise or view pollution, crime rates, etc., and positive accessibility effects. Such a case shall be analyzed in the series of empirical applications presented below.

11.4. Econometric issues and implicit price estimation

As we saw earlier, using LRM involves adhering to certain conditions. The violation of these conditions impacts on the reliability of regression coefficients, that is, on both the hedonic prices and the variance of the estimators. In particular, analysts are regularly confronted with three types of problems: *multicollinearity* among dependent variables, and *heteroskedasticity* and *spatial autocorrelation* that affect the model's residuals. We provide a brief description of these below.

11.4.1. *Multicollinearity between independent variables*

Among the axioms underlying the use of LRM is the mutual independence of the regression model's explanatory variables insofar as they constitute many "vectors" orthogonal to each other. In fact, a certain correlation among explanatory variables in a hedonic model is, for practical purposes, inevitable given the very nature of real estate data. The perverse effects of multicollinearity are only really felt when there is excessive correlation (i.e. $|r| > 0.8$) between two or more variables. Excessive multicollinearity may be referred to as *imperfect*, as is the case, for example, when the number of rooms and number of bedrooms are simultaneously plugged into the regression equation. The parameters of collinear variables will then display significant instability with regard to both their magnitude and sign, while often being at odds with theoretical expectations or common sense. Moreover, the student tests that establish the degree of statistical significance of the coefficients – and thus the reliability of the hedonic prices of the attributes – are being invalidated[5]. The only "sustainable" solution to this problem is to remove the variables at fault – which contribute little if anything to explaining the phenomenon – or, if necessary,

---

5 The presence of excessive multicollinearity mainly affects the regression coefficient's standard error that is used to calculate the *t*-test, thereby making it impossible to assess the degree of statistical significance of the coefficient derived from the analysis and, consequently, to establish whether the latter is statistically different than zero. In some cases, it may even result in the coefficient being biased.

substitute any combination of like variables in their place (for example, the average room area).

For *perfect* multicollinearity, it implies the existence of an exact linear combination between two or more variables (identical vectors) and will not be tolerated by the regression procedure as the X'X matrix cannot be inverted. This occurs, notably, when we uses two or several dummy variables that take exactly the same values for all cases in the regression (e.g. all sold properties have a swimming pool and a garage).

Multicollinearity may be determined by applying an iterative procedure consisting of plugging potentially problematical attributes (described above) into the regression equation one at a time, or alternatively removing them one a time, and analyzing the results. The preferable course of action in general is to calculate the VIF (variance inflation factors) to detect the source of the problem. In this regard, the econometric literature suggests a value of 10 as the limit that should not be exceeded[6] [NET 85]. Based on our experience, it is important to be very careful when the VIF reaches a value of five, since binary variables are less sensitive to this indicator than numerical variables. Lastly, applying a *stepwise regression procedure* automatically eliminates any attribute with marginal contribution to market value that is not sufficiently significant, thereby simplifying the model and avoiding the over-specification trap. This being said, this technique may not be viewed as a substitute for the analyst's judgment with regard to which variables should ultimately enter the model.

11.4.2. *Heteroskedasticity of the residuals*

In addition, for the error terms to be normally distributed, using the OLS regression framework requires that the latter be homoskedastic (i.e. that their variance be constant) and therefore independent from the value that various attributes of the property may assume, particularly the level of real estate prices. In other words, the estimation errors made about high-end real estate properties in the sample must not be more scattered than the ones for the low-end units. Now, this phenomenon is encountered frequently in hedonic modeling, where luxury attributes are characterized by more scattered distribution than standard attributes. Violating this condition translates into a regression model where estimates of low-value units are less reliable because they have been influenced by more luxurious properties.

---

6 To summarize, the VIF shows the extent to which an exogenous variable is explained by the model's set of other explanatory variables. Measured using the following equation: VIF = (1/1- explanatory power of the other independent variables), it assumes a value of 10 if 90% of the other variables are explained by it, five if their explanatory power is 80%, and two if it is only 50%, and so on.

While the presence of heteroskedasticity may be revealed using the White test [WHI 80], there is a range of solutions for countering its effects. These range from transforming the dependent variable – by dividing it, for example, by the living space – or by using a multiplicative functional form, to applying the weighted least-squares procedure that weighs observations as an inverse function of the variance of their error term. That said, prevention remains the best approach: generally, careful segmentation of real estate sub-markets will minimize the import of this problem. The analyst must make sure that the selling price distribution range is not excessive and that the market segments modeled remain relatively homogeneous.

11.4.3. *Spatial autocorrelation of the residuals*

As regression model residuals are in principle generated at random, they should be kept independent of one another. If they are not, these residuals become tainted with autocorrelation errors. Unlike macroeconomic literature, where detection and control of temporal autocorrelation is the main concern, in hedonic modeling where, more than elsewhere, cross-sectional analysis is used, it is the spatial dimension of the autocorrelation that predominates. Thus, a phenomenon affecting economic activity in one area or the behavior of its residents has analogous repercussions on neighboring areas: i.e. *spatial autocorrelation* or *spatial dependence*.

Omnipresent in spatial analysis, it merely reproduces the forces that structure the territory and mirror the great complexity arising from the modalities under which urban phenomena spread (disseminate), often anisotropically, over the territory. For example, there is often heavy spatial dependence between the architectural style of residential properties, their age and certain of their attributes, including lot area. This merely translates the phases of development of the urban fabric and the propensity of residential customers to seek out relatively homogeneous living environments that make up the various sub-markets of the housing market.

Spatial autocorrelation often takes up a non-negligible portion of the explanatory power of hedonic models, which brings to light the paramount role that neighborhood attributes exert on residential values. Mistakes in specifications of the model frequently cause autocorrelation. Thus, omitting a major explanatory variable, which itself is autocorrelated, or adopting an inadequate functional form to describe the relationship between the dependent variable and independent descriptors will impact the level of errors, which become autocorrelated due to this fact.

When there is heavy autocorrelation caused by omitted variables, the regression coefficients are most likely biased and significance tests are thus invalidated because the variance of the coefficients is less efficiently estimated. It is through the analysis of residuals, which are error estimates, that it is possible to obtain information on the latter. This is very useful for acquiring a better understanding of the structuring forces acting on the territory.

Detecting and dealing with spatial autocorrelation is a complex undertaking that involves testing for spatial dependence[7] and using spatial statistics procedures. Among these, trend surface analysis and kriging are meant to capture spatial phenomena that operate locally. The results of these are retrofitted into a regression analysis in the form of interpolated trends [DES 99a]. For its part, Casetti's expansion method [CAS 72] has the benefit of highlighting, with the help of interactive variables, some spatial drifts that capture the emergence of atypical household behaviors relative to the valuation of certain residential attributes, which would otherwise remain undetectable [DES 07].

Lastly, setting up regression procedures makes it possible to substantially refine the modeling of geographical phenomena and their impact on real estate values. This is particularly true of the geographically-weighted regression method [FOT 98] that is based on performing a series of local regressions simultaneously. For each of the model's exogenous variables, these result in establishing the distribution of hedonic prices that varies for each point in space. Other econometric methods are also available for successfully handling spatial autocorrelation, notably the autoregressive methods [ANS 88]. In this regard, Chapter 10, which describes measuring the accessibility of urban services, contains a comparative analysis of the relative performance of various approaches.

11.5. The hedonic approach and measure of externalities: some examples

In the following sections, we present three examples of the application of the hedonic approach, which illustrate both its potential for measuring urban externalities and the issue of specifying a functional form that can adequately capture the effects these externalities exert on residential values.

11.5.1. *Proximity impact of a high-voltage transmission line*

Since the beginning of the 1990s, several studies have investigated the impact the proximity of high-voltage electrical transmission lines (HVTL) can have on residences [COL 90, DEL 92, DES 02, HAM 95, KIN 95]. Unlike most authors who measure the decreasing influence of this infrastructure away from home, Des Rosiers [DES 02a] adopts a microspace approach that emphasizes what happens along the easement. He uses a series of binary variables to describe in detail the

---

7 Moran's index [MOR 50] is by far the most widely applied test for detecting the overall level of spatial autocorrelation.

position and orientation of adjacent properties in the proximity of transmission lines and pylons, so as to capture the degree of visual encumbrance to which homeowners are subjected.

The study, conducted in close collaboration with the real estate unit of the *Groupe TransÉnergie*, a Hydro-Québec subsidiary, used a sample of 507 single-family homes that had been sold (at an average price of $69,000 CAD) between 1991 and 1996 in the area of Ville de Brossard (approximately 69,000 inhabitants in 1996). This area is located in the southern suburbs of the Montreal metropolitan area, along the Saint Laurence River. The area of study, extending 250–500 m in width, is bounded by three major highways and has a 315 kilovolt HVTL, 60 m-wide corridor (easement) down the middle extending about 3 km.

The improved visual appearance of conical steel pylons are 48–55 m high and there are 26 in the study area. They are set 200–350 m apart and the separation between the transmission wires and the ground ranges from 11–20 m. The neighborhood is topographically flat with some tree groves here and there and a bicycle lane along the east side of the easement.

The asymmetry of the line, located 45 m from the easement eastern boundary and only 15 m from its western boundary, was a major advantage in this study as it made the testing of differential impacts of the distance to the adjacent properties possible. In all, 383 properties had a limited, moderate or full view (from the front, side or rear) of the line, and 34 of them were directly adjacent to the HVTL easement. The average distance to the outside boundary of the easement was 248 m. The analysis included some 25 descriptors including:

– the linear distance of each property to the boundary and the easement;

– binary variables of distance in increments of 50–100 m;

– binary variables of control that take into account the location of easement-adjacent properties relative to the position of pylons; and, finally,

– a series of interactive binaries combining the view into the HVTL structure and home orientation.

Functional forms (linear, log-linear, quadratic and gamma) were tested. An overall sample and a series of sub-markets, including the eastern and western areas on either side of the easement, were analyzed.

Global Sample

	HVTL Attribute	Impact %
House facing pylon	FACNGPYL	-9.6
One lot away from pylon	1LOTPYL	+11.6
Two lots away from pylon	2LOTPYL	+8.7
Three lots away from pylon or mid-span location	3LOTMID	-4.7

East Area (45 m Setback to HVTL)

	HVTL Attribute	Impact %
House facing pylon	FACNGPYL	n.s.
One lot away from pylon	1LOTPYL	+15.7
Two lots away from pylon	2LOTPYL	n.s.
Three lots away from pylon or mid-span location	3LOTMID	-7.7

West Area (15 m Setback to HVTL)

	HVTL Attribute	Impact %
House facing pylon	FACNGPYL	-14.0
One lot away from pylon	1LOTPYL	n.s.
Two lots away from pylon	2LOTPYL	+10.3
Mid-span location	MIDSPAN (sig. 0.07)	-7.4

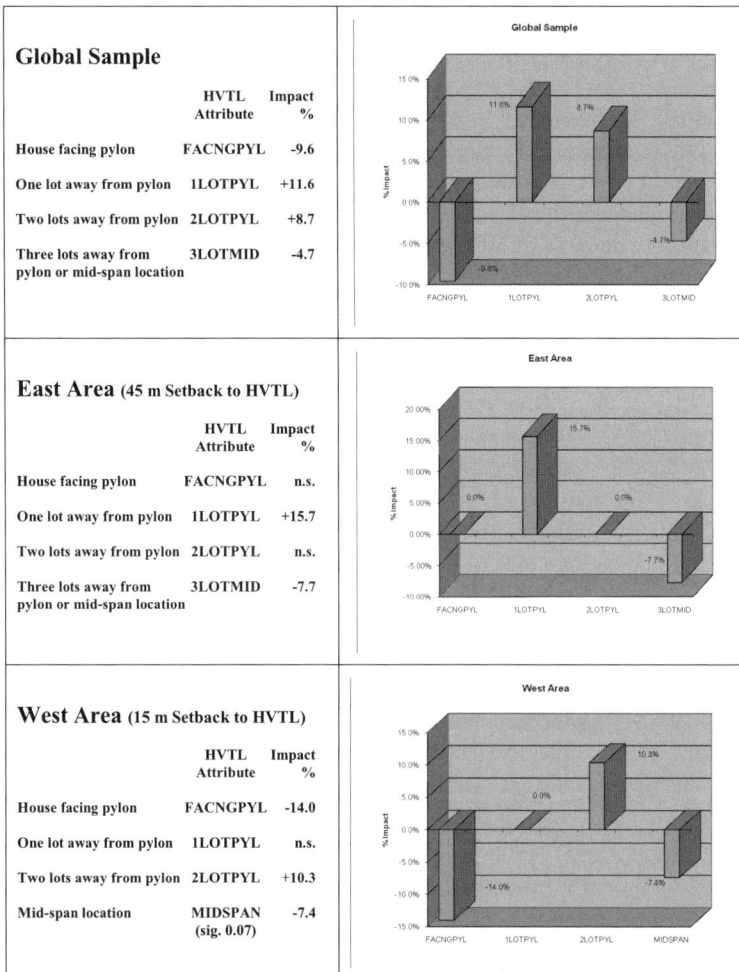

N.B. Percentage price impact reported here are an average of all significant coefficients derived from various functional forms and should therefore be viewed as indicators only. Besides, they reflect "gross" location impacts due to view on pylons and conductors alone.

Figure 11.2. *Impact of HVTL structures on the market value of adjacent properties [DES 02a]*

In general, the explanatory and predictive performances of the models are most satisfying, displaying adjusted R-squared values of 0.90, whereas the relative standard errors of estimate (SEE%) are under 10%. First, the study corroborates earlier research that suggest that negative visual impact on values tend to decrease rapidly over distance and that they are no longer significant beyond 150 m.

Furthermore, "the net visual encumbrance"[8] reaches a maximum for properties located 50–100 m away from the HVTL easement, with their value being reduced by 5–12%.

Given the density of information stemming from the analysis, the only findings discussed here are those relating to the properties adjacent to an HVTL easement, namely the ones obtained with the overall sample and with the east and west areas. These results are illustrated in Figure 11.2 and are summarized as follows:

– The market value of residential properties *adjacent to an HVTL easement and facing a pylon* (FACNGPYL) decreases significantly due to the visual encumbrance arising from this. This decrease, which is about 9.6% of the average price in the overall sample, amounts to 14% in the western area where the residences are only 15 m from the HVTL. In contrast, in the eastern area, characterized by an offset margin of 45 m that tends to neutralize the visual encumbrance, pylons facing residences have no significant impact on the prices of residences.

– On the contrary, *a home located one or two lots away from the pylon* generally benefits from a market premium that reflects the visual clearance created in this manner as well as from the increased privacy its occupants enjoy. Results from the overall sample record an increase in value of 11.6% for the property located right next to the lot affected by a pylon (1LOTPYL) and an increase of 8.7% where the pylon faces the second neighbor (2LOTPYL). The value of a property adjacent to the easement and located one lot away from a pylon sees a 16% increase if it is located in the eastern area; whereas in the western area the potential location premium is cancelled out due to the smaller width of the set off margin with a more marked visual encumbrance. Therefore, when a house is located two lots away from a pylon its value increases and does so by about 10%.

– Lastly, the market value of *a property located three lots away from a pylon or in mid-range* (3LOTMID, MIDSPAN) generally depreciates substantially due to the visual encumbrance caused by the transmission lines in the section of the hydroelectric corridor where the minimal clearance to the ground is low. For the overall model, this depreciation is about 4.7 %, and amounts to 7.7% and 7.4% in the eastern and western areas, respectively.

As can be seen, the preceding analysis enables a highly nuanced interpretation of the impact of proximity of a HVTL on the market value of residential properties located along the boundaries of the easement. More specifically, we can see that, unlike the findings of most previous studies, immediate proximity to a HVTL does not translate into a significant decrease in home prices except when visual field of

---

8 Net visual encumbrance is defined as the difference between the loss of value due to the visual encumbrance and the premium given to properties adjacent to a HVTL easement due to the greater privacy it confers on their occupants (no neighbors facing the backyard).

their occupants is directly obstructed by a pylon or by electrical transmission lines. Otherwise, the benefit of proximity to the easement (better visual clearance and greater privacy) wins out over its disadvantages and commands a market premium that is just as significant.

11.5.2. *Residential values and shopping center proximity*

The second case study presented here illustrates the impact on residential values of proximity to a shopping center, where the emphasis is placed on identification of a functional form appropriate for capturing a phenomenon whose effects are shown to be spatially non-monotonous [DES 96b]. Indeed, the same element – a business or service – may be a benefit or a nuisance for neighborhood residents, depending on the nature of the service in question, its size and its distance from neighboring properties. The net effect observed in the chapter on market values thus reproduces the overlapping of positive effects of accessibility to urban services – which decreases over distance – and negative effects arising from too great a proximity (noise, odors, higher volumes of local traffic, etc.). There are some studies in the hedonic literature on the impact of shopping centers as a source of externalities [COL 85, SIR 94]. These studies find that the size of a shopping center has a positive effect on values due to the time savings that this implies for its customers in their search for products; whereas the price of residences decreases as the distance to the shopping centers increases. However, they are silent on which functional form should be used to capture the non-monotonous nature of the hedonic relationship *vis-à-vis* the distance between the residence and the commercial establishment.

This study was conducted on a sample of more than 4,000 houses (single-storey, stand-alone single-family residences) sold in the area of the ex-QUC (Québec Urban Community) between January 1990 and December 1991. The selling price is used as the dependent variable, whereas there are more than 60 descriptors describing the physical attributes of the property and land as well as the local tax, socioeconomic, time and neighborhood dimensions *vis-à-vis* estimating the "net" effects of the presence of a shopping center. The analysis also incorporates 87 commercial establishments in three size categories: 73 neighborhood centers, nine community centers and five regional centers. Here the number of commercial premises were used to determine the category of each establishment. The distance to the closest shopping center is measured using Euclidean distance or a series of nine binary variables that identify concentric buffer zones varying in width from 100 m to 1 km, up to a distance of 2.5 km from the house. All this is managed with the help of a regional GIS. The average selling price of the residences is $84,000 CAD, whereas the age and average living space area are set at 16 years and 95 m^2. Lastly, the average size of the commercial establishments stands at 15 premises for neighborhood centers, 60 premises for community centers and 220 premises for regional ones.

As with the preceding example, various functional forms (linear, log-linear, quadratic and gamma) are applied to the hedonic price function with the analysis initially covering the overall sample. The models calibrated in this manner exhibit an explanatory power (adjusted R-squared) between 0.865 and 0.890, while the relative predictive performance (SEE%) averages 9%. As for multicollinearity, it is well under control (maximum VIF below five)[9]. With a few exceptions, the coefficients for variables relating to size and distance from the closest shopping center are statistically significant at the 0.01 or 0.001 level and confirm the results of previous research: the size of a shopping center as much as its accessibility have a positive effect on values.

What of the proximity effects? To measure them, we use a *gamma* transformation that has the advantage of being very flexible in terms of adjusting the hedonic function's parameters.

The gamma distribution for a given variable X is a probability density function expressed as follows:

$$f(x) = K * x^{(\alpha-1)} * e^{(-x/\beta)} \text{ for } x > 0 \qquad [11.3]$$
$$= 0 \text{ for } x = 0$$

where α and β are non-null parameters and K is a constant. Depending on the values assigned to α and β, the gamma distribution takes the form of an exponential, chi-squared or even normal distribution. Furthermore, the slope of the curve beyond the optimum distance, i.e. the one that maximizes values, is the steeper the smaller β is. Plugging the gamma function into the hedonic equation results in a generalized gamma formulation as:

$$SALEPRICE = K * INVDSHOP^{(\alpha-1)} * e^{(-INVDSHOP/\beta + \Sigma B_i * Z_i + e)} \qquad [11.4]$$

where *INVDSHOP* stands for the inverse of the distance to the closest shopping center, while B_i stands for the vector of the coefficients to be estimated, and Z_i for the vector of control variables. By applying a logarithmic transformation, we obtain:

$$LnSALEPRICE = LnK + (\alpha-1) * Ln\ INVDSHOP - (1/\beta) * INVDSHOP + \Sigma B_i * Z_i + e \qquad [11.5]$$

---

9 Unless we use the quadratic form for the distance-to-the-center variable, which would by necessity inflate the VIF.

By setting the first derivative of the gamma function at zero, we obtain the measurement of the *optimum* distance (*DSHOP**) to the closest shopping center, that is the one that maximizes the value of a property. We thus can write:

$$\text{d LnSALEPRICE} / \text{d INVDSHOP} = (\alpha\text{-}1) * \text{DSHOP} - (1/\beta) \qquad [11.6]$$

hence:

$$\text{DSHOP}^* = (\alpha\text{-}1) / \beta \qquad [11.7]$$

Furthermore, the expectation of a maximum (value) requires that the second derivative of the function in relation to *INVDSHOP* be negative, which gives us:

$$\text{d}^2 \text{LnSALEPRICE} / \text{d INVDSHOP}^2 < 0 \qquad [11.8]$$

hence:

$$(\alpha\text{-}1) / \text{INVDSHOP}^2 > 0 \qquad [11.9]$$

Since the squared distance is necessarily positive, the condition for maximization is satisfied for all values of α greater than the unit. Using a semi-logarithmic functional form along with a gamma transformation applied to the inverse of the distance to the closest shopping center results in the adjusted R-square rising to almost 0.900, while the model's relative SEE is down to 8.35% (overall sample). Both the size and distance variables display coefficients that are highly significant (p <0.001) and with signs in line with theoretical expectations. As can be seen from Figure 11.3, the overall model – where neighborhood centers largely predominate – suggests an optimal distance from home of about 209 m.

Applying the gamma transformation after segmenting the sample by shopping center size category generates results that are even more interesting, although, as expected, the effect of the size variable is no longer significant. Thus, the optimal distance, estimated to be 215 m in the case of neighborhood centers, goes up to 310 m for community centers and 532 m for regional centers. Thus, the range of influence of the negative externality of proximity tends to increase with center size.

Finally, the lower portion of Figure 11.3 illustrates *ceteris paribus* well, the premium the market assigns to properties depending upon the size category the closest shopping center belongs to. Whereas, in our example, the average value of residences located in the proximity of a neighborhood center is about $87,600, it rises to $94,300 and $100,800 in the cases of residences located near a community center (a differential premium of $6,700) and a regional center ($13,200), respectively. The progressive decline in values stemming from loss of accessibility is also brought to light and suggests a rate of decline that increases (i.e. a steeper slope) with the shopping center hierarchical level.

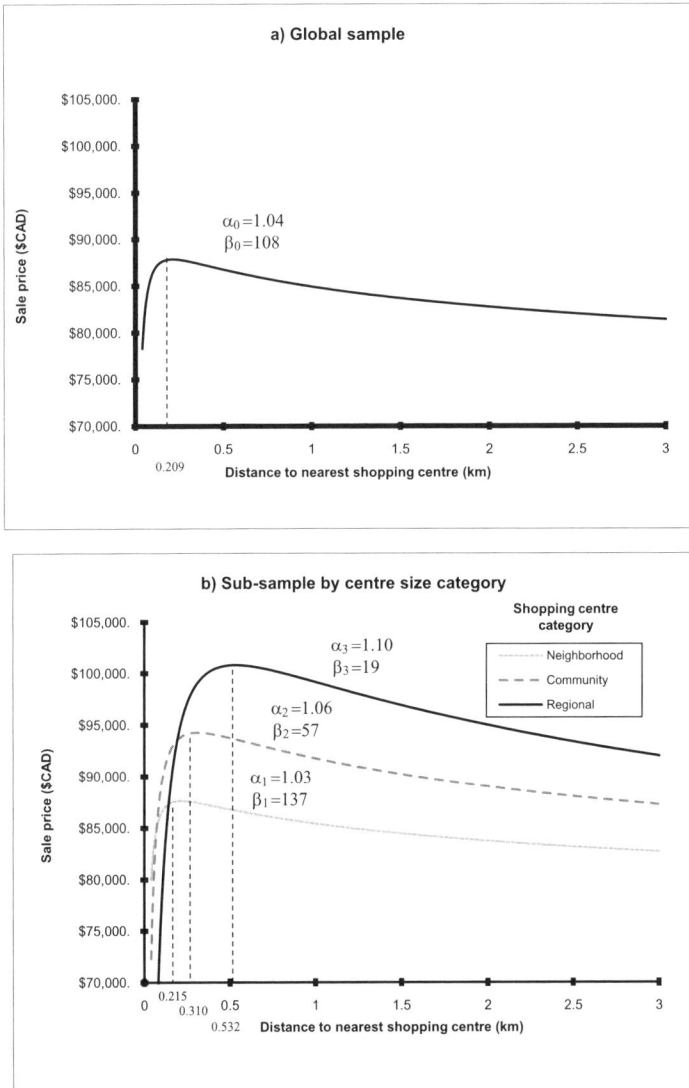

Figure 11.3. *Measuring of the effects of proximity to a shopping center using a gamma transformation [DES 96b]*

11.5.3. *Landscaping and its effect on house values*

We shall end this chapter on the measurement of urban externalities by analyzing the contribution of landscaping to residential value. In this respect, the reader will find a highly sophisticated application of the hedonic approach to measuring the market value of peri-urban landscapes on residential real estate in France in Chapter 12.

Over the last 30 years, several authors have dealt with the impact of trees, green spaces, watercourses and landscaping on real estate values [DES 02b]. A number of analyses on this topic, including more recent ones, use hedonic modeling [DOM 00, LUT 00]. The results of these analyses suggest, in particular, that the presence of trees in the neighborhood of a house increases its value by 2–9%, depending on the type of dwelling and how it is landscaped. The study we present here is based on a field survey carried out in the summer of 2000 of 760 single-family homes (stand-alone houses, single-storey houses and row houses) sold between 1993 and 2000 in the ex-QUC area. The survey consisted of identifying the landscaping attributes characterizing both the properties and their immediate surroundings, i.e. the neighborhood as visible from the property. Almost 250 variables and explanatory factors were used in the analysis [DES 00], including 31 environmental and landscaping attributes recorded by the survey (trees, flower and rock arrangements, shrubbery and flower beds (plantings), patio and balcony arrangements, and visible vegetation density). On average, the selling price of the properties in the sample is $112,000 CAD ($50,000–435,000), the apparent age[10] is 16 years and the living space is 120 m^2. As for landscaping attributes, the tree cover characterizing houses and their immediate surroundings stands at 44% and 46%, respectively, with the difference made up by areas with or without ground vegetation.

Calibration of the hedonic model with a semi-logarithmic functional form is undertaken in three phases. In the first model (Model A), only the physical attributes of the property, the socioeconomic dimensions of the neighborhood and the variables of proximity and accessibility to urban service are included. This initial model contains 18 explanatory variables and generates an adjusted R-square of 0.862 and a SEE of 0.140. All coefficients are significant at the 0.01 or 0.001 level, and their sign and magnitude conform to expectations. Once the basic descriptors[11] are selected, the landscaping attributes are then added on, either individually or interactively with other exogenous variables. *Absolute* interactions (Model B) and *relative* interactions (Model C) are being used successively, with relative

---

10 In contrast with actual building age, apparent age accounts for improvements brought to the house or accelerated structural deterioration.
11 Selection is carried out by the application of a standard regression procedure followed by a backward regression procedure, which considerably reduces the number of descriptors and, hence, limits multicollinearity to a minimum.

interactions calculated on the basis of centered variables. While a lower number of significant coefficients are obtained from the former interactions, they show more stability than those obtained with relative interactions, which, in return, tend to lessen the perverse effects of spatial autocorrelation.

Model B (Absolute interactions)	No. of obs. = 758 No. of explanatory variables = 19		Adjusted R-squared: 0.866 SEE: 0.138		F test: 258.4 Prob.: 0.000
Attributes	*Regression coefficient (B)*	*t-test*	*Prob.*	*Adjustment factor: Exp(B)*	*VIF*
%NbhdTrees * %Age65+	0.0112	6.74	0.000	1.011	2.17
%PropVegetation * Single-storey house	0.0020	8.73	0.000	1.002	2.19
%PropVegetation * Two-storey house	0.0019	7.54	0.000	1.002	2.43
%PropTrees - %NbhdTrees	0.0017	5.50	0.000	1.002	1.40
Hedges	0.0386	3.51	0.000	1.039	1.16
Model C (Relative interactions)	No. of obs. = 758 No. Of explanatory variables = 24		Adjusted R-squared: 0.859 SEE: 0.142		F test: 193.2 Prob.: 0.000
Attributes	*Regression coefficient (B)*	*t-test*	*Prob.*	*Adjustment factor: Exp(B)*	*VIF*
%Property Tree / %Neighborhood Tree	0.0708	5.11	0.000	1.073	2.43
(%Property Tree - %Neighborhood Tree) * Age45-64	-0.0061	-3.46	0.001	0.994	2.45
%Vegetation Proportion centered * Two-storey house	0.0011	2.93	0.003	1.001	1.31
%Neighborhood Tree centered * %Women centered	0.0355	3.56	0.000	1.036	1.25
Visible vegetation density	-0.0225	-2.30	0.022	0.978	1.20
Hedges	0.0355	3.16	0.002	1.036	1.15
Patio	0.1167	2.41	0.016	1.124	1.04
Flower beds	0.0427	2.36	0.019	1.044	1.13

Table 11.1. *The implicit pricing of landscaping attributes*

Both approaches, nevertheless, produce consistent, complementary information on the contribution of landscaping attributes; in particular, the interactions that make it possible to isolate residential behaviors that are "deviant" in relation to local standards. For this reason, results from Models B and C pertaining to these attributes are reported in Table 11.1.

In general, a property that has greater tree cover than the neighborhood is assigned a higher value, even though a negative adjustment is applied to areas with a majority of baby-boomers, i.e. those aged 45–64 (Model C). The importance of tree coverage in the visible environment has a positive impact on prices, even more so in areas with a high percentage of retired persons (Model B) and women (Model C). Furthermore, although trees seem to be favored by most homeowners, single-storey houses and two-storey houses with rich ground vegetation (lawns, flower beds, rock, etc.) command a market premium (Models B and C). It is interesting to observe that houses surrounded by excessive vegetation (i.e. above average) are priced at a discount (Model C), as evidenced by the Payne's work [PAY 73]. Lastly, hedges increase the value of a property by 3.6% (Model C) to 3.9% (Model B), with a premium of 4.4% to 12.4%, respectively, in the case of flowerbeds and patio arrangements (Model C).

With aging of the population, which is particularly marked in Québec City, landscaping is bound to grow in popularity in homeowners' preferences. As has been seen, the hedonic approach enables us to measure, with the necessary level of nuance, the monetary importance these households place on landscaping attributes.

11.6. Conclusion

Multiple urban externalities, both positive and negative, are an essential component of real estate and land values. Their impact is, however, difficult to measure due to the complexity of cross-over influences which, at each point in space overlap or retract according to the nature of the phenomenon studied and profile preferences of the land users. In this chapter, we have shown that it is possible to dissect these influences by using the hedonic approach, on the condition that the analyst is able to develop adequate measurement of the phenomena studied and correctly identify the functional form of the hedonic relationship. Examples relating to Québec's residential sector also demonstrate that it are quite possible, by resorting to adequate methodological devices and adaptations to the linear regression method, to capture the economic impact of urban externalities on home value in a nuanced manner as well as household preferences and their spatial patterns.

11.7. Acknowledgements

The research material presented in this chapter has been developed thanks to the long-term financial support of the Social Sciences and Humanities Research Council of Canada and the Fonds Québécois pour la Recherche sur la Société et la Culture. Databases used have been made available under a series of agreements with the Réseau de Transport de la Capitale, the Québec Urban Community – now merged with Québec City, Hydro-Québec Corporation and the Greater Québec Real Estate Board. The authors also wish to thank the 30 or so contributors and graduate students who, over the past two decades, took part in setting up the georeferenced databases used for this research.

11.8. Bibliography

[ADA 96] ADAIR A.S., BERRY J.N., MCGREAL W.S., "Hedonic modelling, housing submarkets and residential valuation", *Journal of Property Research*, vol. 13, pp. 67-83, 1996.

[BAJ 85] BAJIC V., "Housing market segmentation and demand for housing attributes: Some empirical findings", *AREUEA Journal*, vol. 13, no. 1, pp. 58-75, 1985.

[BAJ 05] BAJARI P., KAHN M., "Estimating housing demand with an application to explaining racial segregation in cities", *Journal of Business and Economic Statistics*, vol. 23, no. 1, pp. 20-33, 2005.

[BEN 90] BENJAMIN J.D., BOYLE G.W., SIRMANSS C.F., "Retail leasing: Determinants of shopping center rents", *AREUEA Journal*, vol. 18, no. 3, pp. 302-312, 1990.

[BEN 05] BENKARD C., BAJARI P., "Hedonic price indexes with unobserved product characteristics, and application to PC's", *Journal of Business and Economic Statistics*, vol. 23, no. 1, pp. 61-75, 2005 .

[BER 79] BERRY B.J.L., BEDNARZ R.S., "The disbenefits of neighbourhood and environment to urban property", in SEGAL D., *The Economics of Neighborhood*, New York, Academic Press, pp. 219-246, 1979.

[BER 95] BERRY S., LEVINSOHN J., PAKES A., "Automobile prices in market equilibrium", *Econometrica*, vol. 63, no. 4, pp. 841-89, 1995.

[CAS 72] CASETTI E., "Generating models by the expansion method: Applications to geographical research", *Geographical Analysis*, vol. 4, pp. 81-91, 1972.

[COL 85] COLWELL P.F., GUJRAL S.S., COLEY C., "The impact of a shopping center on the value of surrounding properties", *Real Estate Issues*, vol. 10, no. 1, pp. 35-39, 1985.

[COL 90] COLWELL P.F., "Power lines and land values", *Journal of Real Estate Research*, vol. 5, no. 1, pp. 117-127, 1990.

[COU 39] COURT A.T., "Hedonic price indexes with automotive examples", *The Dynamics of Automobile Demand*, New-York, General Motors Corporation, pp. 99-117, 1939.

[DEL 92] DELANEY C.J., TIMMONS D., "High voltage power lines: Do they affect residential property value?", *Journal of Real Estate Research*, vol. 7, no. 3, pp. 315-29, 1992.

[DES 96a] DES ROSIERS, F., THÉRIAULT M., "Rental amenities and the stability of hedonic prices: A comparative analysis of five market segments", *Journal of Real Estate Research*, vol. 12, no. 1, pp. 17-36, 1996.

[DES 96b] DES ROSIERS, F., LAGANA A., THÉRIAULT M., BEAUDOIN M., "Shopping centers and house values: an empirical investigation", *Journal of Property Valuation & Investment*, vol. 14, no. 4, pp. 41-63, 1996.

[DES 99a] DES ROSIERS, F., THÉRIAULT M., *House Prices and Spatial Dependence: Towards an Integrated Procedure to Model Neighborhood Dynamics*, Québec, Working Paper #1999-002, Laval University, Faculty of Business Administration, 1999.

[DES 99b] DES ROSIERS, F., BOLDUC A., THÉRIAULT M., "Environment and value: Does drinking water quality affect house prices?", *The Journal of Property Investment and Finance*, vol. 17, no. 5, pp. 444-463, 1999.

[DES 00] DES ROSIERS F., THÉRIAULT M., VILLENEUVE P.-Y., "Sorting out access and neighbourhood factors in hedonic price modeling", *The Journal of Property Investment and Finance*, vol. 18, no. 3, pp. 291-315, 2000.

[DES 01] DES ROSIERS F., LAGANA A., THÉRIAULT M., "Size and proximity effects of primary schools on surrounding house values", *Journal of Property Research*, vol. 18, no. 2, pp. 149-168, 2001.

[DES 02a] DES ROSIERS, F., "Power lines, visual encumbrance and house values: A micro-spatial approach to impact measurement", *The Journal of Real Estate Research*, vol. 23, no. 3, pp. 275-300, 2002.

[DES 02b] DES ROSIERS F., THÉRIAULT M., KESTENS Y., VILLENEUVE, P-Y., "Landscaping and house values: an empirical investigation", *The Journal of Real Estate Research*, vol. 23, no. 1, pp.139-61, 2002.

[DES 05] DES ROSIERS, F., M. THÉRIAULT AND L. MÉNÉTRIER, "Spatial versus non-spatial determinants of shopping center rents: Modeling location and neighborhood-related factors", *Journal of Real Estate Research*, vol. 27, no. 3, pp. 293-319, 2005.

[DES 07] DES ROSIERS F., THÉRIAULT M., KESTENS Y., VILLENEUVE P-Y., "Landscaping attributes and property buyers' profiles: Their joint effect on house prices", *Journal of Housing Studies*, vol. 22, no. 6, pp. 945-964, 2007.

[DOM 00] DOMBROW J., RODRIGUEZ M., SIRMANS C.F., "The market value of mature trees in single-family housing markets", *The Appraisal Journal*, vol. 68, pp. 39-43, 2000.

[EPP 98] EPPLE D., PLATT G.J., "Equilibrium and local distribution in an urban economy when households differ in both preferences and income", *Journal of Urban Economics*, vol. 43, pp. 23-51, 1998.

[FOT 98] FOTHERINGHAM A.S., CHARLTON M.E., BRUNDSON C.F., "Geographical weighted regression: a natural evolution of the expansion method for spatial data analysis", *Journal of Environment & Planning A*, vol. 30, pp. 1905–1927, 1998.

[GOE 01] GOETTLER R., SHACHAR R., "Estimating spatial competition in the network television industry", *RAND Journal of Economics*, vol. 32, no. 4, pp. 624-656, 2001.

[GOO 03] GOODMAN A. C., THIBODEAU T. G., "Housing market segmentation and hedonic prediction accuracy", *Journal of Housing Economics*, vol. 12, pp. 181-201, 2003.

[GOR 80] GORMAN T., "A possible procedure for analysing quality differentials in the egg market", *Review of Economic Studies*, vol. 47, no. 5, pp. 843-856, 1980.

[GRA 88] GRAVES P., MURDOCH J.C., THAYER M.A., WALDMAN D., "The robustness of hedonic price estimation: Urban air quality", *Land Economics*, vol. 64, no. 3, pp. 220-33, 1988.

[GRE 80] GRETHER M.D., MIESZKOWSKI P., "The effects of non-residential land uses on the prices of adjacent housing: Some estimates of proximity effects", *Journal of Urban Economics*, vol. 8, no. 1, pp. 1-15, 1980.

[GRI 60] GRILICHES Z., "Hedonic price indexes for automobiles: An econometric analysis of quality change", *The Price Statistics of the Federal Government*, General Series, National Bureau of Economic Research, vol. 73, pp. 137-196, 1961.

[GRI 71] GRILICHES Z., "Hedonic price indexes revisited", *Price Indexes and Quality Change*, Federal Reserve Board, Massachusetts, pp. 3-15, 1971.

[GUN 83] GUNTERMANN K.L., COLWELL P.F., "Property values and accessibility to primary schools", *Real Estate Appraiser & Analyst*, vol. 49, no. 1, pp.62-8, 1983.

[HAM 95] HAMILTON S.W., SCHWANN G.M., "Do high voltage electric transmission lines affect property value ?", *Land Economics*, vol. 71, no. 4, pp. 436-44, 1995.

[HOC 93] HOCH I., WADDELL, P., "Apartment rents: another challenge to the monocentric model", *Geographical Analysis*, vol. 25, no. 1, pp. 20-34, 1993.

[HOE 97a] HOESLI M., GIACOTTO C., FAVARGER P., "Three new real estate price indices for Geneva, Switzerland", *Journal of Real Estate Finance and Economics*, vol. 15, no. 1, pp. 93-109, 1997.

[HOE 97b] HOESLI M., THION B., WATKINS C., "A hedonic investigation of the rental value of apartments in central Bordeaux", *Journal of Property Research*, vol. 14, pp. 15-26, 1997.

[JUD 91] JUD G.D., WINKLER D.T., "Location and amenities in determining apartment rents: An integer programming approach", *Appraisal Journal*, vol. 59, no. 2, p. 266–75, 1991.

[KES 04] KESTENS Y., THÉRIAULT M., DES ROSIERS F., "The impact of surrounding land use and vegetation on single-family house prices", *Environment & Planning B – Planning and Design*, vol. 31, pp. 539-567, 2004.

[KIN 95] KINNARD W.N. JR, DICKEY S.A., "A primer on proximity impact research: Residential property values near high-voltage transmission lines", *Real Estate Issues*, vol. 20, no. 1, pp. 23-29, 1995.

[LAN 66] LANCASTER K., "A new approach to consumer theory", *The Journal of Political Economy*, vol. 74, no. 2, pp. 132-157, 1966.

[LIN 80] LINNEMAN P., "Some empirical results on the nature of the hedonic price function for the urban housing market", *Journal of Urban Economics*, vol. 8, no. 1, pp. 47-68, 1980.

[LUT 00] LUTTIK J., "The value of trees, water and open space as reflected by house prices in the Netherlands", *Landscape and Urban Planning*, vol. 48, pp. 161-67, 2000.

[MEJ 02] MEJIA L.C., BENJAMIN J.D., "What do we know about the determinants of shopping center sales? Spatial vs. non-spatial factors", *Journal of Real Estate Literature*, vol. 10, no. 1, pp. 3-26, 2002.

[MOR 50] MORAN, P.A.P. "Notes on Continuous Stochastic Phenomena", *Biometrika*, vol. 37, pp. 17–33, 1950.

[NET 85] NETER J., WASSERMAN W., KUTNER M.H., *Applied Linear Statistical Models*, 2nd edition, R. D. Irwin, Homewood, IL, pp. 391-392, 1985.

[NEV 01] NEVO A., "Measuring market power in the ready-to-eat cereal industry", *Econometrica*, vol. 69, no. 2, pp. 307-342, 2001.

[PAY 73] PAYNE B.R., STROM S., "The contribution of trees to the appraised value of unimproved residential land", *Valuation*, vol. 22, no. 2, pp. 36-45, 1975.

[RIC 77] RICHARDSON H.W., "On the possibility of positive rent gradients", *Journal of Urban Economics*, vol. 4, no. 1, pp. 60-68. 1977.

[ROS 74] ROSEN S., "Hedonic prices and implicit markets: Product differentiation in pure competition", *The Journal of Political Economy*, vol. 82, pp. 34-55, 1974.

[SIM 06] SIMONS R.A., SAGINOR J.D., "A meta-analysis of the effect of environmental contamination and positive amenities on residential real estate values", *Journal of Real Estate Research*, vol. 28, no. 1, pp. 71-104, 2006.

[SIR 89] SIRMANS G.S., BENJAMIN J. D., "Determining apartment rent: The value of amenities, services and external factors", *Journal of Real Estate Research*, vol. 4, no. 2, pp. 33–43, 1989.

[SIR 91] SIRMANS G.S., BENJAMIN J. D., "Determinants of market rent", *Journal of Real Estate Research*, vol. 6, no. 3, pp. 357–79, 1991.

[SIR 93] SIRMANS G.S., GUIDRY K.A., "The determinants of shopping centre rents", *The Journal of Real Estate Research*, Vol. 8, No. 1, p. 107-115, 1993.

[SIR 94] SIRPAL R., "Empirical modeling of the relative impacts of various sizes of shopping centres on the value of surrounding residential properties", *Journal of Real Estate Research*, vol. 9, no. 4, pp. 487-505, 1994.

[STO 56] STONE R., *Quantity and Price Indexes in National Accounts*, Paris, Organization for European Economic Cooperation, 1956.

[STR 87] STRASZHEIM M., "The theory of urban residential location", in: Edwin S. MILLS (ed.), *Handbook of Regional and Urban Economics*, Amsterdam, North Holland, 2001-2060, 1987.

[TYR 97] TYRVAINEN L., "The amenity value of the urban forest: an application of the hedonic pricing method", *Landscape and Urban Planning*, vol. 37, pp. 211-222, 1997.

[WHI 80] WHITE, H., "A Heteroskedasticity-Consistent Covariance Matrix Estimator and a Direct Test for Heteroskedasticity", *Econometrica*, vol. 48, no. 4, pp. 817–838, 1980.

[WOR 95] WORZALA E., LENK M., SILVA A., "An exploration of neural networks and its application to real estate valuation", *Journal of Real Estate Research*, vol. 10, pp. 185-201, 1995.

Chapter 12

The Value of Peri-urban Landscapes in a French Real Estate Market

12.1. Introduction

Landscape is a decisive factor in people's environment and quality of life. For example, the beauty of the landscape most certainly guides the behavior of individuals and social groups in their choice of tourist destinations or place of residence. This probably goes some of the way towards explaining urban sprawl. Landscape induces variation in spatial practices in accordance with the representations and consequently the interest it elicits. Beyond this simple observation of fact, attempting to determine the criteria that underpin the value of a landscape is a scientific challenge, the practical repercussions of which concern ordinary citizens and public authorities alike.

This chapter introduces research designed to shed light on the economic valuation of landscapes, or more specifically on the evaluation of that proportion of landscape-related value that is capitalized in the price of real estate transactions regarding the characteristics of the place the real estate is located [CAV 09, JOL 09]. The quality of the environment, and particularly of the landscape, is one of the factors taken into consideration when buying and selling property. Realtors can attest to that, although they are unable to put any precise figure on it. The fact is that the purchase price applies to the property in its entirety, and landscape makes up only a very small part of the whole. Economists classically handle this matter

Chapter written by Thierry BROSSARD, Jean CAVAILHÈS, Mohamed HILAL, Daniel JOLY, François-Pierre TOURNEUX and Pierre WAVRESKY.

through the hedonic pricing method. The scientific approach to ascertaining that proportion of a property price ascribed to landscape is formalized in two parts:

– First, we need to be able to evaluate the quality of a landscape through objective properties that can be measured and quantified. To that end, we employ the resources provided by the geographical analysis of landscapes.

– Second, we must introduce econometric instruments that allow us to emphasize the specific proportion of transaction prices that can be ascribed to the previously-determined landscape variables; here the hedonic pricing method is called upon (see Chapter 11).

This, then, is multidisciplinary research in which landscape and econometric modeling are used in turn. This chapter, which is based on the example of a test zone around Dijon in France, will unfold in three parts. Section 12.2 presents the ingredients of this research work: the study region and its bounds, as well as the geographical and economic data to be mobilized. Section 12.3 covers the development of the analytical and experimentation tools used in the model. The final section sets out the results: maps and graphs aid in describing the landscape and the way it is organized; while tables and commentaries establish the values ascribed to the various constituents of real estate.

12.2. Real estate and landscape data

12.2.1. *The study area*

The outer bound of the area under study is delineated by an access time to Dijon of less than 33 minutes or a road distance of less than 42 km. Its inner bound is that of the urban agglomeration of Dijon, which is composed of the central commune of Dijon (151,500 inhabitants) and the 14 *communes* that make up its suburbs (86,600 inhabitants). This urban agglomeration is excluded, as analytical landscape models used in this work perform poorly for densely built areas with variable building heights. Conversely, the models are well adapted to the analysis of forest or farm landscapes, which may be open or closed, and whose evolution has been dominated for 30 years or so by urban sprawl. We selected 305 *communes* defined in French statistical nomenclature as being peri-urban or belonging to predominantly rural areas. They cover 3,408 km² and comprise 150,800 inhabitants.

The zone is broken down into four major "geographic" groups. In the north-west, the limestone plateaux are home to large cereal farms. Then from west to east, comes a series of three strips:

– first, the Auxois, which is an extensive cattle farming region of grassy valleys criss-crossed by hedgerows and forested hill tops;

– then comes the Arrière-Côte, a limestone plateau incised by dry valleys where the agriculture is diversified (fruit, cereals, livestock); and finally

– there is the floodplain of the Saône River where forested zones and cultivated land are found side-by-side with intensive agricultural productions (market gardening and large farms).

A steep escarpment separates the last two regions and a ribbon of vineyards (Côte d'Or) runs along it producing the renowned Burgundy wine.

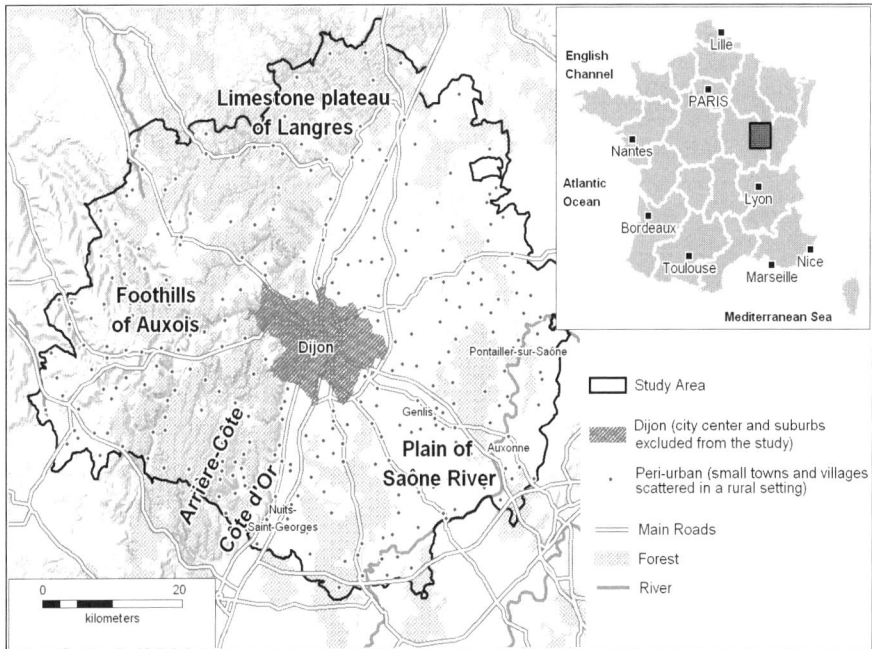

Figure 12.1. *The area under study*

12.2.2. *The real estate market*

The real estate data used were extracted from the lawyers' database (PERVAL) and cover the years 1995–2002. The data give a broad description of the structural characteristics of the transactions (price, surface area, fixtures and fittings, attributes of the seller and the buyer, etc.). In order to pinpoint the transactions within the study area, each item of data was first assigned its geographical coordinates, expressed as latitude and longitude. In total, 6,448 transactions were recorded and georeferenced, with 1,714 transfers relating to developable plots and 4,734 to houses; only the latter are used in the economic section of this research. After removing the houses located in the Dijon agglomeration (as well as a few atypical transactions), the econometric estimation of which we present the results is based on 2,517 observations.

The real estate market is very active close to Dijon (in the suburbs and nearby peri-urban area) and in several small towns or large villages of the peri-urban and rural space. The *communes* that can be reached by the road network to the south-east of Dijon (highway, high traffic roads) also have a very dynamic market. The plot areas increase with distance from Dijon, whereas unitary prices decrease. Blue collar workers represent more than 50% of the buyers in an arc that covers the south-eastern quadrant of the study area. Senior executives and white collar professionals buy in greater numbers in the communes close to Dijon and to the north of the agglomeration.

12.2.3. *Landscape data*

One constraint on the objective and reproducible approach we wish to implement is that landscape must be analyzed everywhere by an equivalent and unchanging process. The resources required were provided by databases from which we extracted the appropriate information. These were digital elevation models (DEMs) from which information on the physical structure of the landscape was derived, and existing databases or satellite images that were used to define the land use, which determines the physiognomy of landscape. The resources of geographical information systems (GISs) were required for managing and matching the databases.

For this purpose, we opted for a matrix mode data format by which the geographical space, which is continuous in nature, is divided up into pixels. Each pixel is defined by its size or resolution, its position in latitude and longitude (georeferencing), and by its attributes or descriptors. The latter refer to nomenclatures that vary in nature: topography, land use, socioeconomic information, etc. These descriptors are either qualitative (e.g. the type of land use) or quantitative (e.g. the altitude of the DEM or the radiometry of the satellite image).

Another constraint arises from the fact that the landscape is organized in space on multiple scales, according to the 2D cartographic surface or the 3D scenic volume. In order to control this scaling effect, it was necessary to constitute four databases with varying levels of precision, both in spatial terms (resolutions at 7, 30, 150 and 1,000 m) and in thematic terms (constitution of appropriate nomenclatures for the various scale levels) [BRO 99].

12.2.3.1. *Digital elevation model (DEM)*

This information is constituted from the initial digital layers supplied by the *Institut Géographique National* (National Geographic Institute), giving the altitude at a ground distance step of 50 or 250 m. Transformations were necessary to bring the four bases to the required resolutions: an interpolation procedure applied to the base at 50 m allowed us to refine the resolution at 7 and 30 m, whereas an aggregation procedure allowed us to enlarge the initial resolutions respectively from 50 to 150 m and from 250 to 1,000 m.

The DEM, now available at four resolutions, served to model the "architecture" of the landscape. We derived many types of indicators from it, such as slopes, orientation, topographical forms or indexes, and the overlooked or overlooking visual configuration.

12.2.3.2. *Land use*

Some land use databases are already available, such as the European Corine Land Cover database whose spatial and thematic characteristics were suitable for providing layers at 150 and 1,000 m. Corine Land Cover data are in vector format. The information is related to spatial entities that are geometrically defined as figures (points, lines or areas) and computer coded as vectors (a set of coordinates that may or may not be interrelated by rules that are specific to each type of figure). Rasterization was therefore required to convert the relevant vector data into the matrix format.

In the absence of satisfactory information that has already been constituted, the development of the 7 and 30 m databases was more involved as we had to use remote sensing resources. Satellite images were acquired from various sensors with complementary performance criteria. The US Landsat satellite provided images whose radiometric properties (seven different spectral bands) are very extensive, being useful for differentiating land cover as much as possible. The Indian IRS satellite, although its radiometric qualities are not as extensive, delivers very high resolution (<6 m) images in panchromatic mode whose definition enables thin features to be identified. A processing protocol was established to make the most of the capabilities of these two satellites; it involved the following sequence of operations:

– Geometric correction rights the images and positions them in conformal projection with the set of databases for the chosen reference framework (in this case the Extended Lambert II). The projection consists of transferring the information from the real spac,e which is curved (the Earth's surface), onto an abstract space, which is plane. They can consequently be more conveniently represented and mapped. The Extended Lambert II is a projection plane that serves as a reference for the whole of France, where it has become a norm.

– The merging of images utilizes certain physical properties of radiometry that allow us to combine the spatial precision of panchromatic high-resolution images and the thematic richness of the images with multiple spectral bands. In this way, merged images are created that simultaneously present the advantages of each type of source.

– Spatial re-sampling is a transformation operation that brings the data to the appropriate 7 and 30 m resolutions.

– Classification is a procedure that allows us to decipher the radiometric information contained in the images thus prepared and translate it in terms of land cover. Well-known image analysis protocols, such as supervised classification, were used again here. They had to be adapted to allow the fine differentiation that we needed, in particular for built areas. This was obviously decisive in the context of this study.

Number	Base 1 7 x 7 m	Base 2 30 x 30 m	Base 3 150 x 150 m	Base 4 1 x 1 km
1	Water	Water	Water	Water
2	Coniferous	Coniferous	Coniferous	Forest
3	Deciduous	Deciduous	Deciduous	Forest
4	Bushes	Bushes	Bushes	Forest
5	Crops	Crops	Crops	Fields
6	Meadows	Meadows	Meadows	Fields
7	Vineyards	Vineyards	Vineyards	Fields
8	Buildings	Buildings	Buildings	Buildings
9	Trading estate	Trading estate	Buildings	Buildings
10	Roads	Roads	/	/
11	Quarries	Quarries	/	/
12	Railways	Railways	/	/
Total	12	12	9	5

Table 12.1. *Land use types for each of the four bases*

The digital layers of land use to which the image processing gave rise were declined in two resolutions for the construction of bases at 7 and 30 m. More generally, each of the four bases, at 7, 30, 150 and 1,000 m, was ascribed its own specific land use nomenclature with a variable number of types depending on the thematic precision that needed to be attained at each level of the scale (see Table 12.1).

12.3. Geographic and econometric models

12.3.1. *Landscape and viewshed*

The term "landscape" is in common use and is meaningful for everyone without us needing to define it. However, as soon as the landscape is posited as an object of scientific investigation, it is necessary to be more specific about what the term covers and what we are seeking to understand. In this chapter, we are interested most of all in the visual dimension of landscape as it presents itself to our view in its diverse configurations. Matters of perception and of individual interpretation are not explicitly taken into account as we are seeking to establish, through modeling, a link between the objectivizable properties of landscape and the price set by the real estate market.

Our field of investigation is thus found to be clearly circumscribed and implies that the terms of analysis be formalized. Geographic surface features may be apprehended visually from two types of viewpoint: from above and from within [ROU 91]. The angle and distance of the view determine the sensorial image that we receive from the landscape. The formalization of the geometric relationships between landscape and view forms the foundation of the objective approach that we are aiming for [BRO 80, BRO 84].

12.3.1.1. *The view from above*

Analysis from a vertical view is generally given precedence by researchers, as it is a way to cover the landscape in its entirety in one swoop. Geographers, ecologists and economists, for example, usually apprehend the landscape as detected from above in order to discover the significant configurations: forms of relief, corridors, borders, ecotones, indicators of compactness and fragmentation, fractal properties, access to jobs, or to goods and services, amenities, etc. [ACH 01, BAS 02, BOL 00, BUR 99, CHE 95, DES 02, KES 04, GEO 97, IRW 02, PAL 03, PAT 02, RAM 04, ROE 04]. In addition, research has been greatly stimulated in this field by the generalization of digital databases and GIS, which are indispensable technical aids. This type of analysis viewed from directly overhead is integrated into our approach.

By using certain ecological analysis tools, we were able to conveniently characterize, as a space property, the environmental amenities or nuisances that signal overall quality of life: the proximity of a park, the configuration of a forest, or even the nearby presence of an industrial facility or highway. Space has to be apprehended vertically here, and we have implemented things so that econometric modeling can apprehend those contextual aspects of landscape. As we based our approach on tried and tested methods, this chapter does not present the details of the protocols followed: we have highlighted them directly in our presentation of the results.

12.3.1.2. *The view from within*

We have focused our attention and investigations on this aspect of landscape, which relates to the ordinary view from the ground and is the foundation of our day-to-day sensory relationship with landscape. Here, we address a distinctive point in our scientific contribution.

The databases, as we have compiled them, allow us to go over the area exhaustively point-by-point and to virtually stop on each pixel in the study zone to examine the landscape. The modeling resources make it possible to approximate the landscape as it appears to an observer positioned at each point. To do this, we combine the information contributed by the digital elevation model and by land use; the volume and visual appearance of the landscape is thus reconstructed. As the parameters that lead to this synthesis of the landscape are well-known and controlled, it is possible to engage in a systematic and multiform exploration of the viewshed, on an objective and reproducible basis, while being aware that the models employed provide only a stylized and simplified reproduction of the actual landscape.

The various terms to be considered and mastered in order to reconstruct the landscape through modeling can be listed as follows:

– The diverging ray method was used to explore the entire horizon around an observation point through angular sampling. All the constraints of the parameters aim to maximize sampling efficiency. That entailed finding the best compromise between the constraints linked to computation time and the need to discriminate fine objects as a function of their distance.

– Along each ray, the pixels that are intersected are classified as seen or unseen by trigonometric calculations. Those calculations take into account the masks due to both the relief and height of the objects that compose land use (trees, houses, structures). The masks formed by objects are processed by assigning a standard height to the corresponding type of land use. The technical means for measuring the actual height of such elevated features exist but they remain complex and expensive. Their use could not be contemplated here.

– Managing the accuracy of the information according to the depth of field is another decisive aspect for the quality of the modeling. The trigonometric reconstruction of the landscape based on information contained in the four databases at different resolutions allowed us to circumvent this technical difficulty. The use of several databases meant a different land use nomenclature was used for each: the diverse types thus defined were identified, counted along each ray and finally added on in order to constitute the quantitative variables that form the basis of analysis of the viewshed.

The formal rules that we have just described address the landscape by simulating the active view of an observer; that is why we associate the notion of active view or scope of view with them [BRO 98]. Yet, the rules that command relationships of intervisibility within the landscape are not univocal; they must be completed by reference to the notion of passive view or subjection to view. Seeing and being seen are two complementary terms, which make up intervisibility, and they are not two symmetrical terms.

Due to the height of certain objects (houses, trees, bushes etc.) and their masking of the view, it is possible to see an object (such as the roof of a house) without being seen from that object (i.e. the ground floor of that house). In terms of subjection to view, each point is characterized by the area from which it is potentially visible. The protocol that guides the analysis of subjection uses the technical principles already outlined for the active view. The difference is that, instead of diverging from a central point, which is visually active, the rays simulating the view converge toward that same point, which is subjected to view.

Thus established, the modeling process allows us to take inventory of the characteristics of the landscape that it is possible to modulate in view of the objectives set. For every point in the study zone, it is possible to define the scope of the panorama that it delivers to our view or to establish whether it is markedly exposed to view from the neighborhood. We have presented a detailed description of how the landscape is organized in the viewfield, in an open view toward the distant horizon or dominated by nearby masks, etc. The analysis can also be broken down field-by-field by paying particular attention to their contents: crops, forests, constructions, road networks, water bodies.

We can infinitely multiply the hypotheses and lines of questioning while trying to cover all the criteria likely to affect the price of real estate. The modeling and formal ordering that it underlies allow us to transcribe, in digital form, those various terms of characterization of the landscape. In this way, we can opt to display them as maps or graphs; or they may be represented in their digital form, since that is how they are transferred as explanatory variables to be subjected to econometric modeling.

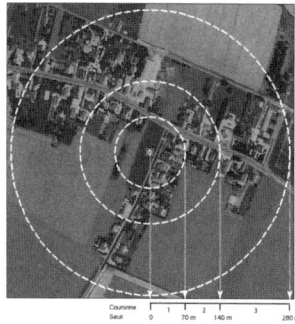

(a) Orthophotograph

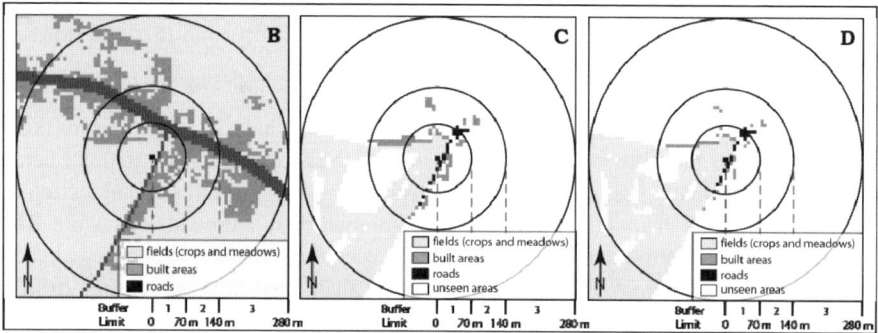

(b) View from above: the
zenith view

(c) View from within: the
scope of the view

(d) View from within:
submission to view

Figure 12.2. *Three landscape models. Each point of real estate transaction is surrounded by elements of the landscape described in terms of land use. Counting the elements allows us to specify their diversity and abundance. In (b) the landscape, presented in a shot taken directly above it, is completely accessible to the view and replicates the structure of orthophotograph (a) despite coarser resolution. In (c) the map shows the segment of landscape seen, in 3D, by an observer located at the center of the circles; the greater part of the surrounding landscape is masked. In (d) all the pixels are marked from which an observer can see the central point.*
Figures (c) and (d) cannot be superposed: thus, "seeing" and "being seen" are two complementary and nonsymmetrical terms

12.3.2. *Landscape and economic value*

12.3.2.1. *Fundamental aspects*

In economics, landscapes are an environmental good belonging to the category of imperfect public goods. While observation of the price of some goods does not pose a very complicated problem when that good is exchanged on the market, specific methods must be developed for non-market goods, such as landscapes. Such goods, when consumed by individuals, may increase their utility, which logically leads us to note that they have a value for those consumers. That value may be declared during direct inquiries (stated preference methods) or it may be revealed indirectly through the observation of an economic agent's behavior on a market, from which the unobserved value of that good is extracted (revealed preference methods). The hedonic pricing method, which we use, refers to the second case. Its interest for us is that, beyond the consumer's personal traits and whatever subjective criteria he/she implements in order to appreciate the beauty of a landscape, the hedonic prices revealed through the prices of the real estate market are objective. The consumer cannot obtain the good for him/herself unless he/she consents to paying that market price [ROS 74].

12.3.2.2. Integration of landscape attributes defined by the geographical approach

The main difficulty in implementing this hedonic method lies in obtaining qualitative and quantitative values for the landscapes. The formalization and instrumentation work, as it was conducted through the geographical approach, allowed us to produce the values in question in the form of qualitative and quantitative variables required by the econometric model. This is because we know that they can be calculated for all points in the space, and therefore, in particular, for the points where real estate transactions are localized. For modeling, the two families of data – one for landscape and the other for real estate transactions – have to be interrelated. For this purpose, we proceeded to precisely locate transaction points (georeferencing) in such a way as to extract the corresponding information from the landscape databases. The analysis protocol of the landscape allows us to propose a great variety of indexes, the choice of which was defined through the following lines of questioning:

– What is the scope of each visual field? We have selected six fields: 0–70 m, 70–140 m, 140–280 m, 280–1,200 m, 1.2–6 km and over 6 km.

– What is the nature of the "landscape objects" seen and at what distance are they located? We were led to classify the forms of land use into a few categories that were meaningful for the economic analysis: forest, farmland, bushes, built area, transportation networks and water bodies.

– Is the household that buys a house exposed to view and from how many places, or is it sheltered from sight?

– How is the landscape around the house structured (diversified, simple, fragmented, partitioned, "mosaic-ed", etc.)? Here, we used indexes commonly used in landscape ecology.

– What is the topography of the site (slope of the plot, dominant exposure, sunshine received, hemmed in)?

– Are there "particular objects" around the house (church, silo, quarry, a Seveso classified industrial site,[1] household garbage incinerator, etc.)?

– What is around the house, even if it cannot be seen (for example, the presence of railroad tracks or roads that cannot be seen but which are the source of noise pollution)?

12.3.2.3. *The econometric model*

The application of the hedonic price method to real estate raises several specific econometric problems, beyond those that may exist in any econometric regression (heteroscedasticity of the residuals and autocorrelations between them, multicolinearity between regressors, etc. – see Chapter 11). Moreover, the estimation of the parameters of household behavior (demand function, demand elasticities) entails other concerns that we omit here, for we did not proceed to the second step of the hedonic method, which is concerned with that aspect. As for the first step to which we confined ourselves, i.e. estimating the hedonic price equation, the two main problems stem from the spatial autocorrelations between the residuals and the endogeneity of certain descriptors. A fundamental postulate of regression methods is violated when one of those two "difficulties" arises, prohibiting the use of the ordinary least-squares method.

Spatial autocorrelation in the residuals may have several causes, the principal one being the existence of omitted characteristics shared by neighboring observations. In this case, the error terms of the model are no longer independent. We have taken the problem into account on two scales:

– the spatial autocorrelations between the residuals of observations belonging to neighboring *communes*[2] (on the *inter-communal* level); and

– the spatial autocorrelations between the residuals of observations of a single *commune* (on the *intra-communal* level).

On the inter-*communal* level, the observations of a single *commune* may share a local policy regarding land (taxation, land zoning), social quality of the neighborhood, enjoyment of local public goods, accessibility to the markets for labor, goods and services, the presence of amenities, and various nuisances and externalities. In addition to this, two neighboring *communes* often have different characteristics in those fields. To take into account those spatial connections that are impossible to introduce into the econometric equation, as they are so infinitely

---

1 The European "Seveso" directive aims to warn of major accidents involving dangerous substances and to limit their consequences for people and the environment. Seveso sites are production sites that are classified as at-risk by this directive.
2 In France, "*communes*" are local municipalities.

diverse, we used the fixed-effects model, while attributing a dummy variable that captures all of those omitted characteristics to each *commune*. By proceeding in this way, the inter-*communal* spatial autocorrelations are eliminated. The Moran index is then calculated to test for correlations among the residuals of observations belonging to a single *commune*. This is calculated with a contiguity matrix in which all transactions less than 200 m away are considered to be neighboring transactions. The statistical test indicates that this index is not significantly different from zero.

Endogeneity of certain covariates may arise here from a simultaneous choice made by households on housing price (explained variable) and certain explanatory variables. For example, the household may make its choice by reasoning simultaneously about the price of the property and its living area (does an additional room justify paying much more?). In that case, that explanatory variable is correlated with the error term, thus violating another postulate of the econometric method. The usual solution, in the place of that endogenous covariate, is to use its projection obtained through an auxiliary econometric equation in which it depends on instruments whose exogeneity has been tested. We use this "instrumental method", which allows us to verify, through a Hausman test, whether the living area of the housing is endogenous or not. The estimation is therefore made by the two-stage least squares (2SLS) method, after the living area has been projected onto instruments (characteristics of the buyers and sellers). The Sargan test indicates whether there is no other endogenous variable.

12.4. Results

12.4.1. *Landscape characterization*

The characterization work is conducted in two ways: first, overall on the scale of the entire zone, which involves the analysis of 127,699,495 points of view (1 per pixel with 7 m long sides); second, they are selectively analyzed by considering only the transaction points (which number 5,956 when we include the land to be built on), or the subgroup of houses (4,050 observations).

12.4.1.1. *Overall characterization according to the map*

The map of the overall view allows us to distinguish the landscapes with respect to the scope of their panoramic view (see Figure 12.3). As this information is the starting point for our investigation protocol, it features in the results, even though the econometric tests have shown that this landscape indicator has a hedonic price of zero, which is an instructive result. In order to surpass this conclusion, we had to proceed iteratively: successive visiting and revisiting allowed us to refine and bring out significant landscape indexes. We also showed that the visual components, such as built areas, deciduous trees, crops and roads, must be considered in the model.

They must always be considered from close up, however, with a variation in the depth of the field: 0–70 m for built areas and deciduous trees, 0–280 m for roads and 70–280 m for crops.

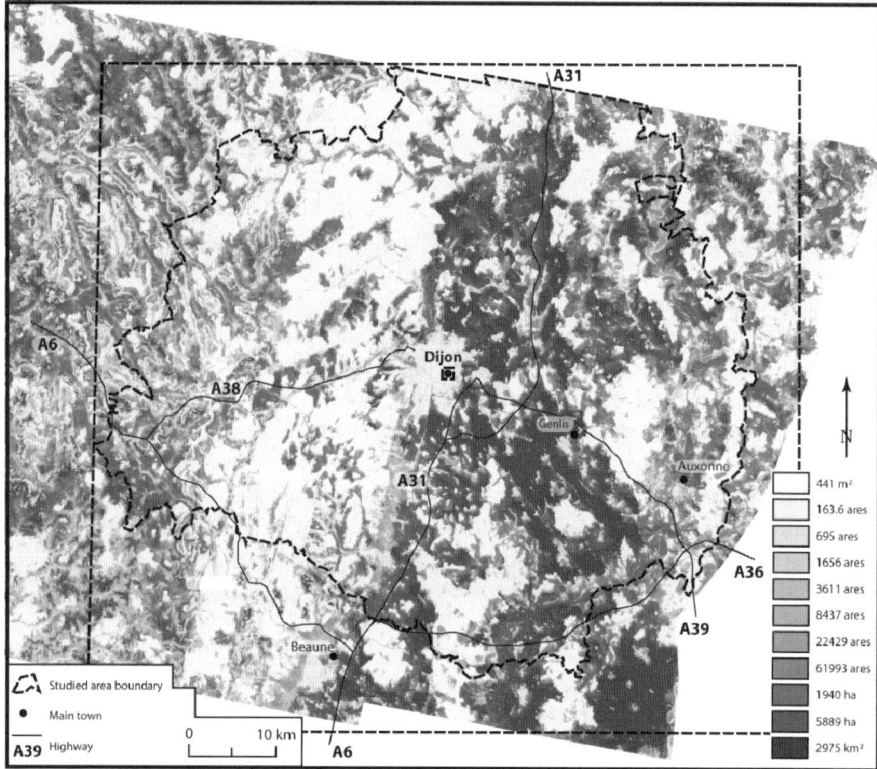

Figure 12.3. *The overall view*

12.4.1.2. *Characterization targeting the transaction points*

Here, we wish to emphasize the way in which the descriptors we selected intervene in the composition of the environment and the views in order to highlight the organizing principles.

Exploring the landscape properties around the transaction points can be organized according to two major ways of apprehending the landscape, which we defined in section 12.3: the view from above and the view from within. Around each transaction point, taking into account the information on the landscape involves defining six concentric rings (or fields) linked to four databases with different resolutions: a base at 7 m for fields 1, 2 and 3; then, respectively, bases at 30, 150 and 1,000 m for fields 4, 5 and 6.

This analysis by field highlights the structuring of the land use around the transaction points: built elements thin out very quickly with distance. Bushes and networks follow a similar tendency from field 3 to fields 4 and 5, whereas forest almost exclusively covers the distant peripheral spaces (see Figure 12.4).

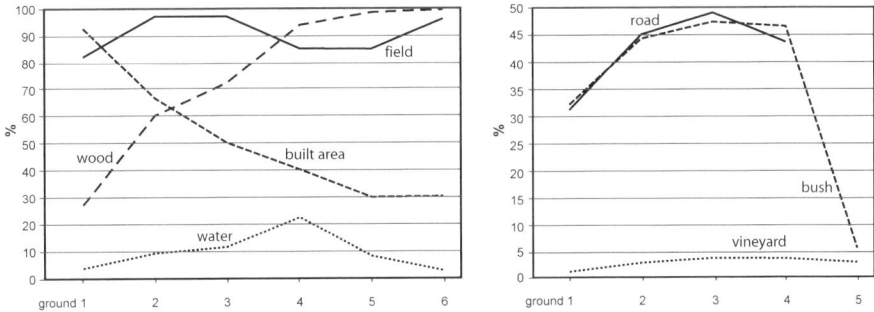

Figure 12.4. *Occurrence of land use themes in the six fields*

The analysis of the landscape, as it is seen from within, brings out the way in which the various categories of land use are distributed in the various visual fields (see Figure 12.5). With a few nuances that distinguish them, we can observe a visual decrease in built areas and bushes as we move from the fore to the background, whereas symmetrically, forests increase their hold. The open spaces, which are mainly farmland, maintain their importance in all the segments of the viewfield. Vineyards and bodies of water are always discreet in the landscape.

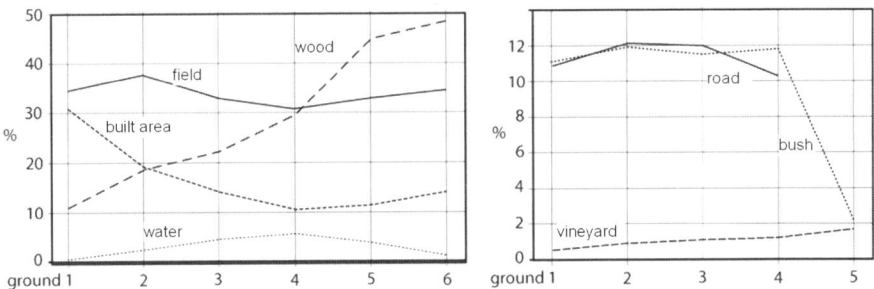

Figure 12.5. *The visual hold of the categories of land use in the six grounds*

In order to establish a sort of susceptibility to the view of various classes of land use, we calculated, by category and by field, the relationship between the number of pixels on the ground and, among them, the number of pixels actually seen in the landscape. We note that an essential proportion of the material elements that compose the geographic space are hidden from view in the landscape. From field 4 (from 280–1,200 m) onward, the landscape that can be seen is made up of elements whose footprint represents <1% of the surface area.

12.4.2. *Hedonic price of non-landscape attributes*

The hedonic price of a square metre of living space, free of any attribute, is worth approximately €1,500, which is a little less than 1.5% of the overall price of a house. One square metre of land of the residential plot, free of any structure, is worth €40–50 near Dijon. That price decreases by 20% when the generalized distance[3] to Dijon increases by 1 km, in such a way that it is no more than about €10 or so at a distance of 12 km, and €2 per square metre when we reach the periphery of the zone under study.

Housing attributes that are identified in the database have hedonic prices that are consistent with the results obtained elsewhere (with only one exception). Bathrooms, attics, swimming pools, basements, parking spaces and cellars have positive hedonic prices, while houses built on several levels or those in a poor state of repair have negative hedonic prices. Apparently, the date of construction has only a very marginal effect on sales prices, which is the exception in relation to most other research (which attributes a high hedonic price to it).

The analysis of the effect of the transaction date shows that half of the approximately 20% increase in real estate values observed between 1995 and June 2002 (the date of the last transfers present in the database) is due to inflation. The 10% progression in inflation-adjusted euros is explained, both by a quantitative effect (living space and plot size) and by a qualitative effect related to the standing of dwelling (modern conveniences have also increased).

---

3 This corresponds to the generalized cost of access to Dijon and takes into account the direct cost of transport (fuel and depreciation, maintenance and vehicle insurance) and cost in time (assessed monetarily from the price of remuneration of an hour's work). Here 1 km costs €0.30 (the average cost allowed by the French fiscal administration for calculating business expenses for personal income tax) and one minute of travel at €0.16 (from estimations by the ministry in charge of infrastructure).

12.4.3. *Hedonic price of landscape attributes*

Table 12.2 indicates the estimated values for the landscape attributes with hedonic prices that are significant at the 5% level.

Field	Landscape attribute	Unit	Value (€)	Value (%)
<70 m	Difference between the number of pixels for a built area seen and the number of pixels of built area from which the location can be seen	+ 100 m²	2,490	2.3
<70 m	Trees seen	+ 100 m²	1,407	1.3
<70 m	Trees unseen	+ 100 m²	356	0.3
<70 m	Number of patches of deciduous trees	+ 1 patch	1,465	1.4
<70 m	Edge length of deciduous tree patches	+ 1 m	-54	0.0
<70 m	Tree patch elongation	+ 1%	327	0.3
70–280 m	Farmland seen	+ 100 m²	43	0.0
70–280 m	Farmland unseen	+ 100 m²	5	0.0
70–280 m	Farmland seen in a developable zone of the local zoning scheme	+ 100 m²	30	0.0
70–280 m	Number of farmland patches	+ 1 patch	186	0,2
<140 m	Transportation networks	+ 100 m²	-105	-0.1
	Highway ribbon (<200 m)	yes	-7,770	-7.2

Table 12.2. *Hedonic price of the statistically-significant landscape attributes. The 100 m² increase in the seen woodland areas <70 ms from a house increases the price of that house by €1,407, or by 1.3% of its price. When the increase is for unseen woodland areas in the same circle, the price increase is €356 (0.3% of the price). A patch of broad-leafed trees is constituted by the set of adjacent points (or pixels) occupied by broad-leafed trees. Its edge is equal to its perimeter*

We have drawn the following main conclusions:

– Households in the study region are characterized by severe nearsightedness: the hedonic price of what they can see beyond a few hundred, or even a few tens of meters is not statistically different from zero. That places a great deal of importance on the urban development undertaken in towns and villages within the immediate proximity of housing. It reduces the role played by farmers as managers of wide-open spaces that offer panoramic views of a far-stretching horizon. The small urban wooded areas, hedges separating houses, and embellishment of the roadways (flowers, trees, etc.) are probably more important than the shape of the fields and types of crops grown by farmers beyond the village limits. However, this conclusion is related to the issue being studied here: the value of the view from a house. It is possible that, on another scale – that of a walk or a bicycle ride – landscape ambiances on smaller scales, measured over distances of several kilometres, acquire significant positive prices. Future research will be done on this topic.

– In this restricted circle of a few tens of meters, or 200–300 m at the very most, it is the view itself from a house that has economic value. Hedonic prices for the simple presence of trees, crops or roads, when they are not seen, come to nought (transportation networks) or prove lower (woodland or farmland: see Table 12.1) than for the actual view of them, taking into account the relief and visual masks.

– We have also raised the dialectic of seeing and being seen, the view being an amenity and subjection to view being a nuisance. In particular, the variable equal to the difference between the number of pixels of built area that we can see and the number of pixels of built area from which we are subjected to view has a positive price (approximately €2,500 for a variation of 100 m² in that variable in a 70 m circle around a transaction).

– Among the objects seen, the wooded formations and farmland have significantly positive hedonic prices while networks, in particular road networks, have a negative price (see Table 12.1). Thus households pay approximately €1,400 to see an additional 100 m² of trees within 70 m of their house (but only €350 for 100 m² of trees present but not seen). For 100 m² of additional farmland they pay barely €50 when it can be seen (only €5 when it cannot be seen), and this in a ring that is not as close as for trees, the results being significant between 70 m and 280 m. These prices are always low, with respect to the price of a house, but they are not zero.

– Some indexes of landscape form (number, shape and arrangement of patches formed by the various land uses), although defined in a rudimentary way by our method, have statistically significant hedonic prices: households prefer complex forms that are fragmented, interlocking or in a mosaic pattern.

– We calculated the price of an obstruction of the landscape, caused, say, by erecting a new building that hides the view. If the viewshed is reduced by 10%, the overall price of the landscape itself decreases, on average, by close to €2,500, which is 2.3% of the price of a house, the median being €1,000. That price is probably lower than the actual value, as it does not take into account the esthetic aspects of landscapes, nor their ecological, recreational and existence value.

– Finally, maps of the value of the landscapes can be established. We present an example, still in the case of 10% reduction of the viewshed (see Figure 12.6). In that way, thanks to the means developed by this multidisciplinary approach, we can generalize the information drawn from the sample of real estate transactions to the entire space. In the example chosen, a new construction has just hidden streets and roads in a few cases, which increases the value of the landscapes, since road networks contribute negatively to the price. In most cases, the light grey to black shading indicates that the reduction of the viewshed is reflected as a loss of value, which is more significant on the periphery of towns or villages (often -2% or greater, for -10% of the landscape that we can see). It is on the margins of the built area cores that the loss relates essentially to woodland or farmland, which contributes positively to landscape values.

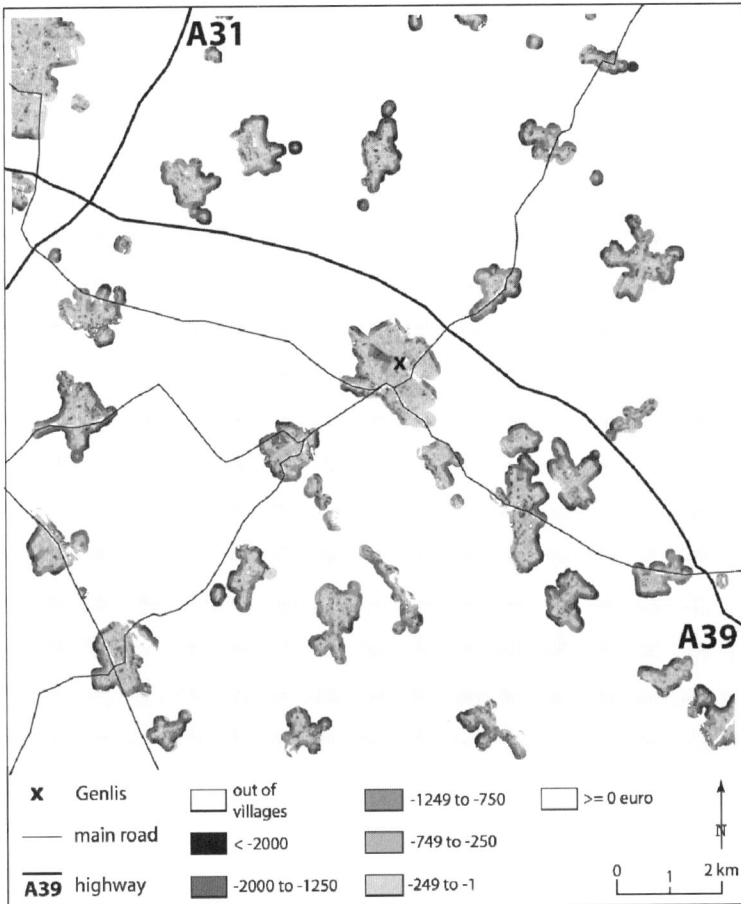

Figure 12.6. *Cost of an obstruction of 10% of the view (in Euros)*

12.5. Conclusion

Geoeconomic models for assessing the price of landscapes are few in number, whereas the two disciplines are complementary for valuing this non-market good which is difficult to evaluate. No doubt, the approach that we adopted provides useful pointers, given the results briefly described above.

Future developments will be necessary, by widening the interdisciplinary collaboration begun here. In particular, certain currently fragile results require more precision (classification of land uses from satellite images seems to have led to an over-estimation of built areas). In terms of econometrics, other models are also contemplated. Comparisons with other regions in France are also underway.

Despite this, even if land use can be refined and transplanted to other regions, our geographic-economic method remains reductive, as it simplifies just what constitutes landscape to the extreme, and only estimates the values of uses related to residential consumption. The fact that it produces significant results, despite those limitations, is encouraging.

We are aware that other types of methods are needed in order to advance the knowledge base in the complex field of the economic valuation of landscapes.

12.6. Acknowledgements

Financial support for this study was received in part from the Conseil Régional de Bourgogne, Conseil Général de Côte-d'Or and Grand Dijon. Real estate data were provided by the PERVAL lawyers'database. Thierry Brossard, Daniel Joly and François-Pierre Tourneux are affiliated with the UMR6049 CNRS-Université de Franche-Comté ThéMA, Besançon. Jean Cavailhès, Mohamed Hilal and Pierre Wavresky are affiliated with the UMR1041 INRA-AgroSup Dijon CESAER, Dijon.

12.7. Bibliography

[ACH 01] ACHARYA G., BENNETT L.L., "Valuing open space and land-use patterns in urban watersheds", *Journal of Real Estate Finance and Economics*, vol. 22, pp. 221-237, 2001.

[BAS 02] BASTIAN C.T., MCLEOD D.M., GERMINO M.J., REINERS W.A., BLASKO B.J., "Environmental amenities and agricultural land values: a hedonic model using geographic information systems data", *Ecological Economics*, vol. 40, pp. 337-349, 2002.

[BLO 88] BLOMQUIST G.C., BREGER M.C., HOEHN J. P., "New estimates of quality of life in urban areas", *American Economic Review*, vol. 78, pp. 89-107, 1988.

[BOL 00] BOLITZER B., NETUSIL N.R., "The impact of open spaces on property values in Portland, Oregon", *Journal of Environmental Management*, vol. 59, pp. 185-193, 2000.

[BRO 80] BROSSARD T., WIEBER J.C., "Essai de formulation systémique d'un mode d'approche du paysage", *Bulletin de l'Association des Géographes Français*, vol. 468-469, pp. 103-111, 1980.

[BRO 84] BROSSARD T., WIEBER J.C., "Le paysage, trois définitions; un mode d'analyse et de cartographie", *L'Espace Géographique*, no. 1, pp. 5-12, 1984.

[BRO 95] BROSSARD T., JOLY D., LAFFLY D., VUILLOD P., WIEBER J.C., "Pratique des systèmes d'information géographique et analyse des paysages", *Revue Internationale de Géomatique*, vol. 4, no. 3-4, pp. 243-256, 1995.

[BRO 98] BROSSARD T., JOLY D., WIEBER J.C., "Analyse visuelle systématique des paysages de cours d'eau par deux approches complémentaires", *Revue Géographique de Lyon*, vol. 73, no. 4, pp. 299-308, 1998.

[BRO 99] BROSSARD T., JOLY D., "Représentation du paysage et échelles spatiales d'information", *Revue Internationale de Géomatique*, vol. 9, no. 3, pp. 359-375, 1999.

[BUR 99] BUREL F., BAUDRY J., *Ecologie du Paysage. Concepts, Méthodes et Applications*, Tec&Doc, 1999.

[CAV 06] CAVAILHÈS J., JOLY D. (eds), *Les Paysages Périurbains et Leur Prix*, Franche-Comté University Presses, 2006.

[CAV 09] CAVAILHÈS J., BROSSARD T., FOLTÊTE J.C., HILAL M., JOLY D., TOURNEUX F.P., TRITZ C., WAVRESKY P., "GIS-based hedonic pricing of landscape", *Environmental and Resource Economics*, vol. 44, no. 4, pp. 571-590, 2009.

[CHE 95] CHESHIRE P., SHEPPARD S., "On the price of land and the value of amenities", *Economica*, vol. 62, pp. 247-267, 1995.

[DES 02] DES ROSIERS F., THÉRIAULT M., KESTENS Y., VILLENEUVE P, "Landscaping and house values: An empirical investigation", *Journal of Real Estate Research*, vol. 23, pp. 139-161, 2002.

[GEO 97] GEOGHEGAN J., WAINGER L.A., BOCKSTAEL N.E, "Spatial landscape indices in a hedonic framework: an ecological economics analysis using GIS", *Ecological Economics*, vol. 23, pp. 251-264, 1997.

[IRW 02] IRWIN E.G., "The effects of open space on residential property values", *Land Economics*, vol. 78, pp. 465-480, 2002.

[JOL 09] JOLY D., BROSSARD, T., CAVAILHÈS J., HILAL M., TOURNEUX F.P., TRITZ C., WAVRESKY P., "A quantitative approach to the visual evaluation of landscape", *Annals of the Association of American Geographers*, vol. 99, no. 2, pp. 292-308, 2009 .

[KES 04] KESTENS Y., THÉRIAULT M., DES ROSIERS F., "The impact of surrounding land use and vegetation on single-family house price", *Environment and Planning B*, vol. 31, pp. 539-567, 2004.

[PAL 03] PALMER J.F., "Using spatial metrics to predict scenic perception in a changing landscape: Dennis, Massachusetts", *Landscape and Urban Planning*, vol. 69, pp. 201-218, 2003.

[PAT 02] PATERSON R.W., BOYLE K.J., "Out of sight, out of mind? Using GIS to incorporate visibility in hedonic property value models", *Land Economics*, vol. 78, pp. 417-425, 2002.

[RAM 04] RAMBONILAZA M., "Evaluation de la demande de paysage: état de l'art et réflexions sur la méthode du transfert des bénéfices", *Cahiers d'Economie et Sociologie Rurales*, vol. 70, pp. 77-101, 2004.

[ROE 04] ROE B., IRWIN E.G., MORROW-JONES H.A., "The effects of farmland, farmland preservation, and other neighborhood amenities on housing values and residential growth", *Land Economics*, vol. 80, no. 1, pp. 55-75, 2004.

[ROS 74] ROSEN S., "Hedonic prices and implicit markets: product differentiation in pure competition", *Journal of Political Economy*, vol. 82, pp. 34-55, 1974.

[ROU 91] ROUGERIE G., BEROUTCHACHVILI N., *Géosystèmes et Paysages: Bilans et Méthodes*, Paris, A. Colin, 1991.

[SMI 02] SMITH V.K., POULOS C., KIM H., "Treating open space as an urban amenity", *Resource and Energy Economics*, vol. 24, pp. 107-129, 2002.

[THO 02] THORSNES P., "The value of a suburban forest preserve: Estimates from sales of vacant residential building lots", *Land Economics*, vol. 78, pp. 626-441, 2002.

[TYR 00] TYRVAINEN L., MIETTINEN A., "Property prices and urban forest amenities", *Journal of Environmental Economics and Management*, vol. 39, pp. 205-223, 2000.

Chapter 13

Conclusion

The completion of this book has allowed us to compile a partial synthesis of the current research into the study of urban dynamics in a Francophone setting. Through the various themes involved and the methodological approaches presented, we can sense the dynamic energy of a burgeoning field. On one hand, it benefits from recent technological advancements (remote sensing, geomatics and computer technology) that provide more control over geographical data. Such data can now be updated more quickly, processed in a more elaborate way (structured in dynamic databases and analyzed with increasingly more sophisticated processes) and extrapolated through the use of simulations in order to model complex systems for studying intersectoral relationships. On the other hand, the formal setting necessary for computerization involves the definition of unequivocal concepts and a reduction of ambiguities. The range of available possibilities encourages researchers to review the basic concepts of urban geography, and evaluate the feasibility and relevance of a paradigm shift (encompassing topographical geography and the geography of people). This consequently allows researchers to weave increasingly strong ties with related disciplines (behavioral psychology, microeconomics, sociology, transportation engineering, management, etc.).

The potential for innovation is great at the conceptual level, as we can now expand the traditional field of interest – geography – which encompasses the study of regions and their exploitation, by taking into account the impact geography has on the individual processes that transform these regions:

– deciding on location;

Conclusion written by Marius Thériault and François Des Rosiers.

– the mobility of individuals;

– households and goods and services;

– choice of location and destination;

– urban planning and development;

– perception of the environment;

– an equilibrium between supply and demand in establishing values, etc.

These phenomena have always been of interest to urban geography, but the lack of sufficient data (generally aggregate) and analysis tools often limited the scope of feasible studies. It is indeed a paradigm shift, because the analysis of the relationship between people (households, businesses, etc.) and this land must be added to the analysis of land use (global systemic vision). We thus gradually move from the functional or virtual landscape to the occupied landscape. It is this transition in progress that surfaces across the mixture of analytical scales and concerns presented in this work.

We believe that, beyond the methodological progress on which it is based, this is the reason for the book's primary originality. Some of the approaches presented in this volume voice our interest in integrating urban subsystems into a multiscalar model capable of considering the wide array of feedback that defines urban dynamics. This model is devised according to the view of the complex system proposed by Jay W. Forrester [FOR 69], and subsequently by George F. Chadwick [CHA 71], but within a multiscalar framework. These new approaches aim to identify the emergent behavior within which sustainable development strategies are sought, to be evaluated when we adequately model and simulate the evolution and adaptation of urban systems. On the other hand, we also seek to detect, in good time, any possible negative overall effects resulting from the aggregation of individual decisions, themselves based on a clear and immediate optimality. The ultimate goal is to make provision for (and not to predict) the evolution of urban systems evaluated from a perspective of social and intergenerational fairness.

Regarding methodology, this work presents several working examples of geomatics, remote sensing, spatial analysis, simulation and statistical modeling used to study urban dynamics. Each chapter focuses on a particular range of themes and methodological tools whose selection and implementation are adapted to the analytical targets. Nevertheless, it is not the availability of the tool that must justify methodology or priority among research themes, being based primarily on social and scientific relevance. We must adapt analysis methods to the subject of study and the hypotheses to be verified. However, when the time comes to plan and carry out the research, a balanced relationship between the two aspects becomes necessary. As with philosophy, the ability to manage concepts is dependent on the necessary vocabulary to express them. In applied science there is an essential link between the potential to verify hypotheses and the means available to test our theories using observable reality.

It is based on this theory/method duality that the empirical validation of ideas can be carried out. This explains why the availability of new analysis tools in urban studies will eventually enable us to improve the quality of research and, possibly, explore new paradigms. In addition to this, the current combination of analyzing general relationships (on an urban system scale) and studying individual coping mechanisms (on a decision-making and transformation process scale) is resulting in concrete repercussions. This is because such a combination of approaches enhances land use planning applications and decision making. Lastly, these new analytical tools, the result of information technologies, are also being heralded as the catalyst of a growing and promising collaboration between fields by gradually substituting the transdisciplinary handling of urban issues to the vertical and often nearsighted compartmentalization of traditional approaches.

The results of applications presented in this volume pave the way to operational tools for urban planning and management. The current practice of land use planning is still very normative: we establish major policies that are then labeled as top priority (e.g. land development plans), which are then broken down into actions (urban planning regulations, zoning plans, etc). Considering the lack of prospective tools required to evaluate long-term impacts, the practice of urban planning remains more of an art than a subject of experimental knowledge. We can of course analyze the consequences of each urban planning decision made (e.g. construction of a highway, factory construction, etc.). When the net effect is known, however, it is too late to react or too costly to repair. Thus, we gain experience by trial and error. In a world that evolves slowly, the accumulated experience should, over the course of generations, make it possible to improve the performance of urban systems. However, our world evolves very quickly; we must adjust to the globalization of economic exchanges that introduce competition among cities. The most efficient cities will grow, while the others will stagnate or regress.

It is in this area that the development of predictive urban land use planning tools becomes a major societal concern for regional development. Their integration is far from complete, but various innovations presented in this volume (process simulations and the production of status indicators to evaluate accessibility, behavior, values, equity, etc.) constitute an essential base on which to nourish the decision-making tools with relevant observable and measurable facts. These facts supplement the people's preference grids when they must establish effective action plans to manage the development of a city. In the long term, we can assume that some of the tools and concepts presented in this volume will become active components in the decision-making process, as a supplement to the political, societal, economic and environmental considerations that any urban land use planning must consider. By continuing the current trend, the field of urban studies will become increasingly apt at evaluating the relevance and potential performance of land use planning policies. Performance will be improved, in particular, through better understanding of the urban system and by better awareness of the internal and external relationships that exist among the system's various elements (infrastructures, people, businesses, environment, etc.).

Lastly, being by nature incomplete, this volume also opens the way towards a research agenda for the next decade. It would be pretentious to want to encompass the whole subject through this work, but we can nevertheless garner useful information from it.

The usefulness and feasibility of studying the elementary processes has been shown, although to harness the full potential it is necessary to cross the disciplinary borders. This will enable us to produce models and set up multisectoral indicators (e.g. to evaluate the impact of an economic measure – an increase in parking prices – on mobility behaviors, and, as a result, on accessibility to services and the property values). As it is a question of developing communities for people, various concepts corresponding to long-term objectives must be dealt with simultaneously: effectiveness, equity, environmental quality, competitiveness, etc. This multiplicity of evaluation criteria requires us to make them comparable without establishing a static exchange rate among objectives, as the legitimacy of this operation in a democratic society falls upon the whole of the community, rather than being the responsibility of a particular scientific community. In order to propose appropriate solutions to complex problems, a transdisciplinary culture must be developed, without compromising the rigor of the analytical procedure or diluting the coherence of each discipline's contribution.

The methodological and epistemological challenges are considerable. Recent technological advances have made it possible to process increasingly large and complex data-processing operations, but the development of novel design comes at a much slower pace, as it often involves a more fundamental revision of the approaches. For example, urban modeling has relied heavily on the statistical approach, and regression in particular, which makes it possible to evaluate the behavior of a dependent variable according to contextual variations. However, beyond its relatively sophisticated variations, this concept of regression is a simple application of differential calculus used to evaluate a median value (or an average probability, etc.).

What about residual variance, also known as residuals? Is a single coefficient sufficient to model the full array of behaviors observed in a population? And what if there were a spatial or temporal drift associated with differences in perception and values among the inhabitants of the various districts of a city? Would it then be reasonable to claim that they all exhibit average behavior? How does one model these behaviors without becoming wrapped up in determinism? Although various possible solutions to these fundamental questions are already the focus of innovative and very promising methodological and conceptual developments, it is only the concerted efforts of the scientific community interested in these territorial issues that will succeed in bringing us closer to uncovering the correct answers.

13.1. Acknowledgements

The completion of this volume was made possible through the excellent collaboration of 33 coauthors. Funding for the programme de soutien aux équipes de recherche, du Réseau Villes, Régions, Monde was received from the Fonds Québécois de la Recherche sur la Société et la Culture and the Social Sciences and Humanities Research Council.

The publication's editors would especially like to thank Marion Voisin and Jean Dubé for their invaluable editing assistance. We would like to express our gratitude to Professor Pierre Dumolard, who offered to produce this work for the Traité IGAT series, and to the staff of Hermès Science Publishing and ISTE, who helped guide us through the editing process. Lastly, the editors wish to acknowledge the efficient collaboration of Gordon Cruise for the English translation and proof reading of chapters in this book.

13.2. Bibliography

[CHA 71] CHADWICK G.F., *System View of Planning: Towards a Theory of the Urban Regional Planning Process*, New York, Pergamon, 1971.

[FOR 69] FORRESTER J.W., *Urban Dynamics*, Waltham (Mass.), Pegasus Communications, 1969.

List of Authors

Philippe APPARICIO
LASER
INRS–Urbanization, Culture
and Society
Montreal
Canada

Arnaud BANOS
Laboratoire Parisgéo
Paris-Sorbonne University
Paris
France

Sandrine BERROIR
Laboratoire Parisgéo
Paris-Sorbonne University
Paris
France

Gjin BIBA
ESAD–CRAD
Laval University
Québec
Canada

Anne-Christine BRONNER
Geography and Development Faculty
Louis Pasteur University
Strasbourg
France

Thierry BROSSARD
Literature and Social Studies Faculty
University of Franche-Comté
Besançon
France

Jean CAVAILHÈS
CESAER
INRA
Dijon
France

François DES ROSIERS
Administration Science Faculty
Laval University
Québec
Canada

Jean DUBÉ
ESAD–CRAD
Laval University
Québec
Canada

Pierre DUMOLARD
MSH-Alpes
Grenoble
France

Christophe ENAUX
Geography and Development Faculty
Louis Pasteur University
Strasbourg
France

Jean-Christophe FOLTÊTE
Literature and Social Science Faculty
University of Franche-Comté
Besançon
France

Cyrille GENRE-GRANDPIERRE
UMR ESPACE
University of Avignon
France

Hélène HANIOTOU
Geography and Development Faculty
Louis Pasteur University
Strasbourg
France

Mohamed HILAL
CESAER
INRA
Dijon
France

Daniel JOLY
Literature and Social Science Faculty
University of Franche-Comté
Besançon
France

Didier JOSSELIN
UMR ESPACE
University of Avignon
France

Yan KESTENS
Department of Public Health
Montreal
Canada

Hélène MATHIAN
Laboratoire Parisgéo
Paris-Sorbonne University
Paris
France

Chryssanthi PETROPOULOU
Geography and Development Faculty
Louis Pasteur University
Strasbourg
France

Thierry RAMADIER
Geography and Development Faculty
Louis Pasteur University
Strasbourg
France

Thérèse SAINT-JULIEN
Laboratoire Parisgéo
Paris-Sorbonne University
Paris
France

Lena SANDERS
Laboratoire Parisgéo
Paris-Sorbonne University
Paris
France

Anne-Marie SÉGUIN
INRS–Urbanization, Culture
and Society
Montreal
Canada

Marius THÉRIAULT
ESAD–CRAD
Laval University
Québec
Canada

Thomas THÉVENIN
University of Burgundy
Dijon
France

Isabelle THOMAS
Department of Geography
Catholic University of Louvain
Louvain-la-Neuve
Belgium

François-Pierre TOURNEUX
Literature and Social Science Faculty
University of Franche-Comté
Besançon
France

Marie-Hélène VANDERSMISSEN
Department of Geography
Laval University
Québec
Canada

Ann VERHETSEL
Department of Transport and
Regional Economics
University of Anvers
Belgium

Paul VILLENEUVE
ESAD–CRAD
Laval University
Québec
Canada

Marion VOISIN
ESAD–CRAD
Laval University
Québec
Canada

Pierre WAVRESKY
CESAER
INRA
Dijon
France

Index